Chambers
Earth Sciences
Dictionary

GENERAL EDITOR:

Professor P. M. B. Walker, CBE, FRSE

CONSULTANT EDITOR:

Dr P. A. Sabine
PhD, DSc, ARCS, FRSA
CEng, FIMM, FRSE, FGS

Chambers

Published 1991 by W & R Chambers Ltd,
43–45 Annandale Street, Edinburgh EH7 4AZ

© W & R Chambers Ltd 1991

All rights reserved. No part of this publication may be reproduced, stored in a retrieval system, or transmitted, in any form or by any means, electronic, mechanical, photocopying, recording or otherwise, without the prior permission of W & R Chambers Ltd.

British Library Cataloguing in Publication Data
Chambers earth science dictionary.
   1. Earth Sciences
   I. Walker, Peter M. B.
   550

ISBN 0 550 13244–9 Hbk
ISBN 0 550 13245–7 Pbk

Cover design by Michael Dancer

Printed in England by Clays Ltd, St Ives plc

Contents

Preface iv
Contributors vi
The Dictionary 1

How the dictionary was made

Chambers Earth Sciences Dictionary was compiled and designed on a COMPAQ 386 personal computer. The original database was made with the INMAGIC library retrieval software from Head Computers Ltd. The text was set using the Xerox VENTURA desktop publishing system and the drawings made with the Micrografx DESIGNER graphics program.

Preface

Chambers Earth Sciences Dictionary is a unique dictionary concerned with the sciences and technologies related to the structure and minerals of the Earth. It has been developed from the database of the acclaimed *Chambers Science and Technology Dictionary* with many new and revised definitions in this group of subjects.

The book contains some 6000 definitions, including 1200 in geology, over a 1000 each in mineralogy and mineral extraction and 150 in crystallography, together with a large number of relevant definitions from chemistry, physics and biology. The aim has been to give a comprehensive treatment of both the basic science and more applied aspects such as mining and oil drilling, with consideration given also to precious stones.

The special articles, placed in separate panels, are a major feature of *Chambers Earth Sciences Dictionary*. These, many of which are illustrated, supply fuller treatments of major topics within the subject. There are, for example, panels on major geological classifications, earthquakes, plate tectonics, drilling rigs and petroleum reservoirs, crystal structure and most of the major fossil groups. There are also panels for those elements important in geology, which, in addition to providing basic information, give their abundances and the minerals in which they are mainly found. Ionizing radiations, their distribution and units of measurement are also described.

These panels, because of the cross references which many include, will be a route into the dictionary for those less familiar with the subject.

Arrangement

The entries are strictly alphabetical with single letter entries occurring at the beginning of each letter. The panels occur either on the same page as the parent entry or on the pages immediately following, with the alphabetical entry stating 'See panel on this page' or 'See panel on p. 0'.

Italic and Bold

Italic is used for:
  (1) alternative forms of, or alternative names for, the headword, usually after 'also' at the end of an entry;
  (2) the expanded form of an abbreviated headword, provided that the expanded form is not found as a headword elsewhere;

(3) terms derived from the headword, often after 'adj.' or 'pl.';
(4) variables in mathematical formulae;
(5) for emphasis.

Bold is used for:

cross-references, either after 'see', 'cf.' etc. or in the **body of the entry**. It is also used after 'Abbrev. for' when the expanded form can be found as a headword elsewhere.

Bold italic is used in the panels to highlight a term explained within that article.

Tables

The main geological divisions will be found as tables either in the 'Geological Column' panel or in other appropriate panels. There are also tables for 'SI units' and 'Fundamental Dynamical Units'. The latter gives comparisons with the older scientific units commonly found in older texts.

Peter M. B. Walker
1991

Contributors

Extensive use has been made of the entries in *Chambers Science and Technology Dictionary* and I would like to thank all those who contributed to individual sections of that dictionary. My particular thanks are due to **Peter Sabine** who not only revised many entries and wrote the articles for the panels in this dictionary but provided excellent rough sketches for the drawings needed to illustrate them.

P.M.B.W.

A

a- Abbrev. in chemistry for asymmetrically substituted.

α– Symbol for (1) substitution on the carbon atom of a chain next to the functional group; (2) substitution on a carbon atom next to one common to two condensed aromatic nuclei; (3) substitution on the carbon atom next to the hetero-atom in a hetero-cyclic compound; (4) a stereo-isomer of a sugar.

α-particle See **alpha particle**.

Å Symbol for Ångström. $1 \text{ Å} = 10^{-10}$ m.

[A] A strong absorption band in deep red of the solar spectrum (wavelength 762.128 nm) caused by oxygen in the Earth's atmosphere. The first of the Fraunhofer lines.

$[\alpha]_\Delta^\tau$ Symbol for the specific optical rotation of a substance at $t°C$, measured for the D line of the sodium spectrum.

aa A term of Hawaiian origin for lava flows with a rough, jagged surface.

Aalenian The oldest stage of the Middle Jurassic. See **Mesozoic**.

abandonment Voluntary surrender of legal rights or title to a mining claim.

Abbe refractometer An instrument for measuring directly the refractive index of liquids, minerals and gemstones.

Abegg's rule Empirical rule that solubility of salts of alkali metals with strong acids decreases from lithium to caesium, i.e. with increase of rel. at. mass, and those with weak acids follow the opposite order. Sodium chloride is an exception to this rule, being less soluble than potassium chloride.

Abegg's rule of eight The sum of the maximum positive and negative valencies of an element is eight, e.g. S in SF_6 and H_2S.

Abel flash-point apparatus A petroleum-testing apparatus for determining the flash-point.

ablation (1) All processes by which snow and ice are lost from a glacier, mainly by melting and evaporation (sublimation). (2) Removal of surface layers of a meteorite or tektite during flight.

abrasion Mechanical wearing away of rocks by rubbing during movement.

abrasion hardness Resistance to abrasive wear, under specified conditions, of metal or mineral.

abrasive A substance used for the removal of matter by scratching and grinding (abrasion), e.g. silicon carbide (carborundum).

absolute age The geological age of a fossil, mineral, rock or event, generally given in years. Preferred synonyms *radiometric age, isotopic age*. See **radiometric dating**.

absolute pressure Pressure measured with respect to zero pressure, in units of force per unit of area.

absolute temperature A temperature measured with respect to **absolute zero**, i.e. the zero of the **Kelvin thermodynamic scale of temperature**, a scale which cannot take negative values. See **kelvin**.

absolute units Those derived directly from the fundamental units of a system and not based on arbitrary numerical definitions. The differences between absolute and international units were small; both are now superseded by the definitions of the SI.

absolute viscosity See **coefficient of viscosity**.

absolute weight The weight of a body in a vacuum.

absolute zero The least possible temperature for all substances. At this temperature the molecules of any substance possess no heat energy. A figure of $-273.15°C$ is generally accepted as the value of absolute zero.

absorbance The logarithm of the ratio of the intensity of light incident on a sample to that transmitted by it. It is usually directly proportional to the concentration of the absorbing substance in a solution. See **Beer's Law**.

absorbed dose Quantity of energy imparted by ionizing radiation to a unit mass of biological tissue. Unit *gray*, symbol Gy (1 Gy = 1 joule/kilogram). See **ionizing radiation: units of measurement**.

absorptiometer An apparatus for determining the solubilities of gases in liquids or the absorption of light.

absorption band A dark gap in the continuous spectrum of white light transmitted by a substance which exhibits selective absorption.

absorption coefficient The volume of gas, measured at s.t.p., dissolved by unit volume of a liquid under normal pressure (i.e. 1 atmosphere).

absorption coefficient (1) At a discontinuity (*surface absorption coefficient*), (*a*) the fraction of the energy which is absorbed, or (*b*) the reduction of amplitude, for a beam of radiation or other wave system incident on a discontinuity in the medium through which it is propagated, or in the path along which it is transmitted. (2) In a medium (*linear absorption coefficient*), the natural logarithm of the ratio of incident and emergent energy or amplitude for a beam of radiation passing through unit thickness of a

absorption plant medium. (The *mass absorption coefficient* is defined in the same way but for a thickness of the medium corresponding to unit mass per unit area.) N.B. *True absorption coefficients* exclude scattering losses, *total absorption coefficients* include them.

absorption plant Plant where oils are removed from natural gas by absorption in suitable oil.

absorption spectrum The system of absorption bands or lines seen when a selectively absorbing substance is placed between a source of white light and a spectroscope. Important in gemmology.

abutment load In stoping or other deep-level excavation, weight transferred to the adjacent solid rock by unsupported roof.

abyssal Refers to the ocean floor environment between ca 4000–6000 m. See **littoral**, **bathyal**.

abyssal deposits Pelagic marine sediments, accumulating in depths of more than 2000 m including, with increasing depth, calcareous oozes, siliceous oozes and red clay (500 m).

abyssal plain A very flat region of the deep ocean floor with a slope of less than 1 : 1000.

acanthite An ore of silver, Ag_2S, crystallizing in the monoclinic system. Cf. **argentite**.

acceleration due to gravity Acceleration with which a body would fall freely under the action of gravity in a vacuum. This varies according to the distance from the Earth's centre, but the internationally adopted value is 9.806 65 m/s^2 or 32.1740 ft/s^2. Abbrev. *g*. See **Helmert's formula**.

acceptor (1) The reactant in an induced reaction whose rate of reaction with a third substance is increased by the presence of the inductor. (2) The atom which accepts electrons in a co-ordinate bond.

accessory minerals Minerals which occur in small, often minute, amounts in igneous rocks; their presence or absence makes no difference to classification and nomenclature.

accessory plates Quartz-wedge, gypsum plate, mica plate. Used with a petrological microscope to help determine the optical character of a mineral as an aid in its examination.

accommodation rig Offshore rig with sleeping, supply and recreational facilities.

accretion The process of enlargement of a continent by the tectonic coalescence of exotic crustal fragments.

ACF diagram A triangular diagram used to represent the chemical composition of metamorphic rocks. The three corners of the diagram are Al_2O_3, CaO and FeO+MgO.

achondrite A type of stony meteorite which compares closely with some basic igneous rocks such as eucrite.

achroite See **tourmaline**.

acicular (1) Needle-shaped. (2) The needle-like habit of crystals.

acid (1) Normally, a substance which (a) dissolves in water with the formation of hydrogen ions, (b) dissolves metals with the liberation of hydrogen gas, or (c) reacts with a *base* to form a *salt*. More generally, a substance which tends to lose a proton or to accept an electron pair. (2) See **acid rock**.

acid cure In extraction of uranium from its ores, lowering of gangue carbonates by puddling with sulphuric acid before leach treatment.

acid drift Tendency of ores, pulps, products, to become acidic through pick-up of atmospheric oxygen through standing.

acidimetry The determination of acids by titration with a standard solution of alkali.

acidity Loosely, the extent to which a solution is acid. See **pH value**. More strictly, the concentration of any species titratable by strong base in a solution.

acidizing an oil well Improving the flow of oil from a limestone formation by pumping acid into it.

acid mine water Water containing sulphuric acid as a result of the breakdown of the sulphide minerals in rocks. Acid mine water causes corrosion of mining equipment, and may contaminate water supplies into which it drains.

acid radical A molecule of an acid minus replaceable hydrogen.

acid rock An igneous rock with >63% quartz.

acid salts Salts formed by replacement of part of the replaceable hydrogen of the acid.

acid sludge Mixture of 'green acid' and 'mahogany soap' salts; reaction product in sulphuric acid treatment of oil refinery heavy fractions, used in froth flotation process for recovering iron minerals.

acid solution An aqueous solution containing more hydroxonium ions than hydroxyl ions, i.e. it has a low **pH value**; one which turns blue litmus red.

acmite A variety of aegirine; also used for the $NaFe^{+3}Si_2O_6$ end-member.

acoustic survey Determination of the porosity of a rock by measuring the time required for a sonic impulse to travel through a given distance.

acritarchs See panel on p. 3.

actinides A group name for the series of

acritarchs, dinoflagellates, pollen and spores

These all occur as microfossils and are of use in establishing geological correlations or in determining the environmental conditions in which these organisms lived. They have resisted destruction but their biological affinities may be uncertain. *Acritarchs* (Figs. 1 & 2) derive their name (Gk. 'origin uncertain') from these doubts. They are minute unicellular micro-organisms only a few tens of micrometres across of unknown or uncertain biological relationship. The group is artificial but some are of algal affinity. They range from the Precambrian to the present day and were most abundant in the Precambrian and Lower Palaeozoic. *Hysterichosphere* is a general term given to a variety of microfossils that have resistant organic shells and range from the Precambrian to the present day. They are spherical or ellipsoidal usually with spines. They are now divided between the dinoflagellates and the acritarchs. *Dinoflagellates* are single-celled micro-organisms of algal or uncertain affinity. They have flagellae; some have siliceous shells, some fossil examples have calcareous shells, and numerous genera, fossil and Recent, have organic shells (Fig. 6). Most post-Palaeozoic hysterichospheres are dinoflagellate *cysts* (*dinocysts*; resting spores). Dinoflagellates range generally from the Jurassic to the present but some Silurian and Permian examples are known.

Spores are also very well adapted to survive extreme conditions. They are a variety of minute, usually unicellular, reproductive bodies or cells (Figs. 3 – 5). *Pollen* are the microspores of seed plants and *palynology* the study of spores and pollen. In many sedimentary rocks these are the only organisms to have resisted destruction and their study is consequently important for stratigraphical correlation. *Pollen analysis* is the study, esp. for Quaternary sediments, of the abundance and distribution in space and time of pollen, to date deposits, and to determine and interpret climatic changes. Used usually for peat and lake-bed studies. *Palynomorphs* are the resistant minute organic bodies including spores, pollen, acritarchs, dinoflagellates and other acid insoluble residues that are left after maceration; they are the objects of palynological study.

Fig. 1. Acritarch, Upper Devonian. × 650.

Fig. 2. Acritarch, Ordovician. × 600.

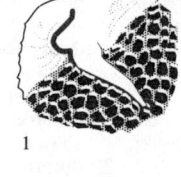
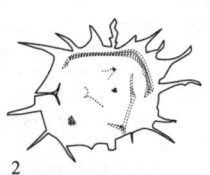

Fig. 3. Spore, Upper Devonian. × 600.

Fig. 4. Spore, Lower Carboniferous. × 350.

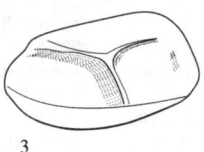
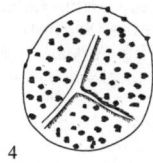

Fig. 5. Spore, Lower Devonian. × 700.

Fig. 6. Organic-walled dinoflagellate, Pleistocene. × 500.

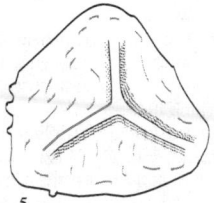

actinium elements of atomic numbers 89–104 (inclusive), which indicates that they have similar properties to actinium.

actinium A radioactive element in the third group of the periodic system. Symbol Ac, at. no. 89, r.a.m. 227; half-life 21.7 years. Produced from natural radioactive decay of the ^{235}U isotope or by neutron bombardment of ^{226}Ra. Gives its name to the actinium (4n+3) series of radio-elements.

actinolite A monoclinic calcium magnesium iron member of the amphibole group, green in colour and usually showing an elongated or needle-like habit; occurs in metamorphic and altered basic igneous rocks.

activated carbon Carbon obtained from vegetable matter by carbonization in the absence of air, preferably in a vacuum. Activated carbon has the property of adsorbing large quantities of gases. Important for gas masks, adsorption of solvent vapours, clarifying of liquids and in medicine.

activated charcoal *Charcoal* treated with acid etc. to increase its adsorptive power.

activating agent See **activator**.

activator Surface-active chemical used in flotation process to increase the attraction to a specific mineral in an aqueous pulp of collector ions from the ambient liquid and increase its aerophilic quality. Also *activating agent*.

active margin A continental margin characterized by earthquakes, igneous activity and mountain building as a result of convergent-plate or transform-plate movements. See **passive margin**.

activity The rate at which transformations occur in a **radionuclide**. Unit *becquerel*, symbol Bq (1 Bq = 1 transformation/second). See **ionizing radiation**.

adamantine See **lustre**.

adamellite A type of granite with approximately equal amounts of alkali-feldspar and plagioclase.

adaptation *Evolutionary adaptation*.The adjustment to environmental demands through the long-term process of natural selection acting on the genotype.

adaptive radiation Evolutionary diversification of species from a common ancestral stock, filling available ecological niches. Also *divergent adaptation*.

adhesion Intermolecular forces which hold matter together, particularly closely contiguous surfaces of neighbouring media, e.g. liquid in contact with a solid. US *bond strength*.

adiabatic Without loss or gain of heat.

adiabatic change A change in the volume and pressure of the contents of an enclosure without exchange of heat between the enclosure and its surroundings.

adinole An argillaceous rock which has undergone albitization during contact-metamorphism.

adit A horizontal passage or tunnel into a mine. See **mining**.

adobe A calcareous clay found in semi-arid plains and basins of the south-western United States and Mexico.

adsorbate A substance, usually in gaseous or liquid solution, which is to be removed by adsorption.

adsorbate Substance adsorbed at a phase boundary.

adsorbent The substance, either solid or liquid, on whose surface **adsorption** of another substance takes place.

adsorption The taking up of one substance at the surface of another.

adsorption catalysis The catalytic influence following adsorption of the reactants, exercised upon many reactions, often upon a specific adsorbent, attributed to the free residual valencies and hence higher reactivity of molecules at an interface (e.g. the ammonia oxidation process).

adsorption chromatography See **chromatography**.

adsorption isotherm The relation between the amount of a substance adsorbed and its pressure or concentration, at constant temperature.

adsorption potential Change of potential in an ion in passing from a gas or solution phase on to the surface of an adsorbent.

adularescence A milky or bluish sheen shown by moonstone.

adularia A transparent or milky-white variety of potassium feldspar, distinguished by its morphology.

advance workings In flattish seams, mining in which the whole face is carried forward, no support pillars being left.

aegirine Green sodium-iron member of the pyroxene group of minerals, essentially $NaFe^{+3}Si_2O_6$. Characteristic of the alkaline igneous rocks. *Acmite* is a brown variety and also used for the pure NaFe end-member.

aegirine-augite Minerals intermediate between **aegirine** and **augite**.

aenigmatite A complex silicate of sodium, iron and titanium; occurs as reddish-black triclinic crystals in alkaline igneous rocks. Also *cossyrite*.

aeolian deposits Sediments deposited by wind and consisting of sand or dust (**loess**).

Aerofall mill Dry grinding mill with large diameter and short cylindrical length in

aerolites   **alabaster**

which ore is mainly or completely ground by large pieces of rock; mill is swept by air currents which remove finished particles.

aerolites A general name for stony as distinct from iron meteorites.

affinity The extent to which a compound or a *functional group* is reactive with a given reagent. See **electron affinity**.

afterburst Delayed further collapse of underground workings after a rockburst.

aftercooler Chamber in which heat generated during compression of air is removed, allowing cool air to be piped underground.

afterdamp The non-flammable heavy gas, carbon dioxide, left after an explosion in a coalmine. The chief gaseous product produced by the combustion of coal-gas. See **black damp, choke damp, fire damp, white damp**.

aftershock A minor earthquake that follows a major one. Many aftershocks may be experienced, decreasing in magnitude.

afwillite Hydrated calcium silicate occurring in natural rocks and set cements.

agalmatolite See **pagodite**.

agate A cryptocrystalline variety of silica, characterized by parallel, and often curved, bands of colour. See **silicon, silica, silicates**.

agglomerate An indurated rock built of angular rock-fragments embedded in an ashy matrix, and resulting from explosive volcanic activity. Occurs typically in volcanic vents.

agglomerating value Index of the binding (sintering) qualities of coal which has been subjected to a prescribed heat treatment.

agglutination The coalescing of small suspended particles to form larger masses which are usually precipitated.

aggregate A mass consisting of rock or mineral fragments.

agitator Tank, usually cylindrical, near bottom of which is a mixing device such as a propellor or airlift pump. Finely ground mineral slurries (the aqueous component perhaps being a leaching solution) are exposed to appropriate chemicals for purpose of extraction of gold, uranium, or other valuable constituents. Types are *Pachuca* or *Brown, Dorr* and *Devereux*.

agmatite A **migmatite** with a brecciated appearance.

agonic line The line joining all places with no magnetic declination, i.e. those where true north and magnetic north coincide on a compass.

air classifier Appliance in which vertical, horizontal, or cyclonic currents of air sort falling ground particles into equal-settling fractions or separate relatively coarse-falling material from finer dust which is carried out. Also *air elutriator*.

air door In mine ventilating system, door which admits air or varies its direction.

airdox US system for breaking coal in fiery mine by use of injected high-pressure air.

air drilling, gas drilling The use of air instead of mud as the cooling and debris removal medium. Faster and easier than mud drilling, it cannot prevent water ingress and emergency mud equipment will then be necessary.

air dry Said of minerals in which moisture content is in equilibrium with that of atmosphere.

aired up Said of an oil plunger pump which no longer sucks because gas or air has filled the suction chamber.

air elutriator See **air classifier**.

air-float table Shaking table in which concentration of heavy fraction in sand-sized feed is promoted by air blown up through the deck, which is porous. Used in desert work. Also *air table*.

air hoist Air winch or other mechanical hoist actuated by compressed air.

air jig In waterless countries, use of pulses of air to stratify crushed ore into heavy and light layers.

airlance Length of piping used to work compressed air into settled sand or to free choked sections of process plant, restoring aqueous flow.

air leg Telescopic cylindrical prop expanded by compressed air, used to support rock drill.

airlift pump An air-operated displacement pump for elevating or circulating pulp in cyanide plants.

air shooting (1) Charging of shot-hole so as to leave pockets of air, thus reducing shatter-effect of blast. (2) In seismic prospecting, explosion in air, above rock formation under examination, as method of propagating seismic wave.

air-swept mill In dry grinding of rock in ball mill, use of modulated current of air to remove sufficiently pulverized material from the charge in the mill.

air table See **air-float table**.

airway Underground passage used mainly for ventilation.

åkermanite The calcium-magnesium end-member, $Ca_2MgSi_2O_7$, of the melilite group of minerals.

alabaster A massive form of gypsum, often pleasingly blotched and stained. Because of its softness it is easily carved and polished, and is widely used for ornamental purposes. Chemically it is a $CaSO_4.2H_2O$. *Oriental alabaster* (also *Algerian onyx, onyx marble*) is a beauti-

5

fully banded form of stalagmitic *calcite*.
alaskite Leucocratic variety of alkali feldspar granite.
albertite A pitch-black solid bitumen of the asphaltite group.
Albian A stage of the Cretaceous System, comprising the rocks between the Aptian stage below and the Cenomanian stage above. See **Mesozoic**.
albite The end-member of the plagioclase group of minerals. Ideally a silicate of sodium and aluminium; but commonly contains small quantities of potassium and calcium in addition, and crystallizes in the triclinic system.
albitization In igneous rocks, the process by which a soda-lime feldspar (plagioclase) is replaced by albite (sodium feldspar).
alexandrite A variety of chrysoberyl, the colour varying, with the conditions of lighting, between emerald green and red.
algae A group of marine plants that includes seaweeds. The group is the only one to have widespread development in the Precambrian, some algae occurring in rocks at least 2700 Ma old.
Algerian onyx Another name for *Oriental alabaster*. See **alabaster**.
alignment Process of orientating e.g. spin axes of atoms, during magnetization and similar operations.
aliquot A small sample of material assayed to determine the properties of the whole, e.g. in process control, the representative fraction whose quantitative analysis gives information on the assay grade. Term often applied to radioactive material. Also *aliquot part*.
alkali (1) A hydroxide which dissolves in water to form an *alkaline* or *basic solution* which has pH>7 and contains hydroxyl ions, OH^-. (2) A prefix given to rocks that contain either feldspathoids or alkali amphiboles or pyroxenes, or normative feldspathoids or acmite. (3) In mineralogy, a prefix for silicate minerals containing alkali metals but little calcium, e.g. the alkali feldspars.
alkali basalt A variety of basalt with normative nepheline.
alkali feldspar Members of the feldspar group of minerals comprising orthoclase, microcline, sanidine, perthite, anorthoclase and the plagioclase end-member albite (Anorthite 0%–5%).
alkali granite Granite containing alkali amphibole or pyroxene. It should not be used as a synonym for alkali feldspar granite.
alkali metals The elements lithium, sodium, potassium, rubidium, caesium and francium, all metals in the first group of the periodic system. In most compounds they occur as univalent ions.
alkalimetry The determination of alkali by titration with a standard solution of acid as in volumetric analysis.
alkaline earth metals The elements calcium, strontium, barium and radium, all divalent metals in the second group of the periodic system.
alkalinity The extent to which a solution is alkaline. See **pH value, acidity**.
allanite A cerium-bearing epidote occurring as an accessory mineral in igneous and other rocks.
alleghanyite A hydrated manganese silicate, crystallizing in the monoclinic system.
allemontite A mineral consisting of a solid solution of antimony and arsenic.
Allen cone Conical tank used for continuous sedimentation of liquids at constant level, the solids being removed from the base of the cone and the clear liquid drawn off from the top.
Allen equation One applied to sedimentation of finely ground particles intermediate between streamline and turbulent in settling mode.

$$\rho = Kr^n p \mu^{2 - n_v n},$$

where ρ is fluid resistance; K, a constant for shape and velocity of fall; r, radius of an equivalent sphere; n, a coefficient of velocity; v; p, density of fluid; and μ the kinematic viscosity b/p, b being absolute viscosity.
alligator See **jaw breaker**.
allochthonous Rock which is exotic to its environment, e.g. a block of limestone that has slid down a submarine slope into a muddy environment or a tectonically-moved block of rock.
allophane Hydrous aluminium silicate, apparently amorphous.
allotriomorphic *Anhedral*. A textural term, used for igneous rocks, describing crystals which show a form related to surrounding previously crystallized minerals rather than to their own rational faces. Cf. **idiomorphic crystals**.
allotropy The existence of an element in two or more solid, liquid or gaseous forms, in one phase of matter, called *allotropes*.
alloy A mixture of metals, or of a metal with a non-metal in which the metal is the major component. Alloys may be compounds, eutectic mixtures or solid solutions.
alluvial mining The exploitation of alluvial or placer deposits. Minerals thus extracted include tin, gold, gemstones, rare

earths, platinum. Term embraces beach deposits, eluvials, riverine and offshore workings.

alluvial values Values which are shown by panning or assay to be recoverable from an alluvial deposit.

alluvium Sand, silt and mud deposited by a river or floods; geologically recent in age.

almandine Iron-aluminium garnet, occurring in mica-schists and other metamorphic rocks. Commonly forms well-developed crystals, often with 12 or 24 faces.

almandine spinel See **ruby spinel**.

alnöite A dark lamprophyre with phenocrysts of mica, olivine and augite in a groundmass of melilite, augite and other minerals.

alpha counter Tube for counting α-particles, with pulse selector to reject those arising from β- and γ- rays.

alpha decay Radioactive disintegration resulting in emission of α-particle. Also *alpha disintegration*.

alpha decay energy The sum of the kinetic energies of the α-particle emitted and the recoil of the product atom in a radioactive decay.

alpha emitter Natural or artificial radioactive isotope which disintegrates through emission of α-rays.

alpha particle Nucleus of helium atom of mass number four, consisting of two neutrons and two protons and so doubly positively charged. Emitted from natural or radioactive isotopes. Often written as α-particle.

Alpine orogeny The fold movements during the Tertiary period which led to the development of the Alps and associated mountain chains.

alstonite A double carbonate of calcium and barium.

altimeter An aneroid barometer used for measuring altitude by the decrease in atmospheric pressure with height. The dial of the instrument is graduated to read the altitude directly in feet or metres, the zero being set to ground or aerodrome level.

altitudes by barometer See **Babinet's formula for altitude**.

alum Hydrated aluminium potassium sulphate and related minerals.

alumina Aluminium oxide. See **corundum**.

aluminium See panel on p. 8.

alumino-silicates Compounds of alumina, silica and bases, with water of hydration in some cases. They include clays, mica, zeolites, constituents of glass and porcelain.

aluminothermic process The reduction of metallic oxides by the use of finely divided aluminium powder. An intimate mixture of the oxide to be reduced and aluminium powder is placed in a refractory crucible; a mixture of aluminium powder and sodium peroxide is placed over this and the mass fired by means of a fuse or magnesium ribbon. The aluminium is almost instantaneously oxidized, at the same time reducing the metallic oxide to metal. This process, also known as the *thermite process*, is used esp. for the oxides of metals which are reduced with difficulty (e.g. titanium, molybdenum). On ignition, the mass may reach a temperature of 3500°C. Magnesium incendiary bombs have thermite as the igniting agent.

alumstone See **alunite**.

alunite Hydrated sulphate of aluminium and potassium, resulting from the alteration of acid igneous rocks by solfataric action; used in the manufacture of alum. Also *alumstone*.

alunogen Hydrated aluminium sulphate, occurring as a white incrustation or efflorescence formed in two different ways: either by volcanic action, or by the decomposition of pyrite in carbonaceous or alum shales.

alvikite A medium- to fine-grained calcite carbonatite.

amalgam The alloy of a metal with mercury. In the treatment of gold ores, the pasty amalgam of gold and mercury obtained from the plates in a mill and containing about ⅓ gold by weight.

amalgamating table Flat sheet of metal to which mercury has adhered to form a thin soft film, used to catch metallic gold as mineral sands are washed gently over it. Also *amalgamated plate*.

amalgamation pan Circular cast-iron pan in which finely-crushed gold-bearing ore or concentrate is ground with mercury, the valuable metal thus being amalgamated before separate retrieval.

amalgam barrel Small ball mill used to regrind gold-bearing concentrates, and then give them prolonged rubbing contact with mercury.

amazonstone, amazonite A green variety of microcline, sometimes cut and polished as a gemstone, and falsely called 'Amazon Jade'.

amber A fossil resin containing succinic acid in addition to resin acid and volatile oils. See **succinite**.

amber mica See **phlogopite**.

amberoid See **ambroid**.

amblygonite Fluorophosphate of aluminium and lithium, a rare white or greenish mineral, crystallizing in the triclinic system and found in pegmatites.

aluminium

US *aluminum*. Symbol Al, atomic number 13, r.a.m. (atomic weight) 26.98, valence 3, one stable isotope, ^{27}Al, atomic radius (6-fold co-ordination) 0.57 Å.

The familiar silver-white metal of everyday use is the commonest metallic element in the Earth's crust (8.8 % by mass) but does not occur as native Al, but always in silicate or other minerals. It is the third commonest element in the crust after oxygen and silicon. In seawater it is only present to the extent of 0.01 ppm.

The minerals in which it occurs are in two main groups: those that formed at relatively high temperatures and those formed at much lower temperatures, often by weathering processes.

Al is found in the rock-forming aluminosilicates, esp. feldspars; in oxides and hydroxides; zeolites; hydrated silicates, esp. micas and clays; and in rarer minerals, e.g. **cryolite**.

The **feldspars** are the most important minerals in the upper **lithosphere**, occurring in acid, intermediate and basic igneous rocks (e.g. granites and basalts), metamorphic rocks (gneisses and schists) and to a lesser extent in other rocks.

Aluminium oxide (*corundum*) is a constituent of metamorphic rocks and some aluminosilicates, e.g. **kyanite**, also occur in these rocks. The mica group of minerals (muscovite, biotite etc.) are hydrated aluminosilicates of alkali metals, magnesium, iron etc. that occur in igneous, metamorphic and sedimentary rocks.

When the primary minerals of aluminium weather, cations are leached out and various clay minerals may be found, e.g **kaolinite**. In tropical weathering hydroxides may be found, e.g. **diaspore**, a principal constituent of many bauxite deposits.

The *zeolites* are hydrated aluminosilicates of the alkalis and alkali earths that occur typically in **amygdales** in basic volcanic rocks and as other late-stage hydrothermal products.

Bauxite is the principal ore of aluminium, which is prepared on a vast scale and has innumerable uses.

ambroid A synthetic amber formed by heating and compressing pieces of natural amber too small to be of value in themselves. Also *amberoid, pressed amber*.

American filter See **disk filter**.

American Petroleum Institute Abbrev. *API*. An association of US companies which represents the interests of the oil industry and sets standards for products such as crude oils using arbitrary units. See **API scale**.

americium Transuranic element. Symbol Am, at.no. 95, half-life 458 years. Of great value as a long life α-particle emitter, free of criticality hazards and γ-radiation, e.g. in laboratory neutron sources.

amesite A variety of septechlorite rich in magnesium and aluminium.

amethyst A purple form of quartz, used as a semi-precious gemstone.

amianthus A fine silky asbestos.

ammonite See **ammonoids**.

ammonoids See panel on p. 9.

amorphous Non-crystalline.

amorphous metal A material with good conductivity, electrical and thermal, and with other metallic properties but with atomic arrangements that are not periodically ordered as in crystalline metal solids, e.g. metallic glass.

amosite A monoclinic amphibole form of asbestos, the name embodying the initials of the company exploiting this material in the Transvaal, viz. the 'Asbestos Mines of South Africa'.

amphiboles An important group of dark-coloured rock-forming silicates, including hornblende, the commonest.

amphibolite A crystalline, coarse-grained rock, containing amphibole as an essential constituent, together with feldspar and

ammonoids

The ammonoids are completely extinct cephalopods that belong to the subclass *Ammonoidea*. They were entirely marine and ranged from the Devonian to the Cretaceous. Ammonoids include the important fossil groups *goniatites* and *ammonites*. The related, more primitive subclass *Nautiloidea* ranged from the Cambrian onwards and includes the *Nautilus* that is found in the Pacific and is used as an ornamental shell.

The ammonites usually have shells coiled in one plane consisting of successive body chambers increasing in size towards the aperture of the shell. The chambers are separated by thin walls or *septae*. Where each septum meets the inner wall of the shell it forms a complicated *suture line* that is of importance in identifying the ammonite and enables them to be used as zonal fossils for the Mesozoic rocks. The goniatites have shells with sutures of angular appearance, and ranged from the Devonian to the Permian. They are used as zonal fossils in the Devonian and Carboniferous.

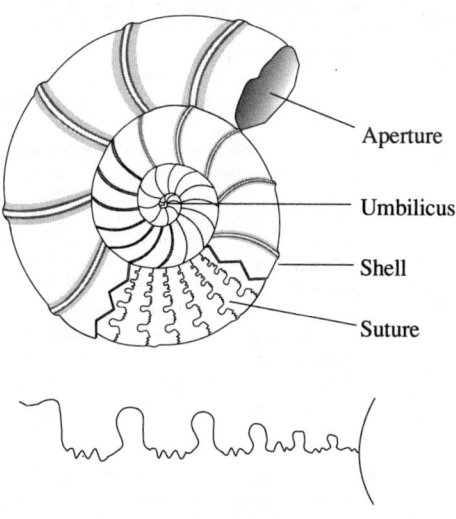

Ammonite from the Trias
Part of the shell removed to show sutures. Below, suture line enlarged. The external margin is at the left.

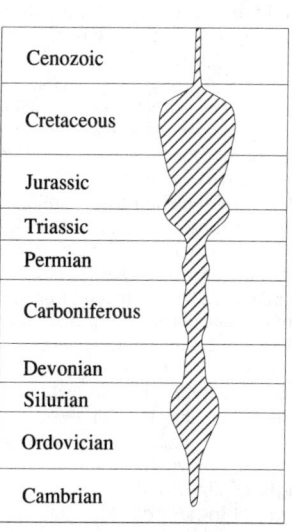

Relative abundance
Cephalopods including ammonites, goniatites, nautiloids.

frequently garnet, e.g. hornblende-schist, formed by regional metamorphism of basic igneous rocks, but not foliated.

amphipathic Descriptive of unsymmetrical molecular group, one end being hydrophilic and the other hydrophobic (wetting and non-wetting).

amphoteric Having both acidic and basic properties, e.g. aluminium oxide, zinc oxide, which form salts with acids and with alkalis.

amygdale, amygdule An almond-shaped infilling (by secondary minerals such as agate, zeolites, calcite etc.) of elongated steam cavities in igneous rocks.

analcime, analcite Hydrated aluminium silicate, a member of the zeolite group, occurring in some igneous and sedimentary rocks, but particularly in cavities in lavas.

analcimite An alkaline volcanic rock composed essentially of analcime, pyroxene and other minerals, with little or no olivine.

analyser The second polarizer (originally a Nicol prism) in a polariscope or petrological microscope. It receives the light which has passed through the object under study; in a petrographical microscope this is a thin section of rock or mineral.

anaphoresis The migration of suspended particles towards the anode under the influence of an electric field.

anatase One of the three naturally occurring forms of crystalline titanium dioxide, of tabular or bipyramidal habit. See **octahedrite**.

anchor string Length of **casing** run into the top of wells and often cemented to prevent a **blowout** outside the casing. It provides fixings for the well-head equipment. Also *surface pipe, top casing*.

andalusite One of several crystalline forms of aluminium silicate; a characteristic product of the contact metamorphism of argillaceous rocks. Orthorhombic. See **chiastolite**.

andesine A member of the plagioclase group of minerals, with a small excess of sodium over calcium: typical of the intermediate igneous rocks.

andesite A fine-grained igneous rock (usually a lava), of intermediate composition, having plagioclase as the dominant feldspar.

andradite Common calcium-iron garnet. Mainly dark brown to yellow or green. See **demantoid, melanite, topazolite**.

angle drilling Technique for drilling at an angle to an existing bore, achieved by special **downhole** equipment, either to straighten a bore, gather oil from a wide area to a production platform or to reach otherwise inaccessible formations, e.g. under a city. Also *deviated drilling*. See **slant rig**.

angle of deviation The angle which the incident ray makes with the emergent ray when light passes through a prism or any other optical device.

angle of dip See **dip**.

angle of incidence The angle which a ray makes with the normal to a surface on which it is incident.

angle of minimum deviation The minimum value of the angle of deviation for a ray of light passing through a prism. By measuring this angle (θ) and also the angle of the prism (α), the refractive index of the prism may be calculated by means of the expression:

$$n = \frac{\sin \frac{1}{2}(\alpha + \theta)}{\sin \frac{1}{2}\alpha}.$$

angle of nip The maximum included angle between two approaching faces in a crushing appliance, such as a set of rolls, at which a piece of rock can be seized and entrained.

angle of reflection The angle which a ray, reflected from a surface, makes with the normal to the surface. The angle of reflection is equal to the *angle of incidence*.

angle of refraction The angle which is made by a ray refracted at a surface separating two media with the normal to the surface. See **refractive index, Snell's law**.

angle of slide Slope at which heaped rock commences to break away.

anglesite Orthorhombic sulphate of lead, a common lead ore; named after the original locality, Anglesey.

Anglian A glacial stage in the late Pleistocene. See **Quaternary**.

Ångström Unit of wavelength for electromagnetic radiation covering visible light and X-rays. Equal to 10^{-10} m. The unit is also used for interatomic spacings. Symbol Å. Superseded by nanometre ($=10^{-9}$ m) but still used widely in crystallography. Named after the Swedish physicist A. J. Ångström (1814-74).

anhedral A term used in petrography to denote a mineral which does not show any crystal faces, i.e. one which is irregular in shape.

anhydrides Substances, including organic compounds and inorganic oxides, which either combine with water to form acids, or which may be obtained from the latter by the elimination of water.

anhydrite Naturally occurring anhydrous calcium sulphate which readily forms gypsum and from which anhydrite plaster is made by grinding to powder with a suitable accelerator.

anhydrous A term applied to oxides, salts etc. to emphasize that they do not contain water of crystallization or water of combination.

anion Negative ion, i.e. atom or molecule which has gained one or more electrons in an electrolyte, and is therefore attracted to an anode, the positive electrode. Anions include all non-metallic ions, acid radicals and the hydroxyl ion. In a primary cell, the deposition of anions on an electrode makes it the negative pole. Anions also exist in gaseous discharge.

anisodesmic structure A structure which gives to a crystal a marked difference between its bond strengths in the intersecting axial planes.

anisomeric Not isomeric.

anisotropic Having some physical property that varies in different directions. In minerals it applies most commonly to the optical properties of crystals: all except those crystallizing in the cubic system are anisotropic.

anisotropic conductivity Body which has a different conductivity for different directions of current flow, electric or thermal.

anisotropic dielectric A material in which electric effects depend on the direction of the applied field, as in many crystals.

anisotropy The property of being anisotropic, applied mainly to conditions shown by rock masses.

ankerite A carbonate of calcium, magnesium and iron.

annabergite Hydrated nickel arsenate, apple-green monoclinic crystals, rare, usually massive. Associated with other ores of nickel. Also *nickel bloom*.

annealing Process of maintaining a material at a known elevated temperature to reduce dislocations, vacancies and other metastable conditions, e.g. steel or glass. In ferrous alloys the metal is held at a temperature above the upper critical temperature for a variable time and then cooled at a predetermined rate, depending on the alloy and the particular properties of hardness, machinability etc. that are needed. The term is usually qualified, e.g. *quench annealing, isothermal annealing, graphitizing*.

Annelida A phylum of worm-like invertebrates which have few skeletal structures and so are rarely found as fossils. Most are known as fossils only from tracks and burrows.

annite The ferrous iron end-member of the biotite series of micas.

annular space The space between the **casing** and the producing or drilling bore.

anomalous dispersion The type of dispersion given by a medium having a strong absorption band, the value of the refractive index being abnormally high on the longer wave side of the band, and abnormally low on the other side. In the spectrum produced by a prism made of such a substance the colours are, therefore, not in their normal order.

anomalous viscosity A term used to describe liquids which show a decrease in viscosity as their rate of flow or velocity gradient increases. Such liquids are also known as *non-Newtonian liquids*.

anorthic system See **triclinic system**.

anorthite The calcium end-member of the plagioclase group of feldspars; silicate of calcium and aluminium, occurring in some basic igneous and other rocks.

anorthoclase A triclinic sodium-rich high-temperature sodium-potassium feldspar; occurs typically in volcanic rocks, and is also known from a syenite, larvikite, from S. Norway, which is widely used for facing buildings.

anorthosite A coarse-grained plutonic igneous rock, consisting almost entirely of plagioclase, near labradorite in composition.

antecedent drainage A river system that has maintained its original course despite subsequent folding or uplift.

anthophyllite An orthorhombic amphibole, usually massive, and normally occurring in metamorphic rocks; magnesium iron silicate, of low aluminium content.

anthracite The highest metamorphic rank of coal. See **rank of coal**.

anthraxolite A member of the asphaltite group.

anthraxylon One of the constituents of coal, derived from the lignin of the plants forming the seam.

anthropoid Resembling Man; pertaining to, or having the characteristics of the Anthropoidea.

Anthropoidea Suborder of the Primates which includes monkeys, apes and humans.

Antian A temperate stage in the Pleistocene. See **Quaternary**.

anticline A type of fold, comparable with an arch, the strata dipping outwards, away from the fold-axis. See **folding**.

antigorite One of three minerals which are collectively known as *serpentine*, hydrated magnesium silicate. It is abundant in the rock type, serpentinite.

antimonial lead ore See **bournonite**.

antimonite See **stibnite**.

antimony Metallic element. Symbol Sb, at.no. 51, r.a.m. 121.75, mp 630°C, rel.d. 6.6. Used in alloys for cable covers, batteries etc.; also as a donor impurity in germanium. Has several radioactive isotopes which emit very penetrating γ-radiation. These are used in laboratory neutron sources. Abundance in Earth's crust 0.2 ppm; it forms stibnite and other sulphide minerals.

antimony glance An obsolete name for **stibnite**.

antiperthite An intergrowth of plagioclase and potassium feldspars with plagioclase as the dominant phase. See **perthite**.

antipodes On a sphere, e.g. the Earth, points on the surface at either extremity of a diameter.

Antonoff's rule The interfacial tension between two liquid phases in equilibrium is equal to the difference of the surface tensions of the two phases.

apatite Phosphate of calcium, also containing fluoride, chloride, hydroxyl or carbonate ions, according to the variety. It is a major constituent of sedimentary phosphate rocks, of the bones and teeth of vertebrate animals (including Man), and it is usually present as an accessory mineral in igneous rocks.

apex The upper edge of a vein reef or lode.

apex law The law entitling the discoverer of an outcrop or exposure of ore to exploit it in depth beyond its lateral boundaries. Also *extra-lateral rights*.

aphanitic The texture of an igneous rock in which the crystals are not distinguishable by the unaided eye.

Aphebian The lower part of the Proterozoic. See **Precambrian**.

API scale *American Petroleum Institute* scale. Scale of relative density, similar to Baumé scale. Degrees API = $(141.5/s) - 131\,s$, where s is the rel.d. of the oil against water at 15°C.

aplite A fine textured, light coloured, igneous rock composed of quartz and feldspar.

apophyllite A secondary mineral occurring with zeolites in amygdales in basalts and other igneous rocks. Composition: hydrated fluorosilicate of potassium and calcium.

apophysis A vein-like offshoot from an igneous intrusion.

Appalachian orogeny A fold period in eastern North America that extended from the Devonian to the Permian. It was caused by *subduction* which led to the closure of the Atlantic Ocean. See **subduction zone**.

apparent expansion, coefficient of The coefficient of expansion when the expansion of e.g. a dilatometer is neglected. See **coefficient of expansion**.

appinite A group name for rocks of variable composition and texture containing conspicuous hornblende in a base of plagioclase with or without orthoclase or quartz.

applied geology Geology studied in relation to human activity.

appraisal well One of a series of wells drilled near a discovery well to determine the size and nature of an oil or gas field. The results will show whether exploitation of the field is economically worthwhile. Also *step-out well*.

apron feeder Short endless conveyor belt, sturdily built of articulated plates, used to draw ore at regulated rate from bottom of stockpile or ore bin.

aprotic Term normally restricted to solvents such as acetonitrile, which have high *relative permittivities* and hence aid the separation of electric charges but do not provide protons.

Aptian A stage of the Cretaceous System lying between the Barremian below and the Albian above. See **Mesozoic**.

aquamarine A variety of beryl, of attractive blue-green colour, used as a gemstone.

aquifer Rock formation containing water in recoverable quantities.

Aquitanian The lowest stage of the Neogene (Miocene). See **Tertiary**.

aragonite The relatively unstable, orthorhombic form of crystalline calcium carbonate, deposited from warm water, but prone to inversion into calcite; also stable at high pressures. See **flos ferri**.

Archaean The oldest rocks of the Precambrian. Usually taken to be older than ca 2500 Ma. See **Precambrian**.

Archimedes' principle When a body is wholly or partly immersed in a fluid it experiences an upthrust equal to the weight of fluid it displaces; the upthrust acts vertically through the centre of gravity of the displaced fluid.

archipelago A sea area thickly interspersed with islands; originally applied to the part of the Mediterranean which separates Greece from Asia.

arenaceous rocks Sedimentary rocks in which the principal constituents are sand grains, including the various sorts of sands and sandstones.

Arenig A series of rocks in the Ordovician System, taking their name from Arenig mountain in N. Wales, where they were originally described by Adam Sedgwick. See **Palaeozoic**.

arenite The general term for any sedimentary rock with sand-sized grains.

arête A sharp-edged precipitous ridge of bare rock, often in mountainous country between two cirques.

arfvedsonite A monoclinic iron-rich alkali-amphibole.

argentiferous Containing silver.

argentite An important ore of silver, having the composition Ag_2S (silver sulphide); crystallizes in the cubic system. Cf. **acanthite**. Also *silver glance*.

argillaceous rocks Sediments of silt or clay-particle size. Common clay minerals are kaolinite and montmorillonite.

argillite A slightly metamorphosed siltstone or mudstone which lacks fissility.

argon Element of atomic number 18, r.a.m. 39.948. A rare gas, it constitutes about 1% by volume of the atmosphere. The isotope argon–40, ^{40}Ar, is formed by the radioactive decay of the isotope potassium–40, ^{40}K. The ratio $^{40}K : ^{40}Ar$ in a rock or mineral is used for the determination of the age in years (abbrev. $K - Ar$ *method*).

argyrodite A double sulphide of germanium and silver, the mineral in which the element germanium was first discovered.

arkose A feldspar-rich, coarse-grained sandstone, derived from the erosion of granites and gneisses.

Armorican orogeny An old name for the Hercynian orogeny.

aromatic compounds Those carbon

arrested crushing compounds that contain at least one benzene ring in the molecule. The ring is hexagonal with a carbon atom at each corner, and is planar with the carbon to carbon bonds intermediate in length between single and double bonds. Aromatic compounds do not undergo addition reactions as other unsaturated compounds do, but show substitution reactions in which hydrogens on the ring are replaced with other molecular species. Heterocyclic compounds such as pyridine also show aromatic properties. Lighter members of the aromatic series have characteristic 'aromatic' odours.

arrested crushing Crushing so conducted that the rock falling through the machine is free to drop clear of the zone of comminution when broken smaller than the exit orifice or set.

Arrhenius theory of dissociation The description of aqueous solutions in terms of acids, which dissociate to give hydrogen ions and bases, which dissociate to give hydroxyl ions. The product of the reaction of an acid and a base is a salt and water. The dissociation of these species gives their solutions the property of conducting electricity.

arsenic Element. Symbol As, at.no. 33, r.a.m. 74.92; valence 3, 5. This element only occurs to the extent of 1.8 ppm in the Earth's crust, but it is very widely distributed being found esp. in sulphide ore deposits. It only occurs to a very minor extent as the native element but is present in a large number of minerals. Arsenic and its compounds are highly toxic. Its uses are mainly in alloys and semiconductors.

arsenical pyrites See **arsenopyrite**.

arsenolite Arsenic oxide, a decomposition product of arsenical ores; occurring commonly as a white incrustation, rarely as octahedral crystals.

arsenopyrite Sulphide of iron and arsenic; the chief ore of arsenic. Also *arsenical pyrites, mispickel*.

artesian Of ground water confined under hydrostatic pressure.

artesian well A well tapping confined ground water that rises in the well under hydrostatic pressure above the level of the water table.

Arthropoda The phylum of marine, freshwater and aerial invertebrates having jointed appendages and a well-developed head. It contains the largest number of known species, and includes **trilobites**, euryterids, **ostracods**, crustaceans and insects. The phylum ranges from the Lower Cambrian to the present day.

artificial disintegration The transmutation of non-radioactive substances brought about by the bombardment of the nuclei of their atoms by high-velocity particles, such as α-particles, protons or neutrons.

artificial radioactivity Radiation from isotopes after high energy bombardment in an accelerator by α-particles, protons and other light nuclei, or by neutrons in a nuclear reactor. Discovered by I. Curie-Joliot and F. Joliot in 1933.

Artinskian A stratigraphical stage in the Lower Permian rocks of Russia and eastern Europe.

asbestos The fibrous form of minerals which are resistant to heat and mineral attack and may be used for making fireproof fabrics, brake linings, electrical and heating insulation etc. Toxic as inhaled dust. Minerals used for asbestos are principally chrysotile (serpentine), and the amphiboles actinolite, amosite, anthophyllite and crocidolite (blue asbestos).

asbolane, asbolite A form of **wad**, i.e. soft earthy manganese dioxide, containing cobalt.

ash (1) The non-volatile inorganic residue remaining after the ignition of an organic material. (2) See **volcanic ash**.

ash curve Graph which shows result of sink-and-float laboratory test in form of relationship between specific gravity of crushed small particles and the ash content at that gravity.

ash flow tuff A tuff deposited by an ash flow or gaseous cloud (*nuée ardente*).

Ashgill The youngest series of the Ordovician. See **Palaeozoic**.

ash tuff A pyroclastic rock in which the average pyroclast size is between 1/16 mm and 2 mm.

asparagus stone Apatite of a yellowish-green colour, thus resembling asparagus.

asphalt (1) A bituminous deposit formed in oil-bearing strata by the removal of the volatiles, occurring in Trinidad, Athabasca (Canada) and elsewhere. (2) A residue in petroleum distillation. (3) A mixture of asphaltic bitumen and rock chippings used in road making. Asphalt is also used extensively for paving, roofing, sealing, paintmaking etc.

asphaltite A group name for the organic compounds albertite, anthraxolite, grahamite, impsonite, libollite, nigrite and uintaite.

assay The quantitative analysis of a substance to determine the proportion of some valuable or potent constituent, e.g. the active compound in a pharmaceutical, or metals in an ore. See **dry assay, wet assay**.

assay balance A balance specially made for weighing the small amounts of matter met with in assaying.

assayer A person who carries out the process of assay. See **dry assay, wet assay.**

assimilation The incorporation of extraneous material in *igneous magma*.

associated gas A gas mixture found associated with crude oil in an underground geological formation. Volume and composition vary widely according to location.

asterism The star effect, with four-, six-, or twelve-rayed stars, seen by reflected light in gemstones, e.g. ruby and sapphire cut *en cabochon*, and produced by rod-like inclusions. See **gems and gemstones.**

asthenosphere The shell of the Earth below the lithosphere. It is identifiable by low seismic wave velocities and high seismic attenuation. It is soft, probably partly molten and a zone of magma generation. It is equivalent to the *upper mantle*. See **Earth.**

astrobleme A circular impact structure on the Earth's crust, caused by a meteorite.

astrogeology The study by geological, geophysical, geochemical and related techniques of the solid bodies of the solar system, excluding the Earth, but including meteorites and tektites.

astrophyllite A complex hydrated silicate of potassium, iron, manganese, titanium and zirconium; occurs in brown laminae in alkaline igneous rocks.

asymmetric atom An atom bonded to three or more other atoms in such a way the arrangement cannot be superimposed on its mirror image. In particular, a carbon atom attached to four different groups. Most *chiral* molecules can be described in terms of specific asymmetric atoms, e.g. the alpha-carbon atoms in amino acids. See **chirality.**

asymmetric system See **triclinic system.**

atacamite A hydrated chloride of copper, widely distributed in South America, Australia, India etc. in the oxidation zone of copper deposits; occurring also at St. Just, Cornwall, UK.

atm Abbrev. for **standard atmosphere.** See **atmospheric pressure.**

atmospheric pressure The pressure exerted by the atmosphere at the surface of the Earth is due to the weight of the air. Its standard value is 1.0132×10^5 N/m^2, or 14.7 lbf/in^2. Variations in the atmospheric pressure are measured by means of the barometer. See **standard atmosphere.**

atoll A coral reef usually forming a circular, elliptical or irregular chain of islets around a shallow lagoon, and surrounded by deep water of the open tropical sea.

atom The smallest particle of an element which can take part in a chemical reaction. See **atomic structure, Dalton's atomic theory.**

atomic bond See **covalent bond.**

atomic disintegration Natural decay of radioactive atoms, as a result of radiation, into chemically different atomic products.

atomic energy Strictly the energy (chemical) obtained from changing the combination of atoms originally in fuels. Now normally applied to energy obtained from breakdown of fissile atoms in nuclear reactors.

atomic heat The product of specific heat capacity and r.a.m. in grams; approx. the same for most solid elements at high temperatures.

atomicity The number of atoms contained in a molecule of an element.

atomic mass unit Abbrev. *u*. Exactly one twelfth the mass of a neutral atom of the most abundant isotope of carbon, ^{12}C. $u = 1.66 \times 10^{-27}$ kg. Before 1960 *u* was defined in terms of the mass of the ^{16}O isotope and *u* was 1.659×10^{-27} kg. See **atomic weight.**

atomic number The order of an element in the periodic (Mendeleev) chemical classification, and identified with the number of unit positive charges in the nucleus (independent of the associated neutrons). Equal to the number of external electrons in the neutral state of the atom, and determines its chemistry. Symbol Z.

atomic orbital A wave function defining the energy of an electron in an atom.

atomic radii Half of the internuclear distance between the nuclei of two identical non-bonded atoms at such a separation that they neither attract nor repel one another.

atomic refraction The contribution made by a mole of an element to the molecular refraction of a compound.

atomic scattering The deflection of radiation, usually electrons or X-rays, by the individual atoms in the medium through which it passes. The scattering is by the electronic structure of the atom in contrast to *nuclear scattering* which is by the nucleus.

atomic structure The chemical behaviour of the various elements arises from the differences in the electron configuration of the atoms in their normal electrically neutral state. Each atom consists of a heavy nucleus with a positive charge produced by a number of **protons** equal to its atomic number. There are an equal number of electrons outside the nucleus to balance this charge. The nucleus also contains electrically neutral **neutrons**; protons and neutrons are collectively referred to as *nucleons*. The principle quantum number indicates the shell to which the orbital belongs and varies from 1 (K-shell) closest to the nucleus to 7 (Q-shell), the most remote. In general, the closer an electron is to the nucleus the greater the coulomb attraction and so the greater the binding energy retaining the electron in the atom. Nuclear binding forces tend to give greatest stability when the neutron number and the proton number are approximately equal. Due to electrostatic repulsion between protons, the heavier nuclei are most stable when more than half their nucleons are neutrons; elements with more than 83 protons are unstable and undergo radioactive disintegration. Those with more than 92 protons are not found naturally on Earth, but can be synthesized in high-energy laboratories. These are the trans-uranic elements which have short half-lives. Most elements exist with several stable **isotopes** and the chemical atomic weight gives the average of a normal mixture of these isotopes.

atomic transmutation The change of one type of atom to another as a result of a nuclear reaction. The transmutation can be produced by high-energy radiation or particles and is most easily produced by neutron irradiation. The change in atomic number means the chemical nature of the atom has been changed, e.g. gold can be transmuted into mercury; the converse of the ancient alchemist's goal.

atomic volume Ratio for an element of the rel. at. mass to the density; this shows a remarkable periodicity with respect to atomic number.

atomic weight *Relative atomic mass* (*r.a.m.*). Mass of atoms of an element in *atomic mass units* on the *unified scale* where $u = 1.66 \times 10^{-27}$ kg. For natural elements with more than one isotope, it is the average for the mixture of isotopes.

attapulgite See **palygorskite**.

Atterburg limits In fine-grained sediments and soils, the empirical moisture-content boundaries between the liquid and plastic states (the *liquid limit*) and between the plastic and semi-solid states (the *plastic limit*).

attrition The wearing down of rocks by friction, esp. that between loose fragments or particles under natural processes.

Au Symbol for **gold** (L. *aurum*).

augen-gneiss A coarsely crystalline rock of granitic composition, containing lenticular, eye-shaped masses of feldspar or quartz embedded in a finer matrix. A product of regional metamorphism.

augite A pyroxene, a complex aluminous silicate of calcium, iron and magnesium, crystallizing in the monoclinic system, and occurring in many igneous rocks, particularly those of basic composition; it is an essential constituent of basalt, dolerite and gabbro.

aureole Area surrounding an igneous intrusion affected by metamorphic changes.

auriferous deposit A natural repository of gold, in the general sense, including gold-bearing lodes and sediments such as sands and gravels, or their indurated equivalents, which contain gold in detrital grains or nuggets. See **banket, lode, placers**.

auriferous pyrite Iron sulphide in the form of pyrite, carrying gold, probably in solid solution.

aurous Containing gold (I).

australites See **tektites**.

authigenic Pertaining to minerals which have crystallized in a sediment during or after its deposition.

automotive gas oil A US term for gas oil used mainly as diesel fuel; same as the UK term *DERV*. Abbrev. *AGO*.

autoradiography Examination of radioactive minerals in thin and polished sections by allowing them to rest on photographic film to record their radioactive emissions.

autoscoper Pneumatic rock drill mounted on long cylinder with extendable ram, so as to be held firmly across opening (*stope*) when packed out by compressed air.

autoxidation (1) The slow oxidation of certain substances on exposure to air. (2) Oxidation which is induced by the presence of a second substance, which is itself undergoing oxidation.

autunite Hydrated calcium uranium phosphate, yellow in colour.

aventurine feldspar A variety of plagioclase, near albite-oligoclase in composition, characterized by minute disseminated particles of red iron oxide which cause fire-like flashes of colour. Also *sunstone*.

aventurine quartz A form of quartz spangled, sometimes densely, with minute inclusions of either mica or iron oxide. Used in jewellery. Sometimes falsely called 'Indian Jade'.

Avogadro number, constant The number of atoms in 12 g of the pure isotope ^{12}C; i.e. the reciprocal of the *atomic mass unit* in grams. It is also by definition the number of molecules (or atoms, ions, electrons) in a **mole** of any substance and has the value 6.0225×10^{23} mol^{-1}. Symbol N_A or L.

Avogadro's law Equal volumes of different gases at the same temperature and pressure contain the same number of molecules.

axes See **crystallographic axes**.
axial plane See **folding**.
axial-plane cleavage **Cleavage** parallel to the axial plane of a fold.
axinite A complex borosilicate of calcium and aluminium, with small quantities of iron and manganese, produced by pneumatolysis and occurring as brown wedge-shaped triclinic crystals.
axis of a fold See **folding**.
axonometry Measurement of the axes of crystals.
azimuthal projection The map projection in which a portion of the globe is projected upon a plane tangent to it, usually at the pole, or a place which is to be the centre of a map.
Azoic That part of the Precambrian without life.
azurite A deep-blue hydrated basic carbonate of copper, occurring either as monoclinic crystals or as kidney-like masses built of closely packed radiating fibres.

B

b- Symbol for (1) substitution on the carbon atom of a chain next but one to the functional group; (2) substitution on a carbon atom next but one to an atom common to two condensed aromatic nuclei; (3) substitution on the carbon atom next but one to the hetero-atom in a heterocyclic compound; (4) a stereo-isomer of a sugar.

β– The intermediate refractive index in a biaxial crystal.

β-particle See **beta particle**.

[B] A Fraunhofer line in the red of the solar spectrum, due to absorption by the Earth's atmosphere. [B] is actually a close group of lines having a head at a wavelength 686.7457 nm.

Ba Symbol for **barium**.

Babinet's compensator A device used, in conjunction with a Nicol prism, for the analysis of elliptically polarized light. It consists of two quartz wedges having their edges parallel and their optic axes at right angles to each other.

Babinet's formula for altitude

$$\text{Altitude} = \frac{32\,(500 + t_1 + t_2)(B_1 - B_2)}{B_1 + B_2} \text{ metres},$$

t_1 and t_2 being the respective temperatures in degrees Celsius, and B_1 and B_2 the barometric heights at sea level and at the station whose altitude is required.

Babo's law The vapour pressure of a liquid is lowered when a non-volatile substance is dissolved in it, by an amount proportional to the concentration of the solution.

back-mixing In a chemical reactor, jet engine or other apparatus through which material flows and thereby undergoes a change in some property, back mixing is said to occur when some of the material is turned back so that it mixes with that which has entered the reactor after it.

back-off (1) Raising the drilling bit or *downhole assembly* for a short distance from the bottom of the hole. (2) To unscrew drilling components.

back pressure The hydrostatic pressure exerted by the fluids in the bore of a well which acts against that of the oil and gas in the strata. Control of this pressure by valves and **chokes** maintains an even and productive flow of oil.

back shift In colliery, the afternoon shift.

back stopes **Overhand** (or **overhead**) **stopes**, worked upslope. Antonym *underhand stopes*.

bacterial leaching See **microbiological mining**.

bacterial recovery See **microbiological mining**.

baddeleyite Zirconium dioxide, found in Brazil, where it is exploited as a source of zirconium.

baffle In hydraulic or rake classifier, a plate set across and dipping into the pulp pool; in mechanically agitated flotation cell, one so set as to reduce centrifugal movement in the upper part of the cell.

bailer Length of piping closed at bottom by check valve, used to remove drilling sands and muds from borehole.

bailiff See **overman** (1).

Bajocian A stage in the Middle Jurassic. See **Mesozoic**.

balas ruby A misnomer for the rose-red variety of the mineral spinel. See **false ruby**.

Balbach process Electrolytic separation of gold and silver from a metal, by making it the anode in a bath of silver nitrate, the silver being discharged at the cathode.

ball clay A fine-textured and highly plastic clay — a reworked china clay — so named because it used to be rolled into balls. Also *potters clay*.

ball mill Mill consisting of a horizontal cylindrical vessel in which a given material is ground by rotation with steel or ceramic balls.

ballstone The name applied to masses of fine unstratified limestone, occurring chiefly in the Wenlock Limestone of Shropshire, and representing colonies of corals in position of growth.

banded structure A structure developed in igneous and metamorphic rocks due to the alternation of layers of different texture and/or composition.

band theory of solids For atoms brought together to form a crystalline solid, their outermost electrons are influenced by a *periodic potential function*, so that their possible energies form *bands* of allowed values separated by bands of forbidden values (in contrast to the discrete energy states of an isolated atom). These electrons are not localized or associated with any particular atom in the solid. This band structure is of fundamental importance in explaining the properties of metals, semi-conductors and insulators.

Banka drill Drill widely used in shallow testing of alluvial deposits. Portable, hand-operated, it consists essentially of an assembly of 5-ft pipes, 4 inch in diameter, which are worked into the ground, the material traversed being recovered by a sand pump or **bailer**.

banket The term originally applied by the Dutch settlers to the gold-bearing conglomerates of the Witwatersrand. It is now used for any metamorphosed conglomerate, containing barren quartz pebbles cemented with siliceous matrix bearing gold.

banking The operations involved in removing full trucks, tubs, or wagons and replacing them by empty ones at the top of a shaft.

bannisterite A hydrated silicate of manganese, crystallizing in the monoclinic system, and occurring in manganese deposits in North Wales and New Jersey.

bar Unit of pressure or stress, 1 bar = 10^5 N/m^2 or pascals = 750.07 mm of mercury at 0°C and lat. 45°. The *millibar* (1 mbar = 100 N/m^2 or 10^3 dyn/cm^2) is used for barometric purposes. (NB Std. atmos. pressure = 1.013 25 bar.) The *hectobar* (1 hbar = 10^7 N/m^2, approx. 0.6475 tonf/in^2) is used for some engineering purposes.

Barbados Earth A siliceous accumulation of remains of Radiolaria, formed originally in deep water and later raised above sea level.

barchan An isolated crescentic sand-dune.

barite See **barytes**.

barium A heavy element in the second group of the periodic system, an alkaline earth metal. Symbol Ba, at.no. 56, r.a.m. 137.34, mp 725°C. In most of its compounds it occurs as Ba^{2+} and is present to the extent of 390 ppm in the Earth's crust. Its ionic radius (1.43 Å) enables it to replace potassium (1.33 Å) especially in feldspar, and it forms various independent minerals including barytes, barium sulphate, and witherite, barium carbonate.

barium feldspar A collective term for barium-bearing feldspars, including celsian and hyalophane.

Barker index A method of identification of crystalline substances from measurements of interfacial angles.

barkevikite A member of the amphibole group, resembling basaltic hornblende but having a higher total iron content and a low ferric-ferrous iron ratio. Occurs in alkaline plutonic rocks, as at Barkevik, Norway.

bar mining Alluvial mining of sandbanks, river bars or submerged deposits.

Barnett effect Magnetization of a ferromagnetic material by rapid rotation of the specimen. Used to measure magnetic susceptibility.

barrel US barrel of 42 US gallons (= 35 Imperial gallons or 159.1 litres), frequently employed as a unit of capacity, esp. of output in the oil industry. Abbrev. *bbl*.

barrels per calendar day Abbrev. *bpcd*. A measure of the output of a production unit, in which the total annual output quoted in **barrels** is divided by 365. Because it includes down time (e.g. for maintenance), the value is less than **barrels per stream day**.

barrels per stream day Abbrev. *bpsd*. The output of a production unit quoted in **barrels** for one day of full operation (stream day). Multiplying the bspd by 365 gives the theoretical maximum annual output. The higher the ratio between bpcd (see previous entry) and bpsd for a particular production unit, the more efficient the unit.

Barremian A stage in the Lower Cretaceous. See **Mesozoic**.

barren Without fossils.

barren solution In chemical extraction of metals from their ores, solution left after these have been removed. Cf. **pregnant solution**.

barrier pillar A pillar of solid coal left in position to protect a main road from subsidence, or as a division, or to protect workings from flooding.

barrier reef A coral-reef developed parallel with the shoreline and enclosing a lagoon between itself and the land. It marks a stage between a fringing reef and an atoll.

Barrovian metamorphism, Barrovian zones Regional metamorphism of the type first described in the Scottish Highlands by G. Barrow in 1893. Zones of increasing metamorphism are characterized by the presence of a series of index minerals: chlorite, biotite, garnet, staurolite, kyanite, sillimanite. This type of metamorphism has since been recognized in many other parts of the world.

Bartonian A stage of the Eocene. See **Tertiary**.

barytes Barium sulphate, typically showing tabular orthorhombic crystals. It is a common mineral in association with lead ores, and occurs also as nodules in limestone and locally as a cement of sandstones. Also *barite, heavy spar*. Used to increase the density of **drilling mud** and thus the **back pressure** during drilling.

barytocalcite A double carbonate of calcium and barium, BaCa(CO$_3$)$_2$, crystallizing on the monoclinic system, and occurring typically in lead veins.

basal conglomerate A first stage of sedimentation resting on a plane of erosion. See **conglomerate**.

basal planes The name applied to the faces

basalt representing the terminating *pinacoid* in all the crystal systems exclusive of the cubic system.

basalt A fine-grained igneous rock, dark colour, composed essentially of basic plagioclase feldspar and pyroxene, with or without olivine; minor amounts of quartz or feldspathoids may be present. In the field, the term is generally restricted to lavas, but many minor intrusions of basic composition show identical characters, and therefore cannot be distinguished in the laboratory. The term probably comes from ancient Egypt. The extrusive equivalent of **gabbro**. See **volcanic rocks**.

basalt glass See **tachylite**.

basaltic hornblende A variety of hornblende with a high ferric-ferrous iron ratio and a low hydroxyl content, occurring chiefly in volcanic rocks. Also *oxyhornblende*.

basanite A basaltic rock containing plagioclase, augite, olivine and a feldspathoid (nepheline, leucite, or analcite). A term of great antiquity, probably Egyptian.

base Generally, a substance which tends to donate an electron pair or co-ordinate an electron. In particular, a substance which dissolves in water with the formation of hydroxyl ions and reacts with acids to form salts.

base exchange The reversible replacement of a cation by another in solution. The superficial physical structure of the solid is not affected. The *ion exchange* may take place in colloids, on surfaces, in crystal lattices, notable zeolites, or in inter-layer crystal lattice sites. Applied in **soil mechanics** to chemical methods used to strengthen clays by replacement of H-ions by Na-ions.

base level The lowest level towards which erosion progresses.

basement (1) A complex of igneous and metamorphic rocks covered by sediments. (2) The crust of the Earth extending downwards to the Mohorovičić discontinuity.

base metals Metals with a relatively negative electrode potential (on IUPAC system). Cf. **noble metals**.

Bashkirian The oldest epoch of the Pennsylvanian period.

basicity The number of hydrogen ions of an acid which can be neutralized by a base.

basic lead carbonate Approximate composition $2PbCO_3.Pb(OH)_2$.

basic lead chromate $PbCrO_4.Pb(OH)_2$. Used as a pigment. Produced when lead chromate is boiled with aqueous ammonia or potassium hydroxide. Also *Austrian cinnabar*.

basic lead sulphate $2PbSO_4.PbO$. A fine powder, obtained by roasting galena (PbS).

basic rocks Igneous rocks with a low silica content. The limits are usually placed at 45% silica, below which rocks are described as ultrabasic, and 52% silica, above which they are described as intermediate. Basic igneous rocks include basalt, the commonest type of lava, and gabbro, its plutonic equivalent.

basic solvent A protophilic solvent, hence one which enhances the acidic (i.e. proton donating) properties of the solute.

basin A large depression in which sediments may be deposited. Alternatively, a gently folded structure in which beds dip inwards from the margin towards the centre.

basin-and-range A structural area of fault-block mountains separated by alluvium-filled basins.

bastite A variety of serpentine, essentially hydrated silicate of magnesium, resulting from the alteration of orthorhombic pyroxenes. Also *schillerspar*.

bastnaesite Fluorocarbonate of lanthanum and cerium. An ore mineral of rare earth elements.

batching sphere When miscible oil fractions are being sent down a pipe line, an inflatable hard rubber sphere, fitting the pipe, is used to separate the fractions.

batholith, bathylith A large body of intrusive igneous rock, frequently granite, with no visible floor.

Bathonian A stage in the Middle Jurassic. See **Mesozoic**.

bathotonic Tending to diminish surface tension.

bathyal Refers to the ocean-floor environment between ca 200 and 4000 m. The three zones of increasing depth are **littoral, abyssal** and **bathyal**. There are numerous definitions of the depth ranges of these zones.

Baumé hydrometer scale The continental Baumé hydrometer has the rational scale proposed by Lunge, in which 0° is the point to which it sinks in water and 10° the point to which it sinks in a 10% solution of sodium chloride, both liquids being at 12.5°C.

Baum jig Pneumatically pulsed **jig** used in coal-washing plants to lift and remove a lighter and low-ash fraction from a denser one containing shale and high-ash material (dirt), which is stratified downward by the effect of pulsed water, and separately withdrawn.

bauxite A residual rock composed almost entirely of aluminium hydroxides formed by weathering in tropical regions. The most important ore of aluminium. See **laterite**.

Baventian A cold stage in the Pleistocene.

See **Quaternary**.

Bayer process A process for the purification of bauxite, as the first stage in the production of aluminium. Bauxite is digested with a sodium hydroxide solution which dissolves the alumina and precipitates oxides of iron, silicon, titanium etc. The solution is filtered and the aluminium precipitated as the hydroxide.

BCC Abbrev. for *Body Centred Cubic*. See **unit cell**.

BCF See **bromochlorodifluoromethane**.

Be Symbol for **beryllium**.

beam A collimated, or approximately unidirectional, flow of electromagnetic radiation (radio, light, X-rays), or of particles (atoms, electrons, molecules). The angular beam width is defined by the half-intensity points.

beater mill In rockbreaking, a mill with swinging hammers, disks or heavy plates, which revolve fast and hit a falling stream of ore with breaking force. Also *hammer mill, disintegrating mill, impactor*.

Becke line A narrow line of light seen under the microscope at the junction of two minerals (or a mineral and the mount) in contact in a microscope section. Used in refractive index determinations.

Beck hydrometer Hydrometer for measuring the relative density of liquids less dense than water. Graduated in degrees Beck, where °Beck = 200 (1− rel.d.).

becquerel The SI unit of radioactivity; 1 becquerel is the activity of a quantity of radioactive material in which 1 nucleus decays per second. Abbrev. *Bq*. Replaces the curie. 1 Bq = 2.7×10^{-11} Ci. See **ionizing radiation**.

becquerelite Hydrated oxide of uranium, an alteration product of uraninite.

bed A packed, porous mass of solid reagent, adsorbent or catalyst through which a fluid is passed for the purpose of chemical reaction.

bed A small rock-unit; the smallest formal lithostratigraphic unit.

bedding A term commonly used for **stratification**.

bedding, bedding plane Surface that separates layers of rock. It is caused by changes in mineralogy, grain size, colour etc.

bedrock Barren formation (seat earth, clay, 'farewell rock') underlying the exploitable part of a mining deposit.

beef Fibrous calcite occurring in veins in sedimentary rocks. Rarely, other minerals with the same structure and occurrence.

beekite A variety of chalcedony, commonly occurring as an incrustation on pebbles.

Beer's law The degree of absorption of light varies exponentially with the thickness of the layer of absorbing medium and the molar concentration in the latter.

Beestonian A cold stage in the Pleistocene. See **Quaternary**.

beetle-stones Coprolitic nodules akin to septaria which, when broken open, give a fancied resemblance to a fossil beetle.

beforsite Medium- to fine-grained dolomite carbonatite, mainly consisting of dolomite and occurring principally in dykes.

behind the pipe Refers to a gas or oil reservoir outside the *casing string*.

beidellite A variety of the montmorillonite group (smectites) of clay minerals.

Beilby layer Flow layer produced by polishing a metal or mineral surface, in which the true lattice structure is modified or destroyed by incipient fusion.

belemnites See panel on p. 21.

bell-metal ore See **stannite**.

Belt, Beltian Series A great thickness (perhaps 12 000 m) of younger Precambrian rocks occurring in the Little Belt Mts., Montana, Idaho and British Columbia. Argillaceous strata predominate, accompanied by algal limestones. Correlated with the Grand Canyon Series in Colorado and the Uinta Quartzite Series in the Uinta Mts. Approximately equivalent of Riphean. See **Precambrian**.

bending of strata See **folding**.

beneficiation See **mineral processing**.

Benioff zone A plane beneath the trenches of the Pacific dipping under the continents; the site of earthquake activity.

benitoite A strongly dichroic mineral, varying in tint from sapphire blue to colourless, discovered in San Benito Co., California. Silicate of barium and titanium.

benmoreite A sodic variety of trachyandesite, consisting of anorthoclase and sodic sanidine.

bentonite A valuable clay, similar in its properties to fuller's earth, formed by the decomposition of volcanic glass, under water. Consists largely of montmorillonite. Used as a bond for sand, asbestos etc.; also in the paper, soap and pharmaceutical industries. Thixotropic properties exploited for altering the viscosity of oil **drilling muds**.

benzene C_6H_6, mp 5°C, bp 80°C, rel.d. 0.879; a colourless liquid, soluble in alcohol, ether, acetone, insoluble in water. Produced from coal tar and coke-oven gas; can also be synthesized from open-chain hydrocarbons. Basis for benzene derivatives. A solvent for fats, resins etc.; very flammable. Benzene is the simplest member of the aromatic series of hydrocarbons. Carcinogenic. Its structure was established by Kekulé in 1858.

belemnites

An extinct cephalopod (*Decapoda*) that in life was related to the octopus (*Octopoda*) in appearance. However, the portion commonly found as a fossil is the *guard*, a solid part often the shape and size of a rifle bullet, as shown in the adjacent figure. They ranged from the Carboniferous to the Eocene.

Typical belemnite

benzol scrubber A device for washing gases and absorbing the benzol contained therein by means of a high-boiling mineral oil.

beraunite Hydrated phosphate of iron; red, found in iron ore deposits.

Bergius process See **hydrogenation** of coal.

Bernoulli's law For a non-viscous, incompressible fluid in steady flow, the sum of the pressure, potential and kinetic energies per unit volume is constant at any point. It is a fundamental law of fluid mechanics.

bertrandite A hydrated beryllium silicate, occurring in pegmatites.

beryl A beryllium aluminium silicate, occurring in pegmatites as hexagonal colourless, green, blue, yellow or pink crystals. Important ore of beryllium, also used as a gemstone. See **aquamarine, emerald**.

beryllium Steely uncorrodible white metallic element (Wöhler, 1828). Gk. and L. *beryl*, the old mineral name. Symbol Be, at. no. 4, r.a.m. 9.0122, mp 1281°C, bp 2450°C, rel.d. 1.93. It is a rare element both cosmically and in the Earth, where its abundance is only 2 ppm. It occurs in a number of minerals, including the gemstone **beryl**, of which **emerald** and **aquamarine** are varieties, and **chrysoberyl**.

beryllonite A rare mineral, found as colourless to yellow crystals. Phosphate of beryllium and sodium.

beta decay Radioactive disintegration with the emission of an electron or positron accompanied by an uncharged antineutrino or neutrino. The mass number of the nucleus remains unchanged but the atomic number is increased by one or decreased by one depending on whether an electron or positron is emitted.

betafite A hydrated, niobate, tantalate and titanate of uranium.

beta particle An electron or positron emitted in beta decay from a radioactive isotope. Also β-*particle*.

beta rays Streams of β–particles.

BHA Abbrev. for **bottom-hole assembly**.

Bi Symbol for **bismuth**.

biaxial The term for crystalline minerals having two optical axes. Minerals crystallizing in the orthorhombic, monoclinic and triclinic systems are biaxial. Cf. **uniaxial**.

bieberite Hydrated cobalt sulphate. Found as stalactites and encrustations in old mines containing other cobalt minerals. Also *cobalt vitriol*.

bilitonites See **tektites**.

binary Consisting of two components etc.

bioclastic limestone A carbonate made up of broken fragments of organic material, esp. shells.

bioherm A reef or mound made up of the remains of organisms growing *in situ*.

biomining See **microbiological mining**.

biostratigraphy Stratigraphy based on the use of fossils, esp. for correlation.

biotite Black mica widely distributed in igneous rocks (particularly in granites) as lustrous black crystals, with a perfect cleavage. Also occurs in mica-schists and related metamorphic rocks. In composition, it is a complex silicate, chiefly of aluminium, iron and magnesium, together with potassium and hydroxyl.

Biot laws The rotation produced by optically active media is proportional to the length of path, to the concentration (for solutions) and to the inverse square of the wavelength of the light.

bipyramid A crystal form consisting of two pyramids on a common base, the one being the mirror-image of the other. Each pyramid is built of triangular faces, 3, 4, 6, 8 or 12 in number. See **pyramid**.

birefringence Double refraction of light in a crystal: the splitting of incident light into two rays vibrating at right angles to each other, and causing two images to appear, e.g.

in calcite. The difference between the greatest and least **refractive index** is a measure of the birefringence.

birnessite Hydrated manganese oxide with sodium and calcium, and enriched in other elements. It is one of the dominant minerals of the deep-sea **manganese nodules**.

bischofite Hydrated magnesium chloride. A constituent of salt deposits, such as those of Stassfurt in Germany; decomposes on exposure to the atmosphere.

bismite The monoclinic phase of bismuth trioxide. Cf. **sillénite**, the cubic phase.

bismuth Element. Symbol Bi, at.no. 83, r.a.m. 208.98, mp 268°C. It is a silvery brittle metal, used as a component of fusible alloys with lead. It is very rare, with only about 0.008 ppm in the Earth's crust, but occurs as native metal and in minerals often in association with gold, sulphur, selenium or tellurium.

bismuth glance See **bismuthinite**.

bismuthinite Sulphide of bismuth, rarely forming crystals. Also *bismuth glance*.

bismuth ochre A group name for undetermined oxides and carbonates of bismuth occurring as shapeless masses or earthy deposits.

bismutite Bismuth carbonate $(BiO)_2CO_3$, a tetragonal mineral.

bisphenoid A crystal form consisting of 4 faces of triangular shape, 2 meeting at the top and 2 at the base in chisel-like edges, at right angles to one another; hence the name, meaning 'double edged'.

bit Cutting end of length of drill steel used in boring holes in rock. See **diamond bit**, **drill bit**, **drilling rig**, **roller bit**.

bittern The residual liquor remaining from the evaporation of seawater, after the removal of the salt crystals.

bitumen (1) All naturally occurring inflammable hydrocarbons, of various compositions and consistencies from liquids to solids: petroleums, tars, asphalts, waxes etc. (2) The non-mineralized substances of coal, lignite etc., and their distillation residues. The heaviest residues of vacuum distillation or long-term weathering of crude oils. Some grades are hardened by oxidation at high temperature to give *blown bitumen*. Used for road surfacing as a component of **asphalt**.

bituminous shale Shale rich in hydrocarbons, which may be recoverable as gas or oil by distillation.

bivalves See panel on p. 23.

black ash *Soda ash.* Impure sodium carbonate, containing some calcium sulphide and carbon. Important pH regulator in flotation process.

black-band iron ore A carbonaceous variety of clay-ironstone, the iron being present as carbonate (siderite); occurs in the English Coal Measures.

black body A body which completely absorbs any heat or light radiation falling upon it. A *black body* maintained at a steady temperature is a full radiator at that temperature, since any black body remains in equilibrium with the radiation reaching and leaving it.

black-body radiation Radiation that would be radiated from an ideal black body. The energy distribution is dependent only on the temperature and is described by *Planck's radiation law*.

black-body temperature The temperature at which a **black body** would emit the same radiation as is emitted by a given radiator at a given temperature. The *black-body temperature* of a carbon-arc crater is about 3500°C, whereas its true temperature is about 4000°C.

black damp Mine air which has lost part of its oxygen as result of fire, and is dangerously high in carbon dioxide. See **fire damp**.

black diamond A variety of cryptocrystalline massive diamond, but showing no crystal form. Highly prized for its hardness as an abrasive. Occurs only in Brazil. Also *carbonado*.

black jack A popular name for the mineral sphalerite or zinc blende.

black lava glass Massive natural glass of volcanic origin, jet black and vitreous.

black lead A commercial form of **graphite**.

black oil Term for heavier, darker petroleum products such as fuel oils. Defines equipment, e.g. tanks and tankers dedicated to these products, which would need cleaning before use for **white oil**.

black opal Includes all opals of dark tint, although the colour is rarely black; the fine Australian blue opal, with flame-coloured flashes, is typical.

black powder Gun powder used in quarry work. Standard contains 75% potassium nitrate, slow 59% and blasting 40%, the balance being charcoal and sulphur.

black smoker A hydrothermal vent, at the crest of an oceanic ridge, producing a plume of large quantities of black fine-grained and very hot sulphide precipitates. There are many varieties of deposits, with iron, copper, manganese and zinc sulphides or oxides common. Cf. **white smoker**, **chimney**.

Blake crusher See **jaw breaker**.

blanket strake Table or sluice with gentle downward slope, with bottom lining of material on which heavy mineral (e.g. metallic gold) is caught. Originally of rough blanket,

bivalves

Members of the class Bivalvia (otherwise Pelecypoda, Lamellibranchia), they are marine, brackish or freshwater molluscs having paired shell valves joined by a hinge and closed by adductor muscles. They usually have calcareous shells and include the mussels, clams, cockles, scallops and oysters. They are of great importance as fossils. The two valves lie on either side of each other, as in the figure, and not dorsally and ventrally as in the brachiopods. The valves may be of similar or unequal size, asymmetric from front to back, and the *beak* or *umbo* points towards the front. The *hinge*, sometimes used in the classification of fossil bivalves, is formed by projections that may be present as *teeth* that alternate in the two valves, along the *hinge-line*. The inner layer of the shell may be a pearly or *nacreous* layer ('mother-of-pearl'). The mode of life of bivalves varies considerably, including the commonest which are free-swimming, but also crawling, burrowing (among them *Teredo*, the shipworm, found fossil in Palaeogene wood), and those attached firmly by a foot to a rock or other object. The shells are appropriately adapted to their different life styles.

The earliest bivalves are from the Cambrian, and many Palaeozoic genera died out before the Mesozoic when they were an important part of the faunas. They remained important in the Tertiary when freshwater bivalves became of greater importance and bivalves are at their acme at the present day. In the Carboniferous they have been used as zonal fossils.

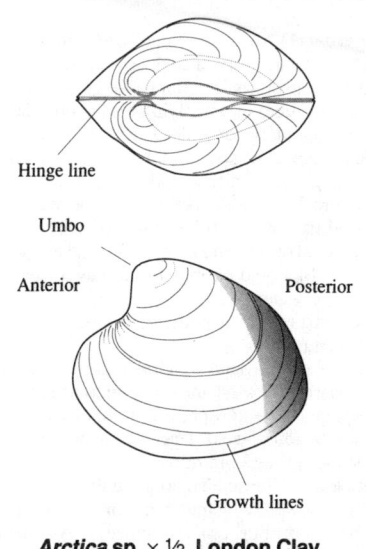

Arctica sp. × ½. **London Clay, Eocene**

now usually corduroy with ribs across flow, or chequered rubber.

blast joint The specially made heavy duty joint of a producing well's **drill string**, able to withstand the abrasive action of the oil entering from a high-pressure formation.

bleed line Used in the **blowout preventer stack** to remove excess gas pressure.

blende, zinc blende See sphalerite.

blind apex Outcrop of mineral vein or lode where ore deposit dies out before reaching the surface (*suboutcrop*), leaving an apparently barren deposit.

blind lode, blind vein A lode which does not outcrop to the surface.

block (1) Rectangular panel of ore defined by drives, raises and winzes, giving all-round physical access for sampling, testing and mining purposes. (2) Crystal, glass or rock fragment of disruptive volcanic origin whose generally angular shape indicates that it was solid when formed, and with an average diameter exceeding 64 mm.

block caving Method of mining in which block of ore is undercut, so that it caves in and the fragments gravitate to withdrawal points.

blocked-out ore See **block** (1).

blocking-out Method of estimating tonnage of mineral reserves in a volume established by drilling, with reference to the grade of the representative sample.

block lava Lava flows with surfaces of rough angular blocks. Similar to **aa** but blocks more regular in shape.

bloodstone Cryptocrystalline silica, a variety of chalcedony, coloured deep green, with flecks of red jasper. Also *heliotrope*.

bloom (1) The efflorescence of altered metallic salt at surface of ore exposure, e.g. cobalt bloom (erythrite). (2) The colour of the fluorescent light reflected from some oils when illuminated. The colour is usually different from that shown by the oil with transmitted light.

blow (1) Ejection of part of the explosive charge (unfired) from hole. (2) Sudden rush of gas from coal seam or ore body.

blowdown stack Container into which refinery vessels can be emptied during emergencies and into which steam or water can be injected to prevent ignition of volatile components.

blower (1) A fissure or thin seam which discharges a quantity of coal gas. (2) An auxiliary ventilating appliance, e.g. a fan or venturi tube, for supplying air to subsidiary working places or to dead ends.

blowhole An aperture near a clifftop through which air, compressed in a sea cave by breaking waves, is forcibly expelled.

blowing a well Old method of temporary removal of pressure at the well-head to allow the tubing and casings to be blown free of debris, water etc.

blowing road In colliery, the main ventilation ingress.

blown sand Sand which has suffered transportation by wind, the grains in transit developing a perfectly spherical form (*millet-seed sand*); grain-size is dependent upon the wind velocity. See **sand dunes**.

blowout The sudden eruption of gas and oil from a well, which then has to be controlled before drilling can recommence. Also *wild well*.

blowout preventer The stack of heavy-duty valves attached to the casing at a well head, designed to control the pressure in the bore when drilling or *working over*. Abbrev. *BOP*. Also *Christmas tree*. See **drilling rig**.

blue asbestos Crocidolite, a fibrous variety of riebeckite. The best-known occurrences are in South Africa.

blue ground *Peridotite, Kimberlite.* Decomposed agglomerate, usually found as a breccia and occurring in volcanic pipes in South Africa and Brazil; it contains a remarkable assemblage of ultramafic plutonic rock fragments (many large) and diamonds.

blue john A massive, blue and white variety of the mineral fluorite, occurring near Castleton, in Derbyshire, and worked for ornamental purposes.

blue lead A name used in the industry for metallic lead, to distinguish it from other lead products such as white lead, orange lead, red lead etc.

blue schist A metamorphic rock formed under conditions of high pressure and relatively low temperature. Characteristic minerals are glaucophane and kyanite.

bluestone, blue vitriol The mineral chalcanthite, hydrated copper sulphate, $CuSO_4.5H_2O$.

boart See **bort**.

body-centred crystal lattice See **unit cell**.

body-centred cubic See **unit cell**. Abbrev. *BCC*.

boehmite Orthorhombic form of aluminium monohydrate, AlO.OH. An important constituent of some bauxites.

boghead coal Coal which is non-banded and translucent, with high yield of tar and oil on distillation. Consists largely of resins, waxes, wind-borne spores and pollen cases. Originated in deeper, more open parts of the coal swamps than ordinary household coals. Essentially a spore coal. Also *parrot coal*. See **torbanite**.

bog iron ore Porous form of **limonite** often mixed with vegetable matter. Found in marshes.

Bohemian garnet Reddish crystals of the garnet *pyrope*, occurring in large numbers in the Mittelgebirge in Bohemia.

Bohemian gemstone A misnomer for the the garnet **pyrope**, the false ruby **rose quartz** and the false topaz **citrine**.

boiling The very rapid conversion of a liquid into vapour by the violent evolution of bubbles. It occurs when the temperature reaches such a value that the saturated vapour pressure of the liquid equals the pressure of the atmosphere.

boiling bed In gas fluidized beds, two separate phases are formed at high gas velocities: gas, containing a relatively small proportion of suspended solids, bubbles through a higher density fluidized phase with the result that the system closely resembles in appearance a boiling liquid, hence *boiling bed*.

boiling point The temperature at which a liquid boils when exposed to the atmosphere. Since, at the boiling point, the saturated vapour pressure of a liquid equals the pressure of the atmosphere, the boiling point varies with pressure; it is usual, therefore, to state its value at the standard pressure of 101.325 kN/m². Abbrev. *bp*.

bolide A brilliant **meteor**, generally one that explodes; a fireball.

bombardment Process of directing a beam of neutrons or high-energy charged particles on to a target material in order to produce

nuclear reactions.
bomb, volcanic See **volcanic bomb**.
bonanza Rich body of ore.
bond Link between atoms, considered to be electrical and arising from electrons as distributed around the nucleus of atoms so bonded. See **chemical bond**.
bond angle The angle between the lines connecting the nucleus of one atom to the nuclei of two other atoms bonded to it, e.g. in water the H-O-H angle is about 105°.
bone Coal containing ash (bone) in very fine layers along the cleavage planes. Also *bony coal*.
bone bed A sediment characterized by abundant fossil bones or bone fragments, scales, teeth, coprolites etc.
bone turquoise Fossil bone or tooth, coloured blue with phosphate of iron; widely used as a gemstone. It is not true turquoise and loses its colour in the course of time. Also *odontolite*.
boninite A highly magnesian siliceous glassy lava containing pyroxene and olivine phenocrysts in a glassy base full of crystallites.
Bonne's projection A derivative **conical projection** in which the parallels are spaced at true distances along the meridians, which are plotted as curves.
boom Device to trap spilled oil floating on water, e.g. at sea, in estuaries. Effective only at slow relative water speeds, ≈ 1 knot.
booster station In long-distance transport by pipeline of oils, other liquids, mineral slurries or water-carried coal, an intermediate pumping station where lost pressure energy is restored.
BOP Abbrev. for **BlowOut Preventer**.
boracite The orthorhombic, pseudocubic form of magnesium borate and chloride, found in beds of gypsum and anhydrite.
borax Hydrated sodium borate, deposited by the evaporation of alkaline lakes.
bord-and-pillar A method of mining coal by excavating a series of chambers, rooms, or stalls, leaving pillars of coal in between to support the roof. Also *room-and-pillar*.
borehole The hole made by a drill for a well; its whole length.
borehole survey Check for deviation from required line of deep borehole. Methods include records of compass, plumbline or gyroscope readings.
boric acid H_3BO_3. A tribasic acid. See **sassolite**.
boring (1) The drilling of deep holes for the exploitation or exploration of oilfields. The term *drilling* is used similarly in connection with metalliferous deposits. (2) A *trace fossil*,
consisting of etchings, grooves etc. caused by fossil animals or plants originally present.
bornite Sulphide of copper and iron occurring in Cornwall and many other localities. Develops a brilliant iridescent red and blue tarnish, hence, also called *erubescite, peacock ore, variegated copper ore*.
borolanite A basic igneous rock occurring near Loch Borolan, Assynt, in N.W. Scotland; it consists essentially of feldspar, green mica, garnet, together with conspicuous rounded white aggregates of 'pseudoleucite', consisting of orthoclase feldspar and altered nepheline.
boron Amorphous yellowish-brown element discovered by Davy, 1808, also Gay-Lussac and Thénard. Symbol B, at.no. 5, r.a.m. 10.811, mp 2300°C, rel.d. 2.5. Can be formed into a conducting metal. Most important in reactors, because of great cross-section (absorption) for neutrons; thus, boron steel is used for control rods. The isotope ^{10}B on absorbing neutrons breaks into two charged particles ^7Li and ^4He which are easily detected, and is therefore most useful for detecting and measuring neutrons. There are numerous minerals in which boron occurs, mainly as borates, or silicates, including tourmaline. They occur in the late stages of magmatic crystallization, in volcanic emanations and in evaporite deposits. Its abundance in the Earth's crust is 9 ppm and in seawater 4.8 ppm.
bort, boart A finely crystalline form of diamond in which the crystals are arranged without definite orientation. Possessing the hardness of diamond, bort is exceedingly tough, and is used as the cutting agent in rock drills.
boss An igneous intrusion of cylindrical form, less than 100 km^2 in area; otherwise like a batholith.
bostonite A fine-grained intrusive igneous rock allied in composition to syenite; essentially feldspathic, and deficient in coloured silicates; type locality, Boston, Mass.
botryoidal Of minerals, formation resembling grapes.
bottom-hole assembly The drilling string attached to the bottom of the **drilling pipe**. It comprises the drill bit and collars used to maintain direction, and may contain several stabilizers and reamers in more difficult conditions, when it is called a *packed-hole assembly*. Abbrev. *BHA*.
bottom-hole pump Electric or hydraulic pump placed at the bottom of a well.
bottomset bed Fine-grained sediment laid down at the front of a growing delta. Cf.

foreset beds, topset beds.
bottom structure See **sole mark**.
boudinage Structure found in sedimentary series subjected to folding. It consists of strike-elongated 'sausages' of more rigid rock, enclosed between relatively plastic rocks.
Bouguer anomaly A gravity anomaly which has been corrected for the station height and for the gravitational effect of the slab of material between the station height and datum. Cf. **free-air anomaly**.
Bouguer law of absorption The intensity p of a parallel beam of monochromatic radiation entering an absorbing medium is decreased at a constant rate by each infinitesimally thin layer db,

$$\frac{-dp}{p} = k\,db,$$

where k is a constant that depends on the nature of the medium and on the wavelength.
boulder The unit of largest size occurring in sediments and sedimentary rocks, the limit between pebble and boulder being placed at 256 mm, although some authorities recognize **cobbles** between pebbles and boulders. Boulders may consist of any kind of rock, may be subangular or well rounded, may have originated in place or have been transported by running water or ice. Accumulations of boulders are *boulder beds*.
boulder clay See **till**.
boule The pear-shaped or cylindrical drop of synthetic mineral, commonly ruby, sapphire or spinel, produced by the Verneuil flame-fusion process.
Bouma cycle A sedimentary succession of five intervals that makes up a complete turbidite deposit. Typically incomplete.
bounce mark A short depression caused by an object (e.g. shell, pebble) bouncing over the surface of a sediment, esp. a turbidite. Also *prod mark*.
boundary layer When a fluid flows past a solid surface, the layer of fluid next to the solid surface is brought to rest, setting up a viscous motion in adjacent layers. The boundary layer is the total thickness of fluid over which the surface exerts a differential effect. It governs the rate of heat, mass, or momentum between the solid surface and the homogeneous bulk of the fluid. The velocity of the layer differs significantly from that of the main fluid stream, and is therefore of considerable importance in heat transfer problems, as in nuclear reactors.
bourne An intermittent or seasonal stream.
bournonite Lead copper antimony sulphide; commonly occurs as wheel-shaped twins, hence known as *cog-wheel ore*, or *wheel-ore*. Also *antimonial lead ore*.
bowenite A compact, finely granular, massive form of serpentine, formerly thought to be nephrite, and used for the same purposes.
bowk A large iron barrel used when sinking a shaft. Also *kibble*.
bowlingite See **saponite**.
box stones Hollow concretions. Nodules of sandstone containing molluscan casts found in the Pleistocene deposits of East Anglia.
BP Before the present. Usually expressed in millions of years (*Ma*).
bp Abbrev. for **boiling point**.
brachiopods See panel on p. 27.
Bragg angle The angle the incident and diffracted X-rays make with a crystal plane when the Bragg equation is satisfied for maximum diffracted intensity.
Bragg equation If X-rays of wavelength λ are incident on a crystal, diffracted beams of maximum intensity occur in only those directions in which constructive interference takes place between the X-rays scattered by successive layers of atomic planes. If d is the interplanar spacing, the Bragg equation, $n\lambda = 2d\,\sin\theta$, gives the condition for these diffracted beams; θ is the angle between the incident and diffracted beams and the planes, and n is an integer. Also applied to electron, neutron and proton diffraction.
braided stream A stream which consists of several channels which separate and join in numerous places. Braided streams occur where the gradient is steep and where seasonal floods are liable to occur. They generally have wide beds filled with loose detritus.
brammallite A variety of illite with sodium as the inter-layer cation.
brattice, brattice cloth A partition for diverting air, for the purpose of ventilation, into a particular working place or section of a mine.
braunite A massive, or occasionally well-crystallized, ore of manganese, Composition $3Mn_2O_3.MnSiO_3$.
Bravais lattices See panel on p. 28.
Brazilian emerald A misnomer for pure-green, deeply coloured variety of tourmaline, occurring in Brazil; used as a gemstone.
Brazilian pebble The name applied to Brazilian quartz or rock-crystal.
Brazilian peridot A misnomer for green crystals of tourmaline or chrysoberyl from Brazil having the typical colour of peridot (olivine).
Brazilian ruby A misnomer for pink topaz, or topaz which has become red after heating, or red tourmaline.

brachiopods

The phylum Brachiopoda is a group of solitary marine invertebrates that vary in size from about 5 mm to 30 cm and have existed, some with little change, from the lower Palaeozoic to the present. They were very important as fossil species but nowadays are much less numerous. Most brachiopods are attached to a rock or other object, but some fossil forms were free. They have two *valves*, dorsal and ventral, differing from each other in size and shape but bilaterally symmetrical, as in the figure. The *dorsal* valve is usually the smaller; the upper one is known as the *brachial* valve. Brachiopods may be distinguished from bivalve molluscs (lamellibranchs) which do not have this bilateral symmetry. The larger, *ventral* valve may have an opening for a muscular stalk or *pedicle* which anchors the animal; this larger valve is the lower and extends to a *beak* or *umbo*. It is also known as the *pedicle* valve.

The Brachiopoda are divided into two classes, the *Inarticulata* and *Articulata*. In the former the two valves are held together with muscles and the mantle which lines the shell, which is usually thin and composed of chitin. *Lingula* is a genus which survives today but ranges back to the Ordovician. The Articulata have valves joined at a *hinge* which consists of two short processes or *teeth* from the ventral valve fitting into sockets in the dorsal valve. The shell is composed of calcareous and chitinous material.

The earliest brachiopods occur in the Lower Cambrian and by the Ordovician were abundant, continuing in the Silurian, when they reached their acme and then into the Devonian. There was some decline in the Carboniferous in which the largest brachiopod, *Productus giganteus* occurs. It is found in the Carboniferous Limestone and reached a foot across. Brachiopods declined in the Permian and Triassic, and after some increase in the Jurassic continued to be less numerous through the Cretaceous to the present day.

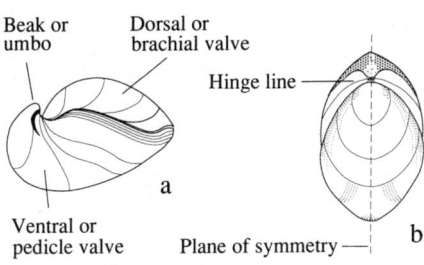

Terebratula sp. Chalk
Shows general characteristics of shell. a, lateral view; b, dorsal view.

Relative abundance

Brazilian sapphire TN for the beautiful clear blue variety of tourmaline mined in Brazil; used as gemstone, but not a true sapphire.

Brazilian topaz True topaz varying in colour from pure white to blue and yellow; mined chiefly in the state of Minas Geraes, Brazil.

bread-crust bomb A type of **volcanic bomb** having a compact and often cracked outer crust and a spongy vesicular interior.

break (1) A jointing plane in a coal seam. (2) Optimum range of size to which ore should be ground before further processing.

Bravais lattices

It was shown by Auguste Bravais in 1845 that, owing to the effects of symmetry, there are only 14 possibly unique crystal lattices. These are shown in the figures below. See **crystal lattice, unit cell**.

1 Cubic primitive (P).
2 Cubic body-centred (I).
3 Cubic face-centred (F).

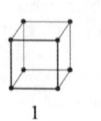 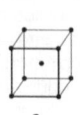 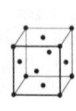
1   2   3

4 Tetragonal primitive (P).
5 Tetragonal body-centred (I).

4   5

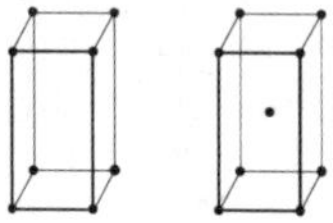

6 Orthorhombic primitive (P).
7 Orthorhombic body-centred (I).
8 Orthorhombic face-centred (F).
9 Orthorhombic C-centred.

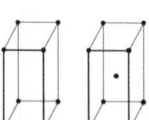 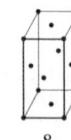 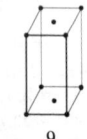
6   7   8   9

10 Rhombohedral primitive (P).
11 Hexagonal primitive (P; C-centred).

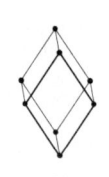 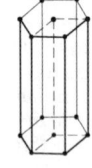
10   11

12 Monoclinic primitive (P).
13 Monoclinic C-centred.
14 Triclinic primitive (P).

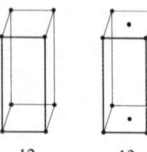 
12   13   14

breaking down, out Unscrewing the drillpipe preparatory to storage.

breakthrough In industrial ion-exchange recovery of metal (e.g. uranium) from solution, the point at which traces of metal begin to arrive in the last of a series of resin-filled stripping columns.

breast (1) The working coalface in a colliery. (2) Underground working face. In flat lodes, breast stopes are those from which detached ore will not gravitate without help.

breathing apparatus Mine rescue equipment in which oxygen is fed to a face mask carried by the wearer, via a demand valve. See **Weg rescue apparatus**.

breccia A coarse-grained clastic rock consisting largely of angular fragments of pre-existing rocks. According to its mode of origin, a breccia may be a *fault-breccia*, a *crush-breccia*, an *intrusion-breccia*, or a *flow-breccia*.

breithauptite Nickel antimonide, occurring as bright coppery-red hexagonal crystals, widely distributed in sulphide ore deposits in small amounts.

breunnerite Variety of **magnesite** containing some iron. Used in manufacture of magnesite bricks.

Brewster angle A plane wave polarized in the plane of incidence is totally transmitted when incident on a plane dielectric boundary at the Brewster angle θ (measured from the normal to the boundary) is given by $\tan\theta = \sqrt{\varepsilon_2/\varepsilon_1}$ where ε_1, ε_2 are the permittivities of the two media. For optical wavelengths, $\tan\theta = n$, the refractive index between the media. Also *polarizing angle*.

brewsterite A rare strontium-barium zeolite.

Brewster law Law relating the Brewster angle (θ) to the refractive index (n) of the medium for a particular wavelength, viz., $\tan\theta = n$. For sodium light incident on particular glass, $n = 1.66$, θ is 51°.

Brewster's bands Interference fringes which are visible when white light is viewed through two parallel and parallel-sided plates, whose thicknesses are in a simple ratio (1:1, 2:1, 1:3 etc.).

brick clay An impure clay, containing iron and other ingredients. In industry, the term is applied to any clay, loam, or earth suitable for the manufacture of bricks or coarse pottery. See **brick earth**.

brick earth Earths used in the manufacture of ordinary bricks; they consist generally of clayey silt interstratified with the fluvioglacial gravels of southern England, frequently exploited in brick manufacture. Also *brickearth*.

bridging Arching of jammed rock so as to obstruct flow of ore; clogging of filtering septum by tiny particles which are individually small enough to pass, but which form such arches.

Bristol diamonds Small lustrous crystals of quartz, i.e. rock crystal, occurring in the Bristol district.

brittle micas A group of minerals (the clintonite and margarite group) resembling the true micas in crystallographic characters, but having the cleavage flakes less elastic. Chemically, they are distinguished by containing calcium as an essential constituent.

brittle silver ore A popular name for **stephanite**.

brochanite A basic sulphate of copper occurring in green fibrous masses, or as incrustations; occurs in the oxidation zone of copper deposits.

brockram A sedimentary rock occurring in the Permian strata in N.W. England; consists of angular blocks which probably accumulated as scree material.

bromine A non-metallic element in the seventh group of the periodic system, one of the halogens. Symbol Br, at.no. 35, r.a.m. 79.909, oxidation states 1, 3, 5, 7, mp − 7.3°C, bp 58.8°C, rel.d. 3.19. A dark red liquid, giving off a poisonous vapour, Br_2. It occurs as a scarce element in the Earth's crust, to the extent of 2.5 ppm, with 65 ppm in seawater. Bromine forms very few minerals and it appears mainly to occur as the bromide ion, sometimes replacing chlorine.

bromochlorodifluoromethane BCF. $CHBrClF_2$. Bp–4°C. Organic substance used as a fire extinguishing fluid, particularly for fires in confined spaces. Low toxicity vapour, 5.7 times as dense as air.

bronzite A form of orthopyroxene, more iron-rich than enstatite and more magnesian than hypersthene; often has metallic sheen, due to the reflection of light from planes of minute metallic inclusions in the surface layers.

bronzitite A rock composed of bronzite with smaller amounts of angite and calcic plagioclase. A common constituent of layered basic igneous intrusions, such as those of the Bushveld, South Africa and Stillwater, Montana.

brookite One of the three naturally occurring forms of crystalline titanium dioxide.

brow The top of the shaft or 'pit'; hence also *pitbrow*.

Brown agitator See **agitator**.

brown coal An intermediate between peat and true coals, with high moisture content, the calorific value ranging from about 4000 to 8300 Btu/lb (9.5 to 20 MJ/kg). Also *lignite*.

brown hematite A misnomer, the material bearing this name being **limonite**, a hydrous iron oxide, whereas true hematite is anhydrous Fe_2O_3.

Brownian movement Small movements of bodies such as particles in a colloid, due to statistical fluctuations in the bombardment by surrounding molecules of the dispersion medium. It may be detected by movement of a galvanometer coil. Also *colloidal movement*. See **colloidal state**.

brucite Magnesium hydroxide, occurring as fibrous masses in serpentinite and metamorphosed dolomite.

brush(ing) discharge Electrical discharge from points along a bar charged to between 18 000 and 80 000 volts to create an electrostatic field through which mineral particles fall and acquire polarity, in the high-intensity separation process.

Bryozoa A phylum of invertebrate animals, the 'moss animals', which are mainly marine. They usually live in colonies (*zoaria*) and are important as fossils from the Ordovician to Recent, but they are rarely found as complete specimens. Also *polyzoa*.

bubble point The temperature at which the first bubble appears on heating a mixture of liquids.

buchite A glassy rock which represents the result of partial fusion and recrystallization at very high temperatures of clay and shale material. It often occurs as xenoliths within igneous rocks.

bucket-dredge System with two pontoons between which a chain of buckets digs through alluvium to mineral-bearing sands and delivers these to concentrating appliances on the pontoon decks. Dredge excavates and floats in a pond in which it traverses the deposit.

buddle A shallow annular pit for concentrating finely crushed, slimed, base-metal ores.

buddle-work Treatment of finely-ground tin-bearing sands by gentle sluicing, in which a heavier fraction of the fed pulp is built up (*buddled*) while the lighter fraction flows to discard. This is continued until a satisfactory concentrate is produced.

buhr mill, burr mill One in which material is ground by its passage between a fixed and a rubbing surface. Types include old-fashioned flour mill, with a circular grindstone rotating above a fixed lower one, radially grooved to facilitate passage of grist from centre to peripheral discharge. Also, rotating cone in fixed casing, material gravitating through the intervening space. Used for softish material, e.g. grain, food processing, and such minerals as soft limestone.

bulk flotation Froth flotation process so applied as to concentrate more than one valuable mineral in one operation.

bulk sample Sample composed of several portions taken from different locations within a bulk quantity of the material under test.

bull wheel The driving pulley for the camshaft of a stamp battery; one on which bull rope of drilling rig is wound.

buoyancy The apparent loss in weight of a body when wholly or partly immersed in a fluid; due to the upthrust exerted by the fluid. See **Archimedes' principle, correction for buoyancy**.

burden (1) Amount of rock to be shattered in blasting between drill hole and nearest free face. (2) See **overburden**.

Burdigalian A stage in the Miocene. See **Tertiary**.

Burgers' vector Translation vector of crystal lattice, representing a displacement which creates a lattice dislocation.

burmite An amber-like mineral occurring in the upper Hukong Valley, Burma, differing from ordinary amber by containing no succinic acid. A variety of **retinite**.

burning Changing the colour of certain precious stones by exposing them to heat.

burnt coal Sooty product of weathering of a coal outcrop.

bustamite A triclinic silicate of manganese, calcium and iron.

butane C_4H_{10}, an alkane hydrocarbon, bp 1°C, rel.d. at 0°C 0.600, contained in natural petroleum, obtained from casing head gases in petroleum distillation. Used commercially in compressed form, and supplied in steel cylinders for domestic and industrial purposes, e.g. Calor gas.

butte A small flat-topped steep-sided hill. See **mesa**.

buttock Coal from which an undercut has been removed, in readiness for bringing it down.

Buxton certification The certification of the suitability of electrical equipment for use in atmosphere in which fire or explosion hazards are present.

bysmalith A form of igneous intrusion bounded by a circular fault and having a dome-shaped top; described by Iddings from the Yellowstone National Park. Cf. **plug**.

bytownite A variety of plagioclase feldspar, containing 70% to 90% of the anorthite molecule; occurs in basic igneous rocks.

C

c Symbol for **concentration**.
c Symbol used for the velocity of electromagnetic radiation *in vacuo*. Its value, according to the most accurate recent measurements, is 2.9979245×10^8 m s^{-1}.
c- Abbrev. for (1) *cyclo-*, i.e. containing an alicyclic ring; (2) **cis-**, i.e. containing the two groups on the same side of the plane of the double bond or ring.
C (1) Symbol for **carbon**. (2) Symbol for **coulomb**. (3) When used after a number of degrees thus: 45°C, the symbol indicates a temperature on the **Celsius scale** or **Centigrade scale**.
C Symbol for (1) **concentration**; (2) (with subscript) **molar heat capacity**: C_p, at constant pressure; C_v, at constant volume.
C- Containing the radical attached to a carbon atom.
Ca Symbol for **calcium**.
cabin A fireman's station underground in a coalmine.
cable tool drilling Method of drilling in which a heavy sharpened tool bit (*churn-*, *percussion-drill*) is reciprocated in a borehole by a steel cable attached to a **walking beam**.
cable tools The variously shaped drilling tools used in **cable tool drilling**.
cadmium White metallic element. Symbol Cd, at.no. 48, r.a.m. 112.40, mp 320.9°C, bp 767°C, rel.d. 8.648. It is a rare element, the Earth's crust containing only 0.16 ppm. Cadmium has considerable affinity for sulphur and, although it forms some independent minerals, it mainly occurs concealed in the crystal lattice of *sphalerite*, zinc sulphide. The smelting of zinc is thus the principal commercial source of the element.
cadmium red line Spectrum line formerly chosen as a reproducible standard of length. Wavelength = 643.8496 nm.
Caerfai The oldest epoch of the Cambrian period.
caesium, cesium Metallic element. Symbol Cs, at.no. 55, r.a.m. 132.905, mp 28.6°C, bp 713°C, rel.d. 1.88. It is a rare element with an abundance of 2.6 ppm in the Earth's crust. Since its ionic radius is large (1.65 Å), it tends to occur in the last stages of magmatic crystallization, forming the pegmatite mineral, *pollucite*. The radioactive ^{137}Cs is used in radiotherapy and as a medical tracer, and the resonance frequency of the natural isotope ^{133}Cs has been used as a standard for the measurement of time.
cage Steel box used in vertical mine shaft to raise and lower men, materials or trucks (trams, tubs). May have two or three decks.
Cainozoic See **Cenozoic**.
cairngorm Smoky-yellow or brown varieties of quartz; named from Cairngorm in the Scottish Grampians, the more attractively coloured varieties being used as semi-precious gemstones. See **smoky quartz**.
caking coal Coal which cakes or forms coke when heated in the absence of air.
calamine A name formerly used in Britain for **smithsonite** and in the US for **hemimorphite**.
calandria Closed vessel penetrated by pipes so that liquids in each do not mix. In evaporating plant the tubes carry the heating fluid and in certain types of nuclear reactor the sealed vessel is called a calandria.
calcarenite A limestone consisting of detrital lime particles (50%) of sand size. See **calcilutite**.
calcareous Containing compounds of calcium, particularly minerals.
calcareous rock Sediment containing a large amount of calcium carbonate (e.g. limestone, chalk, shelly sandstone, calctufa).
calc-flinta A hard fine-grained rock composed of calcium silicate minerals and produced by the contact metamorphism of an impure limestone.
calcilutite A limestone consisting of (50%) clay or silt grade particles of calcite. See **calcarenite**.
calcination The subjection of a material to prolonged heating at fairly high temperatures.
calcination The operation of heating ores to drive off water and carbon dioxide, frequently not distinguished from **roasting**.
calcine Ore, carbonate, mineral or concentrate which has been roasted, perhaps in an oxidizing atmosphere, to remove sulphur as SO$_2$ (*sweet roasting*) or carbon dioxide (*dead roasting*).
calcite The commonest crystalline form of calcium carbonate, showing trigonal symmetry and a great variety of crystal habits. It is the principal constituent of limestone and many marbles, and occurs extensively in other rocks. Also *calcspar*.
calcium See panel on p. 32.
calcium carbonate CaCO$_3$. Very abundant in nature as chalk, limestone and calcite. Almost insoluble in water, unless the water contains dissolved carbon dioxide, when solution results in the form of calcium hydrogen carbonate, causing the temporary hardness

calcium

An alkaline-earth metallic element. Symbol Ca, at.no. 20, r.a.m. 40.08, mp 840°C, bp 1440°C, rel.d. 1.58.

In the Earth's crust, calcium is the fifth commonest element, with an abundance of 4.66%, and in seawater the content is 400 ppm. It occurs extensively in igneous, sedimentary and metamorphic rocks. Its ionic radius 1.06Å, facilitates entry into the isomorphous series of plagioclase feldspars, sodium to calcium aluminium silicates. The feldspars are ubiquitous in most igneous rocks including the granitic and basaltic types which make up much of the Earth's crust. Calcium is also a common constituent of the ferromagnesian silicate minerals, pyroxenes and amphiboles, characteristic of basic and intermediate rocks.

As the primary rocks weather, much of the calcium released to the sea is precipitated as carbonate. It is the main cause of formation of limestone, and if magnesium is present, dolomite. The metamorphic equivalents of these rocks, notably marbles, have major contents of calcium. At a later stage of the sedimentary sequence precipitation from seawater produced evaporite deposits that may contain substantial amounts of the calcium sulphates, gypsum and anhydrite. In all rock types many other primary and secondary minerals of calcium with other elements are also known.

In the biosphere, calcium is of great importance, forming the hard shells of many organisms, principally as the carbonates, calcite and aragonite, and as phosphate in the bones and teeth of vertebrates. Calcium is also a constiuent of all plants.

---

of water.

calcspar See **calcite**.

caldera A large volcanic crater produced by the collapse of underground lava reservoirs or by ring fracture, possibly as a surface expression of **cauldron subsidence**.

Caledonian orogeny The great mountain-building episode of late Silurian to early Devonian time.

Caledonoid direction The direction assumed by the Caledonian (Siluro-Devonian) mountain folds and associated structures in Britain and Scandinavia. Commonly N.E.–S.W., but subject to considerable variations.

calibration The process of determining the absolute values corresponding to the graduations on an arbitrary or inaccurate scale on an instrument.

caliche Concretions of calcium carbonate, sodium nitrate and other minerals occurring in soil in arid regions; due to surface evaporation of subsurface waters.

Californian jade A compact form of green vesuvianite (idocrase) obtained from California, and used as an ornamental stone and in jewellery. Also *californite*.

Californian stamp See **gravity stamp**.

californite See **Californian jade**.

Callovian A stage in the Middle Jurassic. See **Mesozoic**.

calomel *Mercury (I) chloride*, Hg_2Cl_2; found naturally in whitish or greyish masses, associated with cinnabar. Used in physical chemistry as a reference electrode (i.e. a half-cell comprising a mercury electrode in a solution of potassium chloride saturated with calomel).

calorescence The absorption of radiation of a certain wavelength by a body, and its re-emission as radiation of shorter wavelength. The effect is familiar in the emission of visible rays by a body which has been heated to redness by focusing infrared heat rays on to it.

calorie The unit of quantity of heat in the CGS system. The 15°C calorie is the quantity of heat required to raise the temperature of 1 g of pure water by 1°C at 15°C; this equals 4.1855 J. By agreement, the International Table calorie (cal_{IT}) equals 4.186 J exactly, the thermochemical calorie equals 4.184 J exactly. The calorie has now been largely replaced by the SI unit of the joule (J). See **SI units**.

calorific value The number of heat units obtained by the combustion of unit mass of a fuel. The numerical value obtained for the calorific value depends on the units used; e.g. lb-calories/lb, British thermal units (Btu)lb or MJ/kg for solid and liquid fuels, and Btu/ft^3 or MJ/m^3 for gaseous fuels.

calorimeter The vessel containing the liquid used in calorimetry. The name is also applied to the complete apparatus used in measuring thermal quantities.

calorimetry The measurement of thermal constants, such as specific heat, latent heat, or calorific value. Such measurements usually necessitate the determination of a quantity of heat, by observing the rise of temperature it produces in a known quantity of water or other liquid.

calx Burnt lime or quicklime.

Cambrian The lowest division of the **Palaeozoic** era, covering an approximate time span from 570–510 million years. Named after Cambria, the Roman name for Wales. The corresponding system of rocks.

camouflage Describes the relationship between a trace element and a major element whose ionic charge and ionic radius are similar, as a result of which the trace element always occurs in the minerals of the major element but does not form separate minerals of its own. Thus gallium can be considered to be camouflaged by aluminium.

Campanian A stage in the Upper Cretaceous. See **Mesozoic**.

camptonite An igneous rock occurring in minor intrusions, and belonging to the family of lamprophyres. It consists essentially of plagioclase feldspar and brown hornblende, usually *barkevikite*.

Canadian asbestos See **chrysotile**.

Canadian shield The name applied to the vast area of Precambrian rocks which cover 5 million square kilometres in E. Canada.

cancrinite A hydrated silicate of aluminium, sodium and calcium, also containing the carbonate ion. Found in alkaline plutonic igneous rocks. Cf. **vishnevite**.

candela Fundamental SI unit of luminous intensity. If, in a given direction, a source emits monochromatic radiation of frequency 54×10^{12} Hz, and the radiant intensity in that direction is 1/683 watt per steradian, then the luminous intensity of the source is 1 candela. Abbrev. *cd*.

canyon A deep, narrow, steep-sided valley.

Cape asbestos Blue asbestos from South Africa. See **crocidolite**.

Cape diamond A name used in grading diamonds to designate an off-colour stone of a yellowish tint.

Cape ruby A misnomer for the red garnet **pyrope**, obtained in the diamond mines of South Africa. See **false ruby**.

capillarity A phenomenon associated with surface tension, which occurs in fine bore tubes or channels. Examples are the elevation (or depression) of liquids in capillary tubes and the action of blotting paper and wicks. The elevation of liquid in a capillary tube above the general level is given by the formula

$$h = \frac{2T \cos \theta}{\rho g r},$$

where T is the surface tension, θ is the angle of contact of the liquid with the capillary, ρ is the liquid density, g is the acceleration due to gravity and r is the capillary radius.

capillary action See **capillarity**.

capillary condensation Hypothesis that adsorbed vapours can condense under the incidence of capillary forces to form liquid inside the pores of the adsorbate.

capillary pressure The pressure developed by **capillarity**. Mathematically expressed as $p = 2T \cos \theta / r$, where T is the surface tension, θ the angle of contact, and r the radius of the capillary.

capillary pyrite See **millerite**.

capping The fixing of a shackle or a swivel to the end of a hoisting rope. Overburden lying above valuable seam or bed of mineral.

caprock An impervious rock stratum overlying a reservoir formation, thus trapping the oil or natural gas in the reservoir.

Caradoc An epoch of the Ordovician period. See **Palaeozoic**.

carat (1) A standard weight for precious stones. The *metric carat*, standardized in 1932, equals 200 mg (3.086 grain). (2) The standard of fineness for gold. The standard for pure gold is 24 carats; 22 carat gold has two parts of alloy; 18 carat gold 6 parts of alloy. Also *karat*.

carbazole $(C_6H_4)_2NH$; colourless plates: mp 238°C; sublimes readily; contained in coal-tar and crude anthracene oil.

carbenes (1) Reactive uncharged intermediates of formula: CXY, where X and Y are organic radicals or halogen atoms. (2) Such constituents of asphaltic material as are soluble in carbon disulphide but not in carbon tetrachloride.

carbides Binary compounds of metals with carbon. Carbides of group IV to VI metals (e.g. silicon, iron, tungsten) are exceptionally hard and refractory.

carbon See panel on p. 34.

carbonaceous Said of material containing carbon as such or as organic (plant or animal) matter.

carbonaceous chondrite Friable black carbonaceous chondritic stony meteorite.

carbonaceous rocks Sedimentary deposits of which the chief constituent is carbon,

carbon

Amorphous or crystalline (graphite and diamond) element. Symbol C, at.no. 6, r.a.m. 12.011, mp above 3500°C, bp 4200°C. Its allotropic modifications are diamond and graphite. In the Earth's crust its abundance is 180 ppm and in seawater 28 ppm. As CO_2 it also makes up 335 ppm by volume of the atmosphere. In the Earth it occurs mainly in limestone as the carbonate calcite, and with Mg, dolomite; as hydrocarbon in oil and gas, and as coal. It is the basis of all life, living and fossil. The isotope, carbon-14, ^{14}C, is an important source of natural terrestrial radioactivity and is used as a dating method for organic material. See **radiocarbon dating**.

The *carbon cycle* is the exchange and reactions of carbon between the atmosphere, biosphere, hydrosphere and lithosphere as shown in the figure. The major fluxes are the interchange between land plants and the atmosphere; between the atmosphere and the sea, which acts as a huge buffer in the whole system; between marine plant life and the ocean; and oxidation of land and marine humus. Carbon is produced in volcanic gases (as CO_2 and CO) and is abstracted in the formation of sedimentary limestones and dolomites. Industrial activity involving all of the sources contributes to all of the gases.

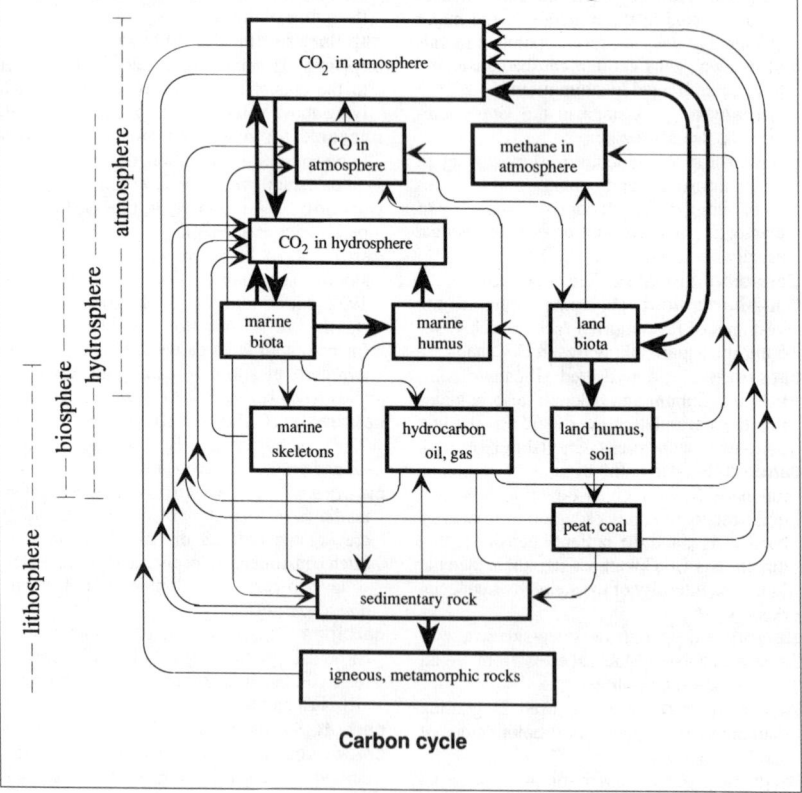

Carbon cycle

derived from plant residues. Under this heading are included peat, lignite, or brown coal, and the several varieties of true coal (bituminous coal, anthracite etc.).

carbonado See **black diamond**.

carbonas Zones of mineralization that have spread into the host rock around mineral veins (commonly tin), and have been rich

enough to mine.
carbonate A compound containing the acid radical of carbonic acid (CO_3 group). Bases react with carbonic acid to form carbonates, e.g. $CaCO_3$, calcium carbonate.
carbonate-apatite A variety of apatite containing appreciable CO_2.
carbonate compensation depth The level in the ocean below which the rate of solution of calcium carbonate exceeds the rate of its deposition, so no limestones etc. are formed. 4000–5000 m in the Pacific, less in the Atlantic. Abbrev. *CCD*.
carbonatite An igneous rock composed largely of carbonate minerals (calcite and dolomite). Carbonatites are invariably associated with alkaline igneous rocks such as nepheline-syenite, and are rich in a number of unusual minerals, esp. those of the rare earth elements.
carbon black Finely divided carbon produced by burning hydrocarbons (e.g. methane) in conditions in which combustion is incomplete. Widely used in the rubber, paint, plastic, ink and other industries. It forms a very fine pigment containing up to 95% carbon, giving a very intense black.
carbon dating *Radiocarbon dating*. Atmospheric carbon dioxide contains a constant proportion of radioactive ^{14}C, formed by cosmic radiation. Living organisms absorb this isotope in the same proportion. After death it decays with a half-life 5.73×10^3 years. The proportion of ^{12}C to the residual ^{14}C indicates the period elapsed since death.
carbon dioxide CO_2. A colourless gas; density at s.t.p. 1.976 kg/m^3, about 1.5 times that of air. Produced by the complete combustion of carbon, by the action of acids on carbonates (e.g. Kipp's apparatus), by the thermal decomposition of carbonates (e.g. lime burning) and during fermentation. It plays an essential part in metabolism, being exhaled by animals and absorbed by plants.
carbonic acid H_2CO_3. A weak acid formed when carbon dioxide is dissolved in water.
carbonic acid gas Carbon dioxide effervescing from liquids which have been saturated with carbon dioxide under pressure. The gas escapes when the pressure is withdrawn.
Carboniferous A geological period extending from approx. 360–290 million years. Divided in the US into the Mississippian and Pennsylvanian periods. In Britain comprises the Carboniferous Limestone, Millstone Grit and Coal Measures. The corresponding system of rocks. See **Palaeozoic**.
carbon-in-pulp The use of carbon as the adsorbent for the gold leached from ore in the **cyaniding** process. This has largely replaced the zinc-dust method and allows commercial recovery down to 2 ppm of gold.
carbonization The destructive distillation of substances out of contact with air, accompanied by the formation of carbon, in addition to liquid and gaseous products. Coal yields coke, while wood, sugar etc. yield charcoal.
carbonization The conversion of fossil organic material to a residue of carbon. Plant material is often preserved in this way.
carbon monoxide CO. Formed when carbon is heated in a limited supply of air, when carbon dioxide is heated in an excess of carbon, when carbon dioxide is passed over some hot metals, or by dehydration of methanoic acid.
carbon value The figure obtained empirically as a measure of the tendency of a lubricant to form carbon when in use.
carbuncle A gem variety of garnet cut *en cabochon*. It has a deep red colour. See **almandine**.
carnallite The hydrated chloride of potassium and magnesium, occurring in bedded masses with other saline deposits, as at Stassfurt, Germany. Such deposits arise from the desiccation of salt lakes. It is used as a fertilizer.
carnelian A translucent variety of chalcedony (silica) of red or reddish-brown colour. Also *cornelian*.
carnotite A hydrated vanadate of uranium and potassium, found as a yellow impregnation in sandstones. It is an important source of uranium.
carrier Non-active material mixed with, and chemically identical to, a radioactive compound. Carrier is sometimes added to carrier-free material.
carstone Brown sandstone in which the grains are cemented by limonite. Also *iron-pan*.
cartesian diver See **diver**.
case-hardening Cement-like surface on porous rock caused by evaporation.
casing (1) Piping used to line a drill hole. (2) The steel lining of a circular shaft.
casinghead The part of the well which is above the surface and to which the casings are attached.
casinghead gas Gas produced as a by-product from an oil well.
cassiterite Oxide of tin, crystallizing in the tetragonal system; it constitutes the most important ore of this metal. It occurs in veins

cast and impregnations associated with granitic rocks; also as 'stream-tin' in alluvial gravels. Also *tinstone*.

cast The impression which fills a natural mould. Most frequently a fossil shell, where the infilling is both a cast of the original animal and an internal mould of the shell. Term also used for impressions of sedimentary structures.

cataclasis The process of rock deformation involving fracture and rotation of mineral grains without chemical reconstitution. Also *kataklasis*.

catalysis The acceleration or retardation of a chemical reaction by a substance which itself undergoes no permanent chemical change, or which can be recovered when the chemical reaction is completed. It lowers the energy of activation.

catalyst A substance which catalyses a reaction. See **catalysis**.

catalytic cracking A process of breaking down the heavy hydrocarbons of crude petroleum, using silica or aluminium gel as a catalyst. See **cracking**.

catalytic poison A substance which inhibits the activity of a catalyst. Also *anticatalyst*.

catalytic reforming Petroleum refining process to improve the **octane number** of light hydrocarbons by reaction with hydrogen at high temperatures (500°C) and pressures over a platinum catalyst. Cf. **hydrofining**.

catophorite See katophorite.

catastrophism The theory that the Earth's geological history has been fashioned by infrequent violent events. See **uniformitarianism**.

catch props In a coalmine, props put in advance of the main timbering for safety, i.e. *watch props* or *safety props*.

cat cracker Refinery vessel in which hydrocarbon fractions are processed in the presence of a catalyst.

cathead Rotating drum around which a rope can be coiled to provide pull for various operations on a drilling rig.

cathetometer An optical instrument for measuring vertical distances not exceeding a few decimetres by moving a small telescope with cross-hairs up and down a scale.

cathodic protection Protection of a metal structure against electrolytic corrosion by making it the cathode (electron receiver) in an electrolytic cell, either by means of an impressed e.m.f. or by coupling it with a more electronegative metal. In ships and offshore structures, corrosion can be prevented by passing sufficient direct current through the seawater to make the metal hull a cathode.

cation Ion in an electrolyte which carries a positive charge and which migrates towards the cathode under the influence of a potential gradient in electrolysis. It is the deposition of the cation in a primary cell which determines the *positive terminal*.

cationic detergents Types of ionic synthetic detergents in which the surface active part of the molecule is the cation, unlike soap and most of the widely used synthetic detergents. Sometimes called *invert soaps*.

catophorite See **katophorite**.

cat's eye A variety of fibrous quartz which shows chatoyancy when suitably cut, as an ornamental stone. The term is also applied to crocidolite when infiltrated with silica (see **tiger's eye, hawk's eye**). A more valuable form is *chrysoberyl cat's eye* (see **cymophane**).

catworks Assemblage of motors and **catheads** which provides power for the many secondary activities on drilling platforms like pipe-hoisting.

Cauchy's dispersion formula

$$\mu = A + \frac{B}{\lambda^2} + \frac{C}{\lambda^{21}} + \ldots$$

An empirical expression for the relation between the refractive index μ of a medium and the wavelength λ of light; A, B and C are the constants for a given medium.

cauldron subsidence The subsidence of a cylindrical mass of the Earth's crust, bounded by a circular fault up which the lava has commonly risen to fill the cauldron. Good examples have been described from Scotland (Ben Nevis and the Western Isles).

caustic lime The residue of calcium oxide, obtained from freshly calcined calcium carbonate; it reacts with water, evolving much heat and producing slaked lime (calcium hydroxide, hydrate of lime, or hydrated lime). Also *quicklime*.

Cavendish experiment An experiment, carried out by Henry Cavendish in 1798, to determine the constant of *gravitation*. A form of torsion balance was used to measure the very small forces of attraction between lead spheres.

cavern A chamber in a rock. Caverns are of varying size and are due to several causes, the chief being the solution of calcareous rocks by underground waters, and marine action.

cavilling The drawing of lots for working places (usually for 3 months) in the coalmine.

caving Controlled collapse of roof in deep mines to relieve pressure. Undercut stoping.
CCD Abbrev. for **Carbonate Compensation Depth**.
CCP Abbrev. for **Cubic Close Packing**.
Cd Symbol for **cadmium**.
cedar-tree laccolith A multiple laccolith, i.e. a series of laccoliths, one above the other, forming part of a single mass of igneous rock. See **dykes and sills**.
celestine Strontium sulphate, crystallizing in the orthorhombic system; occurs in association with rock salt and gypsum; also in the sulphur deposits of Sicily, and in nodules of limestone.
Celite TN for a form of **diatomite** used as an insulating material and filter aid.
cell constant The conversion factor relating the conductance of a conductivity cell to the conductivity of the liquid in it.
cell unit See **unit cell**.
celsian A barium feldspar; barium aluminium silicate. Found in association with some manganese ores.
Celsius scale The SI name for **Centigrade scale**. See **SI units**.
cement The material which binds any loose sediment into a coherent rock. The commonest cements are ferruginous, calcareous and siliceous.
cementstone Argillaceous limestone suitable for the manufacture of cement.
Cenomanian The lowest stage in the Upper Cretaceous. See **Mesozoic**.
Cenozoic The youngest era of the Phanerozoic, covering approx. 65 million years ago to the present day. It includes the **Tertiary** and **Quaternary** periods. Also *Cainozoic*.
Centigrade scale The most widely used method of graduating a thermometer. The fundamental interval of temperature between the freezing and boiling points of pure water at normal pressure is divided into 100 equal parts, each of which is a *Centigrade degree*, and the freezing point is made the zero of the scale. To convert a temperature on this scale to the Fahrenheit scale, multiply by 1.8 and add 32; for the Kelvin equivalent add 273.15. See **Celsius scale**.
centipoise One-hundredth of a **poise**, the CGS unit of viscosity. Symbol cP; $1 \text{ cP} = 10^{-3} \text{ Nm}^{-2}$ s.
centistoke Unit for measuring the kinematic viscosity of oils. Replaces *Redwood* and *Saybolt seconds*, and equivalent to the SI unit, $\text{mm}^2 \text{ s}^{-1}$.
centre of symmetry A point within a crystal such that all straight lines that can be drawn through it pass through a pair of similar points, lying on opposite sides of the centre of symmetry and at the same distance from it. Thus, faces and edges of the crystal occur in parallel pairs, on opposite sides of a centre of symmetry.
centrifugal pump Continuously acting pump with rotating impeller, used to accelerate water or suspensions through a fixed casing to peripheral discharge at the desired delivery height.
centrifuge enrichment The enrichment of uranium isotopes using a high-speed centrifuge. It uses the mass difference between ^{235}U and ^{238}U to effect the separation. The separation factor per stage is much greater than for gaseous diffusion enrichment.
cephalopods Marine molluscs belonging to the class Cephalopoda. They range from the Cambrian to the present day and include the important fossil groups ammonites, goniatites, nautiloids and belemnites. See **Ammonoids**.
ceramics The art and science of non-organic non-metallic materials. The term covers the purification of raw materials, the study and production of the chemical compounds concerned, their formation into components, the study of structure, constitution and properties.
cerargyrite, horn silver Silver chloride. Also a group name for silver halides.
cerium One of the rare earth metals. Symbol Ce, at.no. 58, r.a.m. 140.12, rel.d. at 20°, 6.7, mp 795°C. It is the commonest of the lanthaníde elements, having an abundance in the crust of 66 ppm. Its principal minerals, from which it is extracted, are *bastnaesite* and *monazite*.
cerussite Lead carbonate, crystallizing in the orthorhombic system. A common ore of lead.
cesium See **caesium**.
Ceylon chrysolite See **chrysolite**.
ceylonite See **pleonaste**.
Ceylon peridot A misnomer for the yellowish-green variety of the mineral tourmaline, approaching olivine in colour; used as a semi-precious gemstone.
CGS unit Abbrev. for *Centimetre-Gram-Second* unit, based on the centimetre, the gram and the second as the *fundamental units* of length, mass and time. For most purposes superseded by **SI units**.
chabazite A white or colourless hydrated silicate of aluminium and calcium, found in rhombohedral crystals and belonging to the zeolite group.
chain A series of linked atoms, generally in an organic molecule (catenation). Chains may

chain reaction

consist of one kind of atom only (e.g. carbon chains), or of several kinds of atoms (e.g. carbon-nitrogen chains). There are open-chain and closed-chain compounds, the latter being called ring or cyclic compounds.

chain reaction Chemical or atomic process in which the products of the reaction assist in promoting the process itself such as ordinary fire or combustion, or atomic fission. Hence chain reactions are characterized by an 'induction period' of comparatively slow reaction rate, followed, as the chain-promoting products accumulate, by a vastly accelerated reaction rate.

chalcanthite Hydrated copper sulphate, $CuSO_4.5H_2O$.

chalcedony A microcrystalline variety of quartz with abundant micropores. It occurs filling cavities in lavas and in some sedimentary rocks; flint is a variety found in the Chalk. See **silica**.

chalcocite A greyish-black sulphide of copper, Cu_2S, an important ore of copper. Also *copper glance, redruthite*.

chalcophile Descriptive of elements which have a strong affinity for sulphur and are therefore more abundant in sulphide ore deposits than in other types of rock. Lead is an example.

chalcopyrite, copper pyrites Sulphide of copper and iron, crystallizing in the tetragonal system; the commonest ore of copper, occurring in mineral veins. The crystals are brassy yellow, often showing superficial tarnish or iridescence. Also *cupriferous pyrites*.

chalk A white, fine-grained and soft limestone, consisting of finely divided calcium carbonate and minute organic remains. In N.W. Europe, the Chalk forms the upper half of the Cretaceous system.

chalybite See **siderite**.

chamber process A process for the manufacture of sulphuric acid, in which the reactions between the air, sulphur dioxide and nitric acid gases necessary to produce the sulphuric acid take place in large lead chambers.

chamosite A hydrous silicate of iron and aluminium occurring in oolitic and other bedded iron ores.

characteristic spectrum Ordered arrangement in terms of frequency (or wavelength) of radiation (optical or X-ray) related to the atomic structure of the material giving rise to them.

characteristic X-radiation X-radiation consisting of discrete wavelengths which are characteristic of the emitting element.

chemical constitution

If arising from the absorption of X- or γ-radiation, may be called *fluorescence X-radiation*.

charcoal The residue from the destructive distillation of wood or animal matter with exclusion of air; contains carbon and inorganic matter.

charge Operation of placing explosive, primer, detonator and fuse in drill hole.

Charles's law The volume of a given mass of gas at constant pressure is directly proportional to the absolute thermodynamic temperature. This is also known as *Gay-Lussac's law* and is the equivalent of saying that all gases have the same coefficient of expansion at constant pressure. This is approximately true at low pressures and sufficiently high temperatures as the *ideal gas* behaviour is approached.

charnockite A coarse-grained, dark granite rock, consisting of feldspars, blue quartz and orthopyroxene; it occurs typically in Madras and is named from the rock used for the tombstone of the founder of Calcutta, Job Charnock.

Charnoid direction A N.W.–S.E. direction of folding in the rocks of central and eastern England. Exemplified by the Precambrian rocks of Charnwood Forest.

chatoyancy The characteristic optical effect shown by cat's eye and certain other minerals, due to the reflection of light from minute aligned tubular channels, fibres, or colloidal particles. When cut *en cabochon* such stones exhibit a narrow silvery band of light which changes its position as the gem is turned. See **gems and gemstones**.

Chattian A stage in the Oligocene. See **Tertiary**.

chemical affinity See **affinity**.

chemical bond The electric forces linking atoms in molecules or non-molecular solid phases. Three basic types of bond are usually distinguished: (1) Ionic or electrostatic bonding, in which valence electrons are lost or gained, and atoms which are oppositely charged are held together by coulombic forces. (2) Covalent bonding, in which valence electrons are associated with two nuclei, the resulting bond being described as polar if the atoms are of differing electronegativity. (3) Metallic bonding, in which valence electrons are shared over many nuclei, and electronic conduction occurs.

chemical compound A substance composed of two or more elements in definite proportions by weight, which are independent of its mode of preparation.

chemical constitution See **constitution**.

chemical elements Chemical elements, relevant to the earth sciences, are listed under their names.

chemical energy The energy liberated in a chemical reaction. See **affinity**.

chemical equation A quantitative symbolic representation of the changes occurring in a chemical reaction, based on the requirement that matter is neither added nor removed during the reaction.

chemical kinetics The study of the rates of chemical reactions.

chemically-formed rocks Rocks formed by precipitation of materials from solution in water, e.g. calc-tufa and various saline deposits.

chemical reaction A process in which at least one substance is changed into another.

chemical shift A shift in position of a spectrum peak due to a small change in chemical environment. Observed in the *Mössbauer effect* and in *nuclear magnetic resonance*.

chemical symbol A single capital letter, or a combination of a capital letter and a small one, which is used to represent either an atom or a mole of a chemical element, e.g. the symbol for sodium is Na, for sulphur S etc.

chemiluminescence A process in which visible light is produced by a chemical reaction.

chemisorption Irreversible adsorption in which the absorbed surface is held on the substance by chemical forces.

cheralite A radioactive mineral rich in thorium.

chert A siliceous rock consisting of cryptocrystalline silica, and sometimes including the remains of siliceous organisms such as sponges or radiolaria. It occurs as bedded masses, as well as concretions in limestone. See **silica**.

χ, *Chi*. Symbol for magnetic susceptibility.

χ^2 Symbol for chi-squared. See **chi-squared test**.

chiastolite A variety of andalusite characterized by cruciform inclusions of carbonaceous matter.

Chile nitre, Chile saltpetre A commercial name for **sodium nitrate** (V), $NaNO_3$.

chimney (1) Either an ore shoot or more usually an esp. rich steeply inclined part of the lode. (2) A volcanic pipe. Chimneys include those through which ocean-floor smokers vent.

china clay A clay consisting mainly of **kaolinite**, one of the most important raw materials of the ceramic industry. China clay is obtained from kaolinized granite, for example in S.W. England, and is separated from the other constituents of the granite (quartz and mica) by washing out with high pressure jets of water. Also *kaolin, porcelain clay*.

china stone A kaolinitized granitic rock containing unaltered plagioclase. Also applied to certain limestones of exceptionally fine grain and smooth texture.

chirality Asymmetric molecules which are mirror-images of each other. Quartz displays this property as right-handed and left-handed forms. Also *enantiomorphism*. See **twinned crystals**.

chi-squared test A statistical test used to determine the goodness of fit of observed sample data and expected theoretical population. Symbol χ^2. The chi-squared statistic is also used in contingency table testing.

chitin A resistant horny organic compound that is a common constituent of invertebrate skeletons, e.g. crustaceans and graptolites.

chloanthite Nickel arsenide occurring in the cubic system. A valuable nickel ore, associated with smaltite. Also *cloanthite*.

chlorapatite Chlorophosphate of calcium, a member of the apatite group of minerals. See **apatite, fluorapatite**.

chlorastrolite A fibrous green variety of *pumpellyite*; it occurs in rounded geodes in basic igneous rocks near Lake Superior. When cut *en cabochon*, it exhibits **chatoyancy**.

chlorides Salts of hydrochloric acid. Many metals combine directly with chlorine to form chlorides.

chlorine Element. Symbol Cl, at.no. 17, r.a.m. 35.453, valencies 1−, 3+, 5+, 7+, mp −101.6°C, bp −34.6°C. The second halogen, chlorine is a greenish yellow gas, with an irritating smell and a destructive effect on the respiratory tract. Its occurrence in seawater is 1.9%, far greater than can be accounted for by the weathering of rocks. Most probably this chlorine comes from volcanic gases (Cl_2 and HCl). In the Earth's crust its abundance is 126 ppm. There are numerous minerals containing the chloride ion, including many in evaporite and volcanic deposits, and in the oxidized zones in metalliferous deposits. In igneous rocks, *chlorapatite* is the most significant chlorine-containing mineral. Most commercial chlorine is produced by the electrolysis of concentrated brine or by oxidation of hydrochloric acid.

chlorite A group of minerals, typically green, somewhat resembling the micas, composed of hydrated magnesium, iron and aluminosilicates. They occur as alteration products of igneous rocks, chlorite schists and in sediments.

chloritization The replacement, by alteration, of ferro-magnesian minerals by chlorite.

chloritoid Hydrated iron, magnesium, aluminium silicate, crystallizing in the monoclinic or triclinic systems. Characteristic of low and medium grade regionally metamorphosed sedimentary rocks.

chlorophaeite A mineral closely related to chlorite, dark green when fresh, but rapidly changing to brown, hence the names (Gk. *chloros*, yellowish-green; *phaios*, dun). Described from basic igneous rocks.

chocolate mousse Slang term for the oil (usually crude) and seawater emulsion often produced following an oil spillage. It floats, is very viscous and contaminates anything it touches.

choke damp A term sometimes used for **black damp** (carbon dioxide). More correctly, any mixture of gases which causes choking or suffocation.

chondrite A type of stony meteorite containing *chondrules* — nodule-like aggregates of minerals.

chondrodite Hydrated magnesium silicate, crystallizing in the monoclinic system, and occurring in metamorphosed limestones.

chondrule See **chondrite**.

Chordata See **graptolites**.

Christmas tree Casinghead assembly of a producing well which controls oil or gas flow. Also *Xmas tree*. See **blowout preventer**.

chromatography Method of separating (often complex) mixtures. *Adsorption chromatography* depends on using solid adsorbents which have specific affinities for the adsolved substances. The mixture is introduced on to a column of the adsorbent, e.g. alumina, and the components eluted with a solvent or series of solvents and detected by physical or chemical methods. *Partition chromatography* applies the principle of **countercurrent distribution** to columns and involves the use of two immiscible solvent systems: one solvent system, the *stationary phase*, is supported on a suitable medium in a column and the mixture introduced in this system at the top of the column; the components are eluted by the other system, the *mobile phase*. See **gas-liquid chromatography, paper chromatography**.

chrome iron ore See **chromite**.

chrome spinel Another name for the mineral *picotite*, a member of the spinel group. Cf. **chromite**.

chromic salts Salts in which chromium is in the (III) oxidation state. They are usually green or violet, and are not readily oxidized or reduced.

chromite A double oxide of chromium and iron (generally also containing magnesium), used as a source of chromium and as a refractory for resisting high temperatures. It occurs as an accessory in some basic and ultrabasic rocks, and crystallizes in the cubic system as lustrous grey-black octahedra; also massive. Also *chrome iron ore*.

chromium Metallic element. Symbol Cr, at.no. 24, r.a.m. 51.996, rel.d. at 20°C, 7.138, mp 1900°C. The abundance in the Earth's crust is 122 ppm. The only ore is *chromite*, $FeCr_2O_4$, from which the element is extracted mainly for use in alloy steels, especially stainless steels, and for chromium plating.

chron The timespan of a **chronozone**. See **polarity chron**.

chronostratigraphy The standard hierarchical definition of geological time units, using all possible methods.

chronozone A small non-hierarchical chronostratigraphic unit.

chrysoberyl Beryllium aluminate, crystallizing in the orthorhombic system. The crystals often have a stellate habit and are green to yellow in colour. See **alexandrite**.

chrysoberyl cat's eye See **cymophane**.

chrysocolla A hydrated silicate of copper, often containing free silica and other impurities. It occurs in incrustations or thin seams, usually blue and amorphous.

chrysolite Sometimes applied to the whole of the olivine group of magnesium iron silicates but usually restricted to those richer in the magnesium component. In gemmology, an old name applied to several yellow and greenish-yellow stones.

chrysoprase An apple-green variety of chalcedony; the pigmentation is probably due to the oxide of nickel.

chrysotile A fibrous variety of serpentine once forming the most valuable type of the asbestos of commerce but now little used because of its carcinogenicity. Also *Canadian asbestos*.

churn drill See **cable tool drilling**.

chute (1) An inclined trough for the transference of broken coal or ore. (2) An area of rich ore in an inclined vein or lode, generally of much greater vertical than lateral extent.

cinders Volcanic *lapilli* composed mainly of dark glass and containing numerous vesicles (air or gas bubbles).

cinnabar Mercury sulphide, HgS, occurring as red acicular crystals, or massive; the ore

cinnamon stone See **hessonite**.

CIP Abbrev. for **Carbon-In-Pulp**.

CIPW A quantitative scheme of rock classification based on the comparison of norms; devised by four US petrologists, Cross, Iddings, Pirsson and Washington.

circulation loss Of mud being pumped down a borehole during drilling, it indicates the presence of porous or void conditions *downhole* and is potentially serious.

cirque, corrie A semi-amphitheatre, or 'armchair-shaped' hollow, of large size, excavated in mountain country by, or under the influence of, ice.

cis- A prefix indicating that geometrical isomer in which the two radicals are adjacent in a metal complex or on the same side of a double bond or alicyclic ring (L. 'on this side'). Cf. **trans-**.

citrine Not the true **topaz** of mineralogists but a yellow variety of quartz, which closely resembles it in colour but not in other physical characters; it is of much less value than topaz, and figures under a number of geographical names like *Indian topaz, Spanish topaz*. Also *false topaz*. But see **Brazilian topaz**, the true mineral.

Cl Symbol for **chlorine**.

cladding The material used for the lining of a mine shaft. Traditionally of timber battens or planks, now frequently of precast reinforced concrete slabs.

clan A suite of igneous rock types closely related in chemical composition but differing in mode of occurrence, texture, and possibly in mineral contents.

clarain Bands in coal, characterized by bright colour and silky lustre.

Clarke The average value of a chemical element in the Earth's crust.

classifier A machine for separating the product of ore-crushing plant into 2 portions consisting of particles of different sizes. In general, the finer particles are carried off by a stream of water, while the larger settle. The fine portion is known as the *overflow* or *slime*, the coarse as the *underflow* or *sand*.

clastic rocks Rocks formed of fragments of pre-existing rocks.

clathrate Form of compound in which one component is combined with another by the enclosure of one kind of molecule by the structure of another, e.g. rare gas in 1,4-dihydroxybenzene. See **molecular sieve**.

Clausius-Clapeyron equation Equation showing the influence of pressure on the temperature at which a change of state occurs, and the variation of vapour pressure with temperature.

$$q = T \cdot \Delta v \cdot \frac{dp}{dT},$$

where q is the heat absorbed (latent heat), T is the absolute temperature, Δv is the change in volume.

Claus process A process for recovering elemental sulphur from H_2S, by **desulphurizing**. It is a 2-stage process as defined in the 2 chemical equations. The H_2S is divided into 2 streams. 1st step $2H_2S + 3O_2 = 2H_2O + 2SO_2$. 2nd step $2H_2S + SO_2 = 2H_2O + 3S$. The second step is carried out at a high enough temperature to produce dry sulphur.

clay A fine-textured, sedimentary or residual deposit. It consists of hydrated silicates of aluminium mixed with various impurities. A fine-grained sediment of variable composition having a grain size less than 1/256 mm (4 microns). Clay for use in the manufacture of pottery and bricks must be fine-grained and sufficiently plastic to be moulded when wet; it must retain its shape when dried, and sinter together, forming a hard coherent mass without losing its original shape, when heated to a sufficiently high temperature. See **clay mineral**.

clay ironstone Nodular beds of clay and iron minerals, often associated with the Coal Measure rocks.

clay mineral Any mineral of clay grade but specifically one of a complex group of finely crystalline, metacolloidal or amorphous hydrous aluminosilicates. They have sheet-like lattices (*phyllosilicates*) and most are formed by the weathering of primary silicate minerals. The most common clay minerals belong to the *kaolin, montmorillonite (smectite)* or *illite* groups. See **silicates**.

clay with flints A stiff clay, containing unworn flints, which occurs as a residual deposit in chalk areas, but which is extensively mixed with other superficial deposits.

cleats The main cleavage planes or joint planes in a seam of coal.

cleavage (1) The splitting of a crystal along certain planes parallel to certain actual or possible crystal faces, when subjected to tension. (2) A property of rocks, such as slates, whereby they can be split into thin sheets. Cleavage is produced and oriented by the pressures that have affected rocks during consolidation and earth movements.

cleavelandite A platy variety of **albite**.

cleveite Variety of **pitchblende** containing uranium oxide and rare earths; often occluding substantial amounts of helium.

Cleveland Iron Ore An ironstone consisting of iron carbonate, which occurs in the Middle Lias rocks of North Yorkshire near Middlesbrough. The ironstone is oolitic and yields on the average 30% iron.

clinkstone See **phonolite**.

clinochlore A variety of **chlorite**.

clinohumite One of the four members of the humite group, magnesium silicate with hydroxide, occurring in metamorphosed limestones.

clinopyroxene A general term for the monoclinic members of the pyroxene group of silicates.

clinopyroxenite An ultramafic plutonic rock consisting almost entirely of clinopyroxene.

clino-rhomboidal crystals See **triclinic system**.

clinozoisite A hydrous calcium aluminium silicate in the epidote group. Differs from zoisite in crystallizing in the monoclinic system.

clintonite Hydrated aluminium, magnesium, calcium silicate. One of the brittle micas; differs optically from xanthophyllite.

cloanthite See **chloanthite**.

closed-circuit grinding Size reduction of solids done in several stages, the material after each stage being separated into coarser and finer fractions, the coarser being returned for further size reduction and the finer being passed on to a further stage, so that overgrinding is minimized.

cloud point Standard test on fuel oils, esp. *DERV*, for the temperature at which wax begins to appear (as a white cloudiness) in the fuel as it is cooled. Precipitating wax may lead to blocked filters and fuel lines.

clunch (1) Sear earth below coal seam of marl or shale. (2) A tough fireclay.

CN Abbrev. for **Co-ordination Number**.

Cnidaria A phylum, including the *corals*, previously called *Coelenterata*.

Co Symbol for **cobalt**.

coacervation The reversible aggregation of particles of an emulsoid into liquid droplets preceding flocculation.

coal A general name for firm brittle *carbonaceous rocks*; derived from vegetable debris, but altered, particularly in respect of volatile constituents, by pressure, temperature and a variety of chemical processes. The various types are classified on basis of volatile content, calorific value, caking and coking properties.

coal-cutting machinery Mechanized systems used in colliery to detach coal from its face, and perhaps to gather and load it to transporting device.

Coal Measures The uppermost division of the **Carboniferous**, consisting of beds of coal interstratified with shales, sandstones, and limestones and conglomerates.

coal-oil mixture A stabilized suspension of coal in fuel oil that may be transported as a liquid by pumping through a pipeline, or through jets for burning. Abbrev. *COM*.

coal sizes Sizes officially (UK) recognized are: large coal, over 6 in; large cobbles, 6 to 3 in; cobbles, 4 to 2 in; trebles, 3 to 2 in; doubles, 2 to 1 in; singles, 1 to ½ in; peas ½ to ¼ in; grains, ¼ to ⅛ in.

coaltar The distillation products of the high- or low-temperature carbonization of coal. Coaltar consists of hydrocarbon oils (benzene, toluene, xylene and higher homologues), phenols (carbolic acid, cresols, xylenols and higher homologues), and bases, such as pyridine, quinoline, pyrrole, and their derivatives.

coal washery Cleaning plant where run-of-mine coal is processed to remove shale and pyrite, reduce ash and sort in sizes.

coating and wrapping The process of covering a pipeline with bitumen and winding protective paper around it. Often done by machine just before the welded pipe is lowered into the trench.

cobalt Hard, grey metallic element. Symbol Co, at.no. 27, r.a.m. 58.93, rel.d. 8.90, mp 1495°C, abundance in the Earth's crust 29 ppm. It has a weak affinity for oxygen and sulphur, and is readily soluble in molten iron. There are a number of independent cobalt minerals, e.g. *smaltite* and *cobaltite*. It is an essential trace element in living systems, but toxic in excess.

cobalt bloom See **erythrite**.

cobaltite Sulphide and arsenide of cobalt, crystallizing in the cubic system; usually found massive and compact, with smaltite.

cobalt vitriol See **bieberite**.

cobble (1) A rock fragment intermediate in size between a pebble and a boulder, dia > 64 mm < 256 mm. (2) See **coal sizes**.

cockscomb, cockscomb pyrites A twinned form of **marcasite**.

co-current contact In chemical engineering processes which involve the transfer of heat or mass between two streams A and B, the opposite of **countercurrent contact**. At the start of the process, fresh A is in contact with fresh B, and at each succeeding stage progressively spent reagents are in contact, so that the effective driving force, and hence the overall economy, is lower.

codes of construction

codes of construction Written procedures for design, selection of materials, and manufacturing and operating procedures for production of chemical plant items for onerous conditions. Many are internationally acceptable, e.g. *ASME* (American Society of Mechanical Engineers), *TEMA* (Tubular Exchange Manufacturers Association), *ABCM* (Association of British Chemical Manufacturers).

coefficient of absorption See **absorption coefficient**.

coefficient of elasticity See **elasticity**.

coefficient of expansion The fractional expansion (i.e. the expansion of the unit length, area, or volume) per degree rise in temperature. Calling the coefficients of linear, superficial and cubical expansion of a substance α, β, and γ respectively, β is approximately twice, and γ three times, α.

coefficient of viscosity The value of the tangential force per unit area which is necessary to maintain unit relative velocity between two parallel planes unit distance apart in a fluid; symbol η. That is, if F is the tangential force on the area A and (dv/dz) is the velocity gradient perpendicular to the direction of flow, then $F = \eta A(\delta\varpi/\delta\zeta)$. For normal ranges of temperature, η for a liquid decreases with increase in temperature and is independent of the pressure. Unit of measurement is the **poise**, 10^{-1} Nm^{-2}s in SI units and 1 dyne cm^{-2}s in CGS units.

coelenterate A member of the phylum, Coelenterata. Radially symmetrical aquatic animals including the important fossil corals which range from the Ordovician to the present. See **Cnidaria**.

coesite A high-pressure variety of silica. Found in rocks subjected to the impact of large meteorites; but first made as a synthetic compound.

coffering (1) The operation involved in the construction of dams for impounding water. (2) Shaft lining impervious to normal water pressure.

coffinite Black, hydrous uranium silicate, U(SiO$_4$)$_{1-x}$.(OH)$_{4x}$. See **uranium**.

cohesion The attraction between molecules of a liquid which enables drops and thin films to be formed. In gases the molecules are too far apart for cohesion to be appreciable.

cold pinch Emergency closing of a ruptured pipe line by flattening it with hydraulic pincers.

colemanite Hydrated calcium borate, crystallizing in the monoclinic system; occurs as nodules in clay found in California and else-

colorimetric analysis

where.

collar (1) Concrete platform from which shaft linings are suspended at top-entry end of drill hole. (2) Heavy components placed above the drill in a *drill string* to stabilize the drill and to smooth torsional shock.

collector agent In froth-flotation, a chemical which is adsorbed by one of the minerals in an ore pulp, causing it to become hydrophobic and removable as a mineralized froth. Also *promoter*.

colligative properties Those properties of solutions which depend only on the concentration of dissolved particles, ions and molecules, and not on their nature. They include depression of freezing point, elevation of boiling point and osmotic pressure.

colloid From Gk. *kolla*, glue. Name originally given by Graham to amorphous solids, like gelatine and rubber, which spontaneously disperse in suitable solvents to form lyophilic sols. The term currently denotes any colloidal system.

colloidal movement See **Brownian movement**.

colloidal mud Thixotropic mixture of finely divided clays with baryte and/or bentonite, used in the drilling of deep oil bores. See **drilling mud**.

colloidal state A state of subdivision of matter in which the particle size varies from that of true 'molecular' solutions to that of coarse suspensions, the diameter of the particles lying between 1 nm and 100 nm. The particles are charged and can be subjected to electrophoresis, except at the **iso-electric point**. They are subject to Brownian movement and have a large amount of surface activity.

colloid mill Mill with very fine clearance between the grinding components, operating at high speed, and capable of reducing a given product to a particle size of 0.1 to 1 μm.

collophane A cryptocrystalline variety of *apatite*.

Colorado ruby An incorrect name for the fiery-red garnet (pyrope) crystals obtained from Colorado and certain other parts of US.

Colorado topaz True topaz of a brownish-yellow colour is obtained in Colorado, but quartz similarly coloured is sometimes sold under the same name.

colorimeter An instrument used for the precise measurement of the hue, purity and brightness of a colour.

colorimetric analysis Analysis of a solution by comparison of the colour produced by a reagent with that produced in a standard

colour index

solution.

colour index A number which represents the percentage by volume of dark-coloured heavy silicates in an igneous rock, and is thus a measure of its leucocratic, mesocratic, or melanocratic character.

colours of thin films When light is reflected from a thin film, such as a soap bubble or a layer of oil on water, coloured effects are seen which are due to optical interference between light reflected from the upper surface and that from the lower. The colours are not bright unless the film is less than about 10^{-3} mm thick.

colour temperature That temperature of a black body which radiates with the same dominant wavelengths as those apparent from a source being described.

columbite Niobate and tantalate of iron and manganese. When the Nb content exceeds that of Ta, the ore is called columbite. See **tantalite**.

columnar structure A form of regular jointing, produced by contraction following crystallization and cooling in igneous rocks, esp. those of basic composition. The columns are generally roughly perpendicular to the cooling surface.

comagmatic assemblage Refers to those igneous rocks with a common set of chemical, mineralogical and textural features, which suggest a derivation from a common parent magma.

combustion The chemical union of oxygen with gas that is accompanied by the evolution of light and rapid production of heat (exothermic).

comendite A variety of rhyolite with phenocrysts of quartz, alkali feldspar, sodic pyroxenes and amphiboles.

comminution Size-reduction by breaking, crushing, or grinding, e.g. in ore-dressing.

comparison prism A small right-angled prism placed in front of a portion of the slit of a spectroscope or spectrograph for the purpose of reflecting light from a second source of light into the collimator, so that two spectra may be viewed simultaneously.

compensator (1) A glass plate used in various optical interferometers to achieve equality of optical path length. (2) An apparatus in a polarizing microscope that measures the phase difference between the two components of polarized light, e.g. a *Berek compensator*. (3) A plate of variable thickness of optically active quartz used to produce elliptically polarized light of a given orientation.

components The individual chemical substances present in a system. See **phase rule**.

concordant intrusion

composition The nature of the elements present in a substance and the proportions in which they occur.

composition of atmosphere Dry atmospheric air contains the following gases in the proportions (by weight) indicated: nitrogen, 75.5; oxygen, 23.14; argon, 1.3; carbon dioxide, 0.05; krypton, 0.028; xenon, 0.005; neon, 0.00086; helium, 0.000056. There are variable trace amounts of other gases incl. hydrogen and ozone. Water content, which varies greatly, is excluded from this analysis.

compound See **chemical compound**.

compound fault A series of closely spaced parallel or subparallel faults.

compressibility The reciprocal of the bulk modulus.

compression waves See **earthquakes**.

Compton's rule An empirical rule that the melting point of an element in kelvin is equal to half the product of the r.a.m. and the specific latent heat of fusion.

CONCAWE Abbrev. for *CONservation Clean Air and Water - western Europe*. An association of oil refining companies operating in Europe that provides a service to international bodies, e.g. the EC, and publishes the results of environmental studies.

concentrate The products which have been concentrated so that a relatively high content of mineral has been obtained, and which are ready for treatment by chemical methods.

concentrating table Supported deck, across or along which mineralized sands are washed or moved to produce differentiated products according to the gravitational response of particles of varying size and/or density. Stationary tables include strakes, sluices, buddles; moving tables include shaking tables (e.g. *Wilfley table*), vanners and rockers.

concentration (1) Number of molecules or ions of a substance in a given volume, generally expressed as moles per cubic metre or cubic decimetre. (2) The process by which the concentration of a substance is increased, e.g. the evaporation of the solvent from a solution.

concentration plant *Concentrator mill, reduction works, washing, cleaning plant*. Buildings and installations in which ore is processed by physical, chemical and/or electrical methods to retain its valuable constituents and discard as tailings those of no commercial interest. See **mineral processing**.

concordant intrusion An igneous intrusion that lies parallel to the bedding or foliation of the country rock which it

concretion

intrudes. See **dykes and sills**.

concretion Nodular or irregular concentration of siliceous, calcareous, or other materials, formed by localized deposition from solution in sedimentary rocks.

condensation The union of two or more molecules with the elimination of a simpler group, such as H_2O, NH_3 etc.

condensed nucleus A ring system in which two rings have one or more (generally two) atoms in common, e.g. naphthalene, phenanthrene, quinoline.

condensed system System in which there is no vapour phase. The effect of pressure is then practically negligible, and the *phase rule* may be written $P+F = C+1$.

condenser (1) Apparatus used for condensing vapours obtained in distillation. In laboratory practice usually a single tube, either freely exposed to air or contained in a jacket in which water circulates. (2) A system of lenses below the stage of a polarizing microscope arranged to illuminate the object in a manner suitable for the type of observation required.

conditioning In froth flotation, the treatment of mineral pulp with small additions of chemicals designed to develop specific aerophilic or aerophobic qualities on surfaces of different mineral species as a prelude to their separation.

conduction of heat The transfer of heat from one portion of a medium to another, without visible motion of the medium, the heat energy being passed from molecule to molecule. See **thermal conductivity**.

conductor See **top casing**.

cone classifier Large inverted cone into which ore pulp is fed centrally from above. Coarser material settles to bottom discharge and finer overflows peripherally.

cone-in-cone structure Cones stacked inside one another, occurring in sedimentary rocks; usually of fibrous calcite, sometimes of other minerals.

cone sheets Minor intrusions which occur as inwardly inclined sheets of igneous rock and have the form of segments of concentric cones. See **dykes and sills**.

conformable strata An unbroken succession of strata. See **non-sequence, unconformity**.

conglomerate A coarse-grained clastic sedimentary rock composed of rounded or subrounded fragments larger than 2 mm in size (e.g. pebbles, cobbles, boulders).

Coniacian A stage in the Upper Cretaceous. See **Mesozoic**.

conical drum Winding drum, to which

constituents

hoisting rope of cage or skip is attached. This shape aids in smooth acceleration from rest to full speed where the rope reaches the flat central part of the compound drum (the full diameter between the cones).

conical projections In these, the projection is on to an imaginary cone touching the sphere at a given standard parallel. There are equal-area and two standard (*Gall's projection*) versions, as in the cylindrical type. Best for middle latitudes.

coning and quartering Production of representative sample from a large pile of material such as ore, in which it is first formed into a cone by deposition centrally, and then reduced by removal, one shovelful at a time, into four separate piles drawn alternately from four peripheral opposed points. Two are then discarded and the process (perhaps with intermediate size reduction) is repeated until a manageable hand sample is obtained.

conjugate solutions If two liquids A and B are partially miscible, they will produce at equilibrium two conjugate solutions, the one of A in B, the other B in A.

connate water Water trapped in a rock at the time of its deposition.

conodont One of a large number of microscopic phosphatic fossils, of doubtful affinity, having a tooth-like appearance. They ranged from the Cambrian (or earlier) to the Triassic and are useful for stratigraphical correlation.

conoscopic observation The investigation of the behaviour of doubly refracting crystal plates under convergent polarized light. Interference pictures obtained are important in explaining crystal optical phenomena.

consanguinity Term applied to rocks having a similarity or community of origin, which is revealed by common peculiarities of mineral and chemical composition and often also of texture.

consequent drainage A river system directly related to the geological structure of the area in which it occurs. See **drainage patterns**.

conservation laws Usually refers to the classical laws of the separate conservation of mass, of energy and of atomic species, which are sufficiently accurate for most chemical reactions.

consolidation Of strata, the drying, compacting and induration of rocks, as a result of pressures operating after deposition.

constituents All the substances present in a system.

constitution Structural distribution of atoms and/or ions composing a regularly co-ordinated substance. Includes percentage of each constituent and its regularity of occurrence through the material.

constitutional ash Ash resulting from combustion of coal and derived from siliceous matter in the coal-forming plants. 'Fixed ash' as distinct from entrained impurity of 'free ash'.

constitutional formula A formula which shows the arrangement of the atoms in a molecule.

constitutional water See **water of hydration**.

constringence Inverse of the dispersive power of a medium. Ratio of the mean refractive index diminished by unity to the difference of the refractive indices for red and violet light.

contact angle (1) The angle between the liquid and the solid at the liquid-solid-gas interface. It is acute for wetting (e.g. water on glass) and obtuse for non-wetting (e.g. water on paraffin wax). (2) The angle between a bubble of air and the chemically clean, polished and horizontal surface of a specimen of mineral to which it clings, measured between that surface and the side of the bubble. This forms an index to the floatability of the species under prescribed conditions, in which chemicals are added to the water and change in angle observed.

contact aureole See **aureole**.

contact metamorphism The alteration of rocks caused by their contact with, or proximity to, a body of igneous rock.

contact vein A vein occurring along the line of contact of two different rock formations, one of which may be an igneous intrusion.

contaminated rock Igneous rock whose composition has been modified by the incorporation of other rock material.

continent One of the Earth's major land masses, including the dry land and continental shelves.

continental crust That part of the Earth's crust which underlies the continents and continental shelves. It is approximately 35 km thick in most regions but is thicker under mountainous areas. Sedimentary rocks predominate in its uppermost part and metamorphic rocks at depth, but the detailed composition of the lower crust is uncertain. Cf. **oceanic crust**. See **Earth**.

continental deposit A rock formed under subaerial conditions or in water not directly connected with the sea. See **aeolian deposits, glacial deposits**.

continental drift A hypothesis put forward by Wegener in 1912 to explain the distribution of the continents and oceans, and the undoubted structural, geological and physical similarities which exist between continents. The continents were believed to have been formed from one large land mass and to have drifted apart. See **plate tectonics**.

continental rise That part of the continental margin between the **continental slope** and the **abyssal plain**. It is characterized by a relatively gentle slope.

continental shelf The gently sloping offshore zone, extending usually to about 200 m depth.

continental slope The relatively steep slope between the continental shelf and the more gentle rise from the abyssal plain.

continuous distillation An arrangement by which a fresh distillation charge is continuously fed into the still in the same measure as the still charge is distilled off. The contrary process is known as *batch distillation*.

continuous extraction Extraction of solids or liquids by the same solvent, which circulates through the extracted substance, evaporates and is condensed again, and continues the same cycle over again; or by exhaustive extraction with solvents in counter-current arrangement.

continuous spectrum Spectrum which shows the continuous non-discrete changes of intensity with wavelengths or particle energy.

contour fringes Interference fringes formed by the reflection of light from the top and bottom surfaces of a thin film or wedge. The fringes correspond to optical thickness. Also *Fizeau fringes*. See **Newton's rings**.

convection The very slow mass movement of subcrustal material; believed to be the mechanism that drives tectonic plates.

convergence zone In **plate tectonics** the zone where moving plates collide and area is lost by shortening and crustal thickening or by subduction and destruction of the crust. The site of earthquakes, mountain building, trenches and vulcanism.

convergent evolution The tendency of unrelated species to evolve similar structures, physiology or appearance due to the same selective pressures, e.g. the eyes of Vertebrates and Cephalopods.

co-ordination compound A compound generally described from the point of view of the central atom to which other atoms, called ligands, are bound or co-ordinated. Normally, the central atom is a (transition)

co-ordination number The number of atoms or groups (ligands) surrounding the central (nuclear) atom of a complex ion or molecule. Abbrev. *CN*.

copalite, copaline Synonyms for **Highgate resin**.

copper A bright reddish metallic element. Symbol Cu, at.no. 29, r.a.m. 63.55, mp 1083° C. Its abundance in the Earth's crust is 68 ppm. As the native element it forms thin sheets, but it mainly occurs in a large number of other minerals. These are primarily sulphides, sulphates and other oxidized minerals, and hydrates. It is an essential trace element for living organisms but toxic in excess.

copperas Iron sulphate, $FeSO_4.7H_2O$. See **melanterite**. *White copperas* is goslarite, $ZnSO_4.7H_2O$.

copper glance A popular name for **chalcocite**.

copper nickel See **niccolite**.

copper pyrites See **chalcopyrite**.

coprecipitation The precipitation of a radio-isotope with a similar substance, which precipitates with the same reagent, and which is added in order to assist the process.

coprolite Fossilised excreta of animals. Generally composed largely of calcium phosphate.

coquimbite Hydrated ferric sulphate, crystallizing in the hexagonal system, occurring in some ore deposits and also in volcanic fumaroles such as those of Vesuvius.

coquina A limestone made up of coarse shell fragments, usually of molluscs.

coral See panel on p. 48.

coral reef A calcareous bank formed of the calcareous skeletons of corals and algae which live in colonies. The various formations of coral reefs are known as *atolls*, *barrier reefs* and *fringing reefs*.

coral sand A sand made up of calcium carbonate grains derived from eroded coral skeletons, often found in deep water on the seaward side of a coral reef.

cordierite Magnesium aluminium silicate with some iron, typically occurring in thermally metamorphosed rocks and in some gneisses. Often shows cyclic twinning. *Iolite* is a name often used for gem varieties and the mineral is also sometimes called *dichroite* from its strong blue to colourless dichroism.

core (1) The central part of the Earth at a depth below 2900 km. It is believed to be composed of nickel and iron. (2) Cylindrical rock section cut by rotating hollow drill bit in prospecting, sampling, blasting.

core sample Sample from a borehole to give information on the rock formation at side or bottom. Usually a few inches in diameter.

cornelian See **carnelian**.

cornstone Concretionary limestone common in the Old Red Sandstone and Permo-Triassic rocks of Britain. Characteristic of soils formed in arid conditions.

corrasion The work of vertical or lateral cutting performed by a river by virtue of the abrasive power of its load. See **rivers, geological work of**.

correction for buoyancy In precision weighing it is necessary to correct for the differences in the buoyancy of the air for the body being weighed and for the weights. The correction to be added to the value w, of the weights (in grams) is: $1.2w(1/D-1/\delta)$ mg, where D and δ are the densities of the body and the weights respectively in g cm^{-3}.

correlation The linking together of strata of the same age occurring in separate outcrops.

corrie See **cirque**.

corrosion (1) The slow wearing away of solids, esp. metals, by chemical attack. (2) Petrologically, the modification of crystals formed early in the solidification of an igneous rock by the chemical action of the residual magma. (3) Geomorphologically, erosion by chemical processes.

corundum Oxide of aluminium, crystallizing in the trigonal system. It is next to diamond in hardness, and hence is used as an abrasive. The clear blue variety is *sapphire* and the clear red *ruby*. See **white sapphire**.

cosmic rays Highly penetrating rays from outer space. The *primary cosmic rays* which enter the Earth's upper atmosphere consist mainly of protons with smaller amounts of helium and other heavier nuclei. Cosmic rays of low energy have their origin in the Sun, those of high energy in galactic or extragalactic space, possibly as a result of supernova explosions. Collision with atmospheric particles results in *secondary cosmic rays* and particles of many kinds, including neutrons, mesons and hyperons.

cosmogenic Said of an isotope capable of being produced by the interaction of cosmic radiation with the atmosphere or the surface of the Earth.

cossyrite See **aenigmatite**.

costeaning Prospecting by shallow pits or trenches designed to expose lode outcrop.

cotton ball See **ulexite**.

coral

Coral is a general name given to a group of bottom-dwelling, sessile marine invertebrates that belong to the class Anthozoa of the phylum Coelenterata (Cnidaria). They secrete external calcareous skeletons that are found abundantly in many rocks from the Ordovician onwards and they are common in warm modern seas. They occur as solitary individuals or grow into colonies, and have been used as zone fossils in the Carboniferous Limestone Series. The whole skeleton of a coral is the *corallum*; in colonial corals, a skeleton of the compound coral formed by an individual coral *polyp* is a *corallite*. The *Tabulate* corals range from the Ordovician to the Permian. They are simple colonial types with numerous and well-developed transverse plates (*tabulae*) and rudimentary septa (Fig. 1). The Zoantheria is a subclass that has two important orders, the Rugosa or Tetracoralla, and the Scleractinia or Hexacoralla which include most post-Palaeozoic and living corals, and range from the Trias to the reef-building corals of the present day. The Rugosa have solid corallite walls and septa with fourfold divisions (Fig. 2). They range from simple to complex colonial types. The Scleractinia have septa in sets of six (Fig. 3).

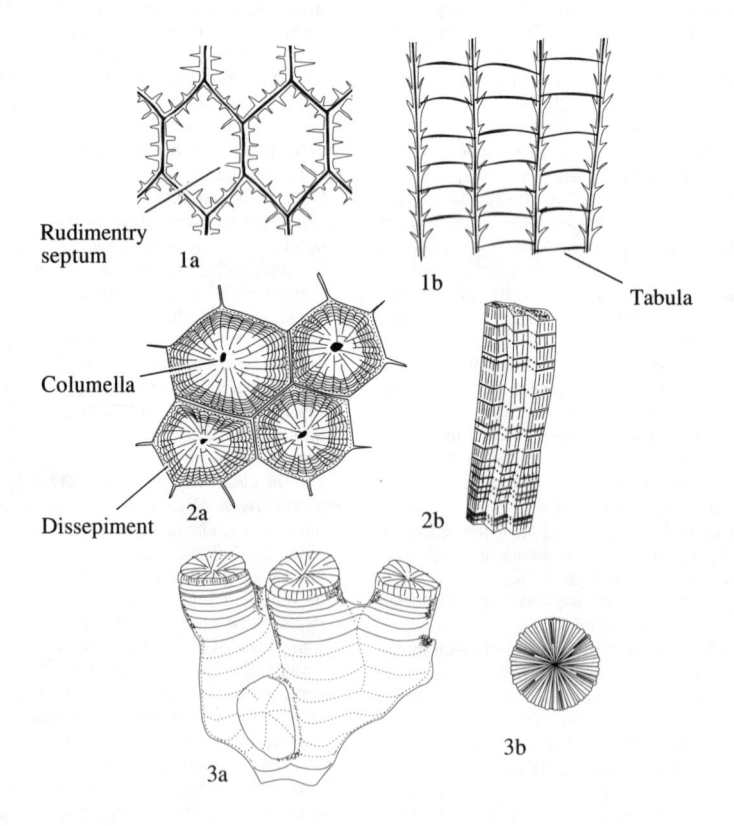

Fig. 1. *Favosites* sp., Silurian. a, horizontal section × 5; b, vertical section × 3.
Fig. 2. *Lithostrotion* sp., Carboniferous Limestone. a, horizontal section × 2; b, vertical view × ⅓.
Fig. 3. *Thecosmilia* sp., Jurassic. × ½. a, general view; b, septal view.

coulomb SI unit of electric charge, defined as that charge which is transported when a current of one ampere flows for one second. Symbol C.

Coulomb's law Fundamental law which states that the electric force of attraction or repulsion between two point charges is proportional to the product of the charges and inversely proportional to the square of the distance between them. The force also depends on the permittivity of the medium in which the charges are placed. In SI units, if Q_1 and Q_2 are the point charges a distance d apart, the force is

$$F = \frac{1}{4\pi\varepsilon} \frac{Q_1 Q_2}{d^2},$$

where ε is the permittivity of the medium. The force is attractive for charges of opposite sign and repulsive for charges of the same sign.

countercurrent contact In chemical engineering processes involving the transfer of heat or mass between two streams A and B, as in liquid extraction, the arrangement of flow so that at all stages the more spent A contacts the less spent B, thus ensuring a more even distribution and greater economy than with **co-current contact**.

countercurrent distribution A repetitive distribution of a mixture of solutes between two immiscible solvents in a series of vessels in which the two solvent phases are in contact. The components are distributed in the vessels according to their partition coefficients.

countercurrent treatment Arrangement used in chemical extraction of values from ore, in washing rich liquor away from spent sands. The stripping liquid enters 'barren' at one end of a typical layout and the rich ore pulp at the other. They pass countercurrent through a series of vessels, the pulp emerging stripped after its final wash with the new 'barren' liquor, and the liquor leaving at the far end, now rich with dissolved values and 'pregnant'.

country rock The rock forming the walls of a reef or lode.

coupling Short tube internally threaded at both ends for joining two lengths of tube. See **slip joint**.

covalency The union of 2 or more atoms by the sharing of one or more pairs of electrons.

covalent bond A chemical bond in which two or more atoms are held together by the interaction of their nuclei with one or more pairs of electrons.

covellite, covelline Sulphide of copper, CuS. The colour is indigo-blue or darker. Also *indigo copper*.

Cr Symbol for **chromium**.

cracking Breaking down heavier crude oil fractions to lighter fractions by heat, pressure and the use of catalysts in the refinery.

crag A rough, steep, precipitous projecting rock. Also *craig* (Scottish).

Crag A local type of shelly and sandy rocks which have been deposited in relatively shallow water; found in the Pliocene of East Anglia.

crag-and-tail A hill consisting partly of solid rock shaped by ice action, with a *tail* of morainic material banked against it on the lee-side, e.g. the Castle Rock, Edinburgh.

crane barge In offshore drilling, a special vessel for handling heavy loads in the supply and maintenance of drilling platforms.

crater A more or less circular depression generally caused by volcanic activity, occasionally by meteoritic activity.

craton A part of a continent that has attained crustal stability; typically Precambrian or Lower Palaeozoic in age. Also *kraton*.

creep Gradual rising of the floor in a coal-mine due to pressure. See **crush**.

creosote oil A coal-tar fraction, boiling between 240°C and 270°C. The crude creosote oil is used as raw material for producing tar acids etc., or used direct as a germicide, insecticide, or disinfectant in various connections (e.g. soaps, sheep dips, impregnation of railway sleepers etc.).

crest The highest part of an anticlinal fold. A line drawn along the highest points of a particular bed is called the *crest line*, and a plane containing the crest lines of successive folds is called the *crest plane*. See **folding**.

Cretaceous The youngest period of the Mesozoic era covering an approx. time span from 145–65 million years ago. The corresponding system of rocks; the **Chalk** is its most striking deposit. See **Mesozoic**.

crevasse A fissure, often deep and wide, in a glacier or ice sheet.

crib (1) An interval from work underground for croust, bait, snack, downer, piece, chop, snap, bite or tiffin. (2) A job. (3) A form of timber support.

crinanite A basic igneous rock, consisting of intergrown crystals of feldspar, titanaugite, olivine and analcite. Similar to **teschenite**.

crinoid See **echinoderms**.

cristobalite The high-temperature form of SiO_2. It is found in volcanic rocks and is stable from 1470°C to 1713°C, but exists at lower temperatures. Another, low temperature form is not stable above atmospheric

critical angle

pressure and is a constituent of *opal*.

critical angle See **gems and gemstones**.

critical coefficient Additive property, a measure of the molar volume. It is defined as the ratio of the critical temperature to the critical pressure.

critical humidity Humidity at which the equilibrium water vapour pressure of a substance is equal to the partial pressure of the water vapour in the atmosphere, so that it would neither lose nor gain water on exposure.

critical point The critical temperature of a substance at which the pressure and volume have their critical values. At the critical point the densities (and other physical properties) of the liquid and gaseous states are identical.

critical solution temperature The temperature above which two liquids are miscible in all proportions.

critical volume The volume of unit mass (usually 1 mole) of a substance under critical conditions of temperature and pressure.

crocidolite The blue asbestos of South Africa, a fibrous variety of the amphibole riebeckite. Long coarse flexible spinning fibre with a high resistance to acids. No longer in use because of its carcinogenic properties. See **Cape asbestos, tiger's eye**.

crocoite, crocoisite Lead chromate, bright orange-red in colour.

Cromerian A temperate stage in the late Pleistocene. See **Quaternary**.

crop See **outcrop**.

cross bedding Internally inclined planes in a rock related to the original direction of current flow. Also *current bedding, false bedding*.

crosscut In metal mining, a level or tunnel driven through the country rock, generally from a shaft, to intersect a vein or lode. See **mining**.

crossed Nicols Two Nicol prisms arranged with their principal planes at right angles, in which position the plane-polarized light emerging from one Nicol is extinguished by the other. Similarly for *polarizers*.

Crowe process Method of removing oxygen from cyanide solution before recovery of dissolved gold, in which liquor is exposed to vacuum as it flows over trays in a tower.

crucible A refractory vessel or pot in which metals are melted. In chemical analysis, smaller crucibles, made of porcelain, nickel or platinum, are used for igniting precipitates, fusing alkalies etc.

crude oil See **petroleum**.

crump Rock movement under stress due to underground mining, possibly violent.

crystal dislocation

crush The broken condition of pillars of coal in a mine due to pressure of the strata. See **creep**.

crush breccia A rock consisting of angular fragments, often recemented, which has resulted from the faulting or folding of pre-existing rocks. See **crush conglomerate, fault breccia**.

crush conglomerate A rock consisting of crushed and rolled fragments, often recemented; it has resulted from the folding or faulting of pre-existing rocks.

crusher Machine used in earlier stages of comminution of hard rock. Typically works on dry feed as it falls between advancing and receding breaking plates. Types include **jaw breaker** or *jaw crusher*, **gyratory breaker, rolls, stamp**.

crust The outermost layer of the lithosphere consisting of relatively light rocks. Continental crust consists largely of granitic material; oceanic crust is largely basaltic.

crust of the Earth See **Earth**.

cryogenic Term applied to low-temperature substances and apparatus.

cryogenics The study of materials at low temperatures.

cryolite Sodium aluminium fluoride, used in the manufacture of **aluminium**. Also *Greenland spar*.

cryptocrystalline The texture of a rock in which the crystals are too small to be recognized under a petrological microscope.

crystal (1) Homogenous solid body of chemical element, compound or isomorphous mixture having a regular atomic lattice arrangement that may be shown by crystal faces. See **Bravais lattices, crystal lattice, unit cell**. (2) Old name for **quartz**.

crystal anisotropy In general, directional variations of any physical property, e.g. elasticity, thermal conductivity etc., in crystalline materials. Leads to existence of favoured directions of magnetization, related to lattice structure in some ferromagnetic crystals.

crystal axes The axes of the natural co-ordinate system formed by the crystal lattice. These are perpendicular to the natural faces for many crystals. See **uniaxial, biaxial**.

crystal boundaries The surfaces of contact between adjacent crystals in a metal. Anything not soluble in the crystals tends to be situated at the crystal boundaries, but in the absence of this, the boundary between two similar crystals is simply the region where the orientation changes.

crystal dislocation Imperfect alignment between the lattices at the junctions of

crystal lattice

A crystal lattice is the three-dimensional repeating array of points used to represent the structure of a crystal. The lattices were classified into fourteen groups by Bravais. See **Bravais lattices**, **unit cell**.

The crystal lattice extends indefinitely in all directions. Thus, for example, in a unit cell of sodium chloride, NaCl (*halite, rock salt*) any sodium or chloride ion can be considered to be at the corner of a unit cell. The pattern is cubic, Na^+ ions alternating with Cl^- ions. In the left-hand figure their sizes are not drawn to scale and either the white or black circles can be considered to be Na^+ or Cl^-.

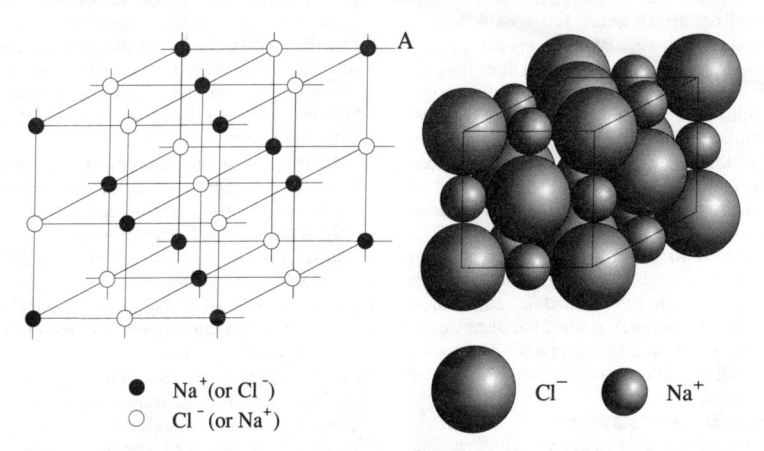

- ● Na^+ (or Cl^-)
- ○ Cl^- (or Na^+)

Cl^- Na^+

The ion marked A above is at the corner of the cell shown, so it is shared by a total of 8 adjacent cells (7 not shown). Four of these are below it and four above, occurring both to the right and left of the ion. To each of these cells it can be considered as contributing ⅛ Na^+ (or Cl^-). Each of these unit cells has eight such corners so each cell has a complete ion, Na^+ or Cl^-, contributed by the eight corners. There are four NaCl formula units in the unit cell.

The radii of the sodium and chloride ions are 0.98 Å and 1.81 Å respectively so the real arrangement of the ions more closely resembles the *spacefilling* drawing shown on the right.

---

small blocks of ions ('mosaics') within the crystal. The resulting mobility and opportunity for realignment of the molecules is of importance in crystal growth, plastic flow, sintering etc.

crystal face One of the bounding surfaces of a crystal. In the case of small, undistorted crystals, each face is an optically plane surface. A *cleavage face* is the smooth surface resulting from cleavage; in such minerals as mica, the cleavage face may be almost a plane surface, diverging only by the thickness of a molecule.

crystal goniometer An instrument for measuring angles between crystal faces.

crystal indices See **Miller indices**.

crystal lattice See panel on this page.

crystalline Having a crystal structure.

crystalline form The external geometrical shape of a crystal.

crystalline overgrowth The growth of one crystal round another, frequently observed with isomorphous substances. Cf. **cubic system**.

crystalline rocks Rocks consisting wholly, or chiefly, of mineral crystals. They are usually formed by the solidification of molten rock, by metamorphic action or by precipitation from solution.

crystalline solid A solid in which the atoms or molecules are arranged in a regular manner, the values of certain physical properties depending on the direction in which they are measured. When formed freely, a

crystallites

crystalline mass is bounded by plane surfaces (faces) intersecting at definite angles. See **X-ray crystallography**.

crystallites (1) Very small, often imperfectly formed crystals. (2) Minute bodies occurring in glassy igneous rocks, and marking a stage in incipient crystallization.

crystallization Slow formation of a crystal from melt or solution.

crystalloblastic texture The texture of metamorphic rocks resulting from the growth of crystals in a solid medium.

crystallographic axes Lines of reference intersecting at the centre of a crystal. Crystal (or morphological) axes, usually three in number, by their relative lengths and attitude, determine the system to which a crystal belongs.

crystallographic notation Brief method of writing down the relation of any crystal face to certain axes of reference in the crystal.

crystallographic planes Any set of parallel and equally spaced planes that may be supposed to pass through the centres of atoms in crystals. As every plane must pass through atomic centres, and no centres must be situated between planes, the distance between successive planes in a set depends on their direction in relation to the arrangement of atomic centres.

crystallographic system Any of the major units of crystal classification embracing one or more symmetry classes.

crystal structure Structure consisting of the whole assemblage of rows and patterns of atoms, which have a definite arrangement in each crystal.

crystal systems A classification of crystals based on the intercepts made on the crystallographic axes by certain planes. See **cubic, tetragonal, hexagonal, orthorhombic, monoclinic** and **triclinic systems; Bravais lattices**.

crystal texture The size and arrangement of the individual crystals in a crystalline mass.

crystal tuff A tuff with crystal fragments more abundant than either lithic or vitric fragments.

Cs Symbol for **caesium**.

Cu Symbol for **copper**.

cubic close packing Stacking of spheres formed by stacking close-packed layers in the sequence ABCABC. The unit cell of such an arrangement is a face-centred cube, with four atoms per cell. This structure is adopted by many metals, e.g. Cu, Ag and Au. Abbrev. *CCP*. See **unit cell**.

cubic crystal lattice See **crystal lattice**.

cubic system The crystal system which has the highest degree of symmetry; it embraces such forms as the cube and octahedron. Also *isometric system*. See **Bravais lattices**.

cubic zirconia Artificial diamond simulant, difficult to detect.

cuesta A hill or ridge with steep slope on one side and a gentle dip-slope on the other. Synonym *escarpment*. Common term in US.

culm The name given to the rocks of Carboniferous age in the south-west of England, consisting of fine-grained sandstones and shales, with occasional thin banks of crushed coal or 'culm'.

cummingtonite Hydrated magnesium iron silicate, a monoclinic member of the amphibole group, occurring in metamorphic rocks. Differs from *grunerite* in having magnesium in excess of iron.

cupferron Ammonium-nitroso-β-phenyl-hydrazine. Reagent used in the colorimetric detection and estimation of copper.

cupid's darts See **flèches d'amour**.

cupola A dome-shaped offshoot rising from the top of a major intrusion.

cupric *Copper (II)*. Containing divalent copper. Copper (II) salts are blue or green when hydrated and are stable.

cupriferous pyrites See **chalcopyrite**.

cuprite Oxide of copper, crystallizing in the cubic system. It is usually red in colour and often occurs associated with native copper; a common ore.

cupro-uranite Synonym for *torbernite*. See **uranium**.

cuprous *Copper (I)*. Containing monovalent copper. Soluble copper (I) salts generally disproportionate to copper (0) and copper (II) salts.

curb Framework fixed in rock of mine shaft to act as foundation for brick or timber lining.

curie Unit of radioactivity: 1 curie is defined as 3.70×10^{10} decays per second, roughly equal to the activity of 1 g of ^{226}Ra. Abbrev. *Ci*. Now replaced by the becquerel (Bq): $1\ Bq = 2.7 \times 10^{-11}\ Ci$.

Curie point, Curie temperature (1) The temperature above which a ferromagnetic material becomes paramagnetic. Also *magnetic transition temperature*. (2) The temperature (*upper Curie point*) above which a ferroelectric material loses its polarization. (3) The temperature (*lower Curie point*) below which some ferroelectric materials lose their polarization.

Curie's law For paramagnetic substances, the magnetic susceptibility is inversely proportional to the absolute temperature.

Curie-Weiss law At the Curie temperature, θ, a ferromagnetic material becomes paramagnetic. Well above this temperature its paramagnetic susceptibility is $\chi = C/(T-\theta)$, where T is the absolute temperature and C is the Curie constant.

current Rate of flow of charge in a substance, solid, liquid or gas. Conventionally, it is opposite to the flow of (negative) electrons, this having been fixed before the nature of the electric current had been determined. Practical unit of current is the **ampere**.

current bedding See **cross bedding**.

customs plant A crushing or concentrating plant serving a group of mines on a contract basis. It buys ore according to valuable content and complexity of treatment, and relies for profit on sale of products.

cut A petroleum fraction.

cutback Bitumen that has been diluted with suitable solvents, e.g. kerosine, to make it liquid and easier to handle.

cutter loader Coal-cutting machine which both severs the mineral and loads it on to a transporting device such as a face conveyor.

cyanicide Any constituent in ore or chemical product made during treatment of gold-bearing minerals by cyanidation, which attacks or destroys the sodium or calcium cyanide used in the process.

cyanides Salts of hydrocyanic acid.

cyaniding The process of treating finely ground gold and silver ores with a weak solution of sodium cyanide, which readily dissolves these metals. The precious metals were formerly recovered by precipitation from solution with zinc. Currently a number of methods have proved more economical, including adsorption on resins and carbon. See **resin-in-pulp, carbon-in-pulp**.

cyanite See **kyanite**.

cycle of erosion The hypothetical course of development followed in landscape evolution; it consists of the major stages of youth, maturity and old age.

cyclic compounds Closed-chain or ring compounds consisting either of carbon atoms only (carbocyclic compounds), or of carbon atoms linked with one or more other atoms (heterocyclic compounds).

cyclic test See **locked test**.

cyclone Conical vessel used to classify dry powders or extract dust by centrifugal action. See **hydrocyclone**.

cyclonite Hexogen, cyclotrimethylene trinitramine $(CH_2)_3(N.NO_2)_3$, a colourless, crystalline solid, mp 200°–202°C, odourless, tasteless, non-poisonous, soluble in acetone, prepared by oxidative nitration of hexamethylene tetramine. Used, generally with TNT, as an explosive.

cyclosilicates Silicate minerals whose atomic structure contains rings of SiO_4 groups, e.g. **beryl**. See **silicates**.

cyclothem A series of beds formed during one sedimentary cycle. Particularly associated with coal-bearing rocks.

cymophane A variety of the gem-mineral *chrysoberyl* which exhibits chatoyancy. Also *chrysoberyl cat's eye* or *Oriental cat's eye*.

D

d- Abbrev. for *dextrorotatory*.

δ- Substituted on the fourth carbon atom of a chain.

D Symbol for **deuterium**.

D Symbol for (1) **angle of deviation**; (2) electric flux density (displacement); (3) **diffusion coefficient**.

Δ Prefixed symbol for a double bond beginning on the carbon atom indicated.

dacite A volcanic rock intermediate in composition between andesite and rhyolite, the volcanic equivalent of a granodiorite. See **volcanic rocks**.

Dalradian Series Very thick and variable succession of sedimentary and volcanic rocks which have suffered regional metamorphism. Occurring in the Scottish Highlands approximately between the Great Glen and the Highland Boundary fault. Referred to the Precambrian to Lower Palaeozoic in age.

dalton The atomic mass unit.

Dalton's atomic theory Theory stating that matter consists ultimately of indivisible, discrete particles (atoms), and atoms of the same element are identical; chemical action takes place as a result of attraction between these atoms, which combine in simple proportions. It has since been found that atoms of the chemical elements are not the ultimate particles of matter, and that atoms of different mass can have the same chemical properties (**isotopes**). Nevertheless, this theory of 1808 is fundamental to chemistry. See **atomic structure**.

Dalton's law See **law of multiple proportions**.

Dalton's law of partial pressures The pressure of a gas in a mixture is equal to the pressure which it would exert if it occupied the same volume alone at the same temperature.

dam (1) A retaining wall or bank for water or tailings. (2) An air-tight barrier to isolate underground workings which are on fire.

danburite A rare accessory mineral, occurring in pegmatites as yellow orthorhombic crystals. Chemically, danburite is a calcium borosilicate. $CaB_2Si_2O_8$.

Danian The lowest stage of the Palaeogene (Palaeocene). Some regard it as Cretaceous. See **Tertiary**.

dannemorite A rare manganese-rich monoclinic amphibole, the name being used for the manganese-rich, iron-rich end-member and differing from **tirodite** in having more iron than magnesium.

daphnite A variety of chlorite very rich in iron and aluminium.

darcy A unit used to express the permeability coefficient of a rock, for example in calculating the flow of oil, gas or water. More commonly used is the *millidarcy* (mD), one-thousandth of a darcy (D).

dark red silver ore See **pyrargyrite**.

datolite Hydrated calcium borosilicate occurring as a secondary product in amygdales and veins, usually as distinct prismatic white or colourless monoclinic crystals.

daughter product A nuclide that originates from the radioactive disintegration of another *parent* nuclide.

Davy lamp The name of the safety lamp invented by Sir Humphrey Davy in 1815.

deactivation The return of an activated atom, molecule, or substance to the normal state. See **activation (2)**.

dead crude oil Crude oil that has been stabilized so that free gas is absent. Cf. **live crude oil**.

dead ground Ground devoid of values: ground not containing veins or lodes of valuable mineral: a barren portion of a coal seam. Also *deads*.

dead oil Crude oil without dissolved gas.

deads (1) Same as **dead ground**. (2) Waste (*back fill*) used to support roof.

dead time Time after ionization during which a detector cannot record another particle. Reduced by a *quench* as in Geiger-Müller counters. When the dead time of a detector is variable a fixed electron dead time may be incorporated in subsequent circuits. Also *insensitive time*.

debris flow A mass movement involving rapid flow of various kinds of debris, esp. in mud. May be associated with earthquakes and volcanic eruptions (e.g. South America) or with excessive rainfall on unstable material (e.g. Aberfan disaster). See **mud flow**.

Debye and Scherrer method A method of X-ray crystal analysis applicable to powders of crystalline substances or aggregates of crystals.

Debye-Hückel theory A theory of electrolytic conduction which assumes complete ionization and attributes deviations from ideal behaviour to inter-ionic attraction.

decade Any ratio of 10:1. Specifically the interval between frequencies of this ratio.

decant To pour off the supernatant liquor when a suspension has settled.

decay law If for a physical phenomenon, the rate of decrease of a quantity is proportional to the quantity at that time, then

decay product

the decay law is an exponential, i.e.

$$\alpha = \alpha_0 \, e^{-\lambda t},$$

where λ is the *decay constant* and α_0 is the value at time $t = 0$, e.g. activity in a radioactive substance, sound intensity in an enclosure, discharge of capacitor.

decay product See **daughter product**.

decay time Time in which the amplitude of an exponentially decaying quantity reduces to e^{-1} (36.8%) of its original value.

décollement Detachment structure of strata due to deformation, with independent styles of deformation above and below a discontinuity, the *plane of décollement*. The rocks above the dislocation commonly show complex folding or thrusting.

decomposition The breaking down of a substance into simpler molecules or atoms.

decrepitation The crackling sound made when crystals are heated, caused by internal stresses and cracking.

decussate texture The random arrangement of prismatic or tabular crystals in a rock.

dedolomitization The recrystallization of a dolomite rock or dolomitic limestone consequent on contact metamorphism; essentially involving the breaking down of the dolomite into its two components, $CaCO_3$ and $MgCO_3$. The former merely recrystallizes into a coarse calcite mosaic; but the latter breaks down further into MgO and CO_2. See **forsterite-marble**.

deep-sea deposits Pelagic sediments accumulating in depths of more than 2000 m. They include, in order of increasing depth, calcareous oozes, siliceous oozes and red clay. Terrigenous material is absent.

deerite A black, monoclinic hydrous silicate of iron and manganese.

defect Lattice imperfection which may be due to the introduction of a minute proportion of a different element into a perfect lattice, e.g. indium into germanium crystal, to form an intrinsic semi-conductor for a transistor. A 'point' defect is a 'vacancy' or an 'interstitial atom', while a 'line' defect relates to a dislocation in the lattice. Often accounts for the colour of gemstones.

defect structure Intense localized misalignment or gap in the crystal lattice, due to migration of ions or to slight departures from stoichiometry. The resulting opportunity for mobility is important for semi-conductors, catalysis, photography, rectifiers, corrosion etc. Cf. **dislocation**.

deflagration Sudden combustion, generally accompanied by a flame and a crackling sound.

deflation The winnowing and transport of dry loose material, esp. silt and clay, by wind.

deflocculate To break up agglomerates and form a stable colloidal dispersion.

defoaming agent Substance added to a boiling liquid to prevent or diminish foaming. Usually hydrophobic and of low surface tension, e.g. silicone oils.

deformation A general term used to describe the structural processes that may affect rocks after their formation. Includes folding and faulting.

degassing The removal of entrained gas from drilling mud. This is important because the gas may seriously reduce the hydrostatic pressure available in the borehole.

degree The unit of temperature difference. It is usually defined as a certain fraction of the fundamental interval, which for most thermometers is the difference in temperature between the freezing and boiling points of water. See **Centigrade scale, Fahrenheit scale, Kelvin thermodynamic scale of temperature**.

degree of dissociation The fraction of the total number of molecules which are dissociated.

degree of ionization The proportion of the molecules or 'ion-pairs' of a dissolved substance dissociated into charged particles or ions.

degrees of freedom (1) The number of variables defining the state of a system (e.g. pressure, temperature) which may be fixed at will. See **phase rule**. (2) Number of independent capacities of a molecule for holding energy, translational, rotational and vibrational. A molecule may have a total of $3n$ of these, where n is the number of atoms in the molecule.

dehydration (1) The removal of H_2O from a molecule by the action of heat, often in the presence of a catalyst, or by the action of a dehydrating agent, e.g. concentrated sulphuric acid. (2) The removal of water from crystals, tars, oils etc., by heating, distillation or by chemical action.

de-ionized water Water from which ionic impurities have been removed by passing it through cation and anion exchange columns.

delessite An oxidized variety of chlorite, relatively rich in iron.

deliquescence The change undergone by certain substances which become damp and finally liquefy when exposed to the air, owing to the very low vapour pressure of their saturated solutions, e.g. calcium chloride.

delta The more or less triangular area of river-borne sediment deposited at the mouth of rivers heavily charged with detritus. A delta is formed on a low-lying coastline, particularly in seas of low tidal range and little current action, and in areas where subsidence keeps pace with sediment deposition. The Nile delta is a good example.

deltaic deposit The accumulations of sand, silt, clay and organic matter, deposited as **topset**, **foreset** and **bottomset beds**. A good active example is the Mississipi delta and in the geological record, the Millstone Grit in England.

demagnetization (1) *Removal* of magnetization of ferromagnetic materials by the use of diminishing saturating alternating magnetizing forces. (2) *Reduction* of magnetic induction by the internal field of a magnet, arising from the distribution of the primary magnetization of the parts of the magnet. (3) *Removal* by heating above the Curie point. (4) *Reduction* by vibration. (5) In the **dense-media process** for separating ores, using ferro-silicon, and passage of the fluid through an a.c. field to deflocculate the agglomerated solid.

demantoid Bright-green variety of the garnet andradite, essentially silicate of calcium and iron.

demulsification number The resistance to emulsification by a lubricant when steam is passed through it; indicated in the minutes and half minutes required for the separation of a given volume of oil after emulsification.

dendrite A tree-like crystal formation.

dendritic markings Tree-like markings, usually quite superficial, occurring on joint-faces and other fractures in rocks, frequently consisting of oxide or manganese, or of iron. Less frequently the appearance is due to the inclusion of a mineral of dendritic habit in another mineral or rock, e.g. chlorite in silica as in 'moss agate'.

dendroclimatology Science of reconstructing past climates from the information stored in tree trunks as annual radial increments of growth (rings). Also *dendrochronology*, in which timber is dated by matching growth ring patterns to standard climatic periods.

dense-media process *Heavy-media* or sink-float process. Dispersion of ferrosilicon or other heavy mineral in water separates lighter (floating) ore from heavier (sinking) ore.

densitometer Any instrument for measuring the optical transmission or reflecting properties of a material, in particular the optical density (absorbance) of exposed and processed photographic images.

density The mass of unit volume of a substance, expressed in such units as kg/m^3, g/cm^3 or lb per cubic foot. See **relative density**.

denudation The laying bare (L. *nudus*, naked) of the rocks by chemical and mechanical disintegration, and the transportation of the resulting rock debris by wind or running water. Ultimately denudation results in the degradation of the hills to the existing base level. The process is complementary to sedimentation, the amount of which in any given period is a measure of the denudation.

depleted uranium Sample of uranium having less than its natural content of ^{235}U.

depletion allowance Analogous to a depreciation allowance, it recognizes that the resources of a well or mine are being depleted as they are produced.

depolymerization The breakdown of a complex material into simple units, either with or without incorporation of any external reagent.

deposit See **deposition**.

deposition The laying down or placing into position of sheets of sediment (often referred to as *deposits*) or of mineral veins and lodes. It is synonymous with *sedimentation* in the former sense.

depressing agent *Wetting agent*. Agent used in froth flotation to render selected fraction of pulp less likely to respond to aerating treatment.

depression of freezing point A solution freezes at a lower temperature than the pure solvent, the amount of the depression of the freezing point being proportional to the concentration of the solution, provided this is not too great. The depression produced by a 1% solution is called the *specific depression* and is inversely proportional to the molecular weight of the solute. Hence the depression is proportional to the number of moles dissolved in unit weight of the solvent and is independent of the particular solute used.

depression of land Depression relative to sea level may be caused in many ways, including sedimentary consolidation, the superposition of large masses of ice, the migration of magma, or changes of chemical phase at depth. It is generally recognized by the marine transgression produced but is often difficult to distinguish from eustatic changes in sea level. See **drowned valleys**.

deputy Man of responsibility in a coalmine. See **fireman**.

Derbyshire spar A popular name for the mineral *fluorite* or *fluorspar*.

derived fossils Fossils eroded from an older sediment and redeposited in a younger sediment.

derrick Steel structure at the well head that supports the drilling assembly, including drill bit and pipe, and the raising and lowering mechanism. See **drilling rig, kelly**.

descloizite Hydrated vanadium-lead-zinc ore with the general formula $PbZn(VO_4)OH$. Important source of vanadium. Occurs in the oxidation zone of lead-zinc deposits.

desert rose A cluster of platy crystals, often including sand grains, formed in arid climates by evaporation. Typically barytes or gypsum. Also *rock rose*.

de-sliming Removal of very fine particles from an ore pulp, or classification of it into relatively coarse and fine fractions.

desmine See **stilbite**.

desorption Reverse process to **adsorption**. See **outgassing**.

destructive distillation The distillation of solid substances accompanied by their decomposition. The destructive distillation of coal results in the production of coke, tar products, ammonia, gas etc.

desulphurizing The process of removing sulphur from hydrocarbons by chemical reactions. Many are known by proprietary process names, e.g. Appleby-Frodingham, Claus, Houdry.

detergents Cleaning agents (solvents, or mixtures thereof, sulphonated oils, abrasives etc.) for removing dirt, paint etc. Commonly refers to soapless detergents, containing surfactants which do not precipitate in hard water; sometimes also a protease enzyme to achieve 'biological' cleaning and a whitening agent.

detonating fuse Fuse with core of TNT which burns at some 6 km/s and is itself set off by detonator.

detrital mineral A mineral grain derived mechanically from a parent rock. Typically such minerals are resistant to weathering, e.g. diamonds, gold and zircon. See **heavy mineral**.

detrition The natural process of rubbing or wearing down strata by wind, running water or glaciers. The product of detrition is *detritus*.

detritus See **detrition**.

deuteration The addition of or replacement by deuterium atoms in molecules.

deuterium Isotope of element hydrogen having one neutron and one proton in nucleus. Symbol D, when required to be distinguished from natural hydrogen, which is both ^1H and ^2H. This heavy hydrogen is thus twice as heavy as ^1H, but similarly ionized in water. Used in isotopic 'labelling' experiments (e.g. mechanism of esterification).

development Opening of ore body by access shafts, drives, crosscuts, raises and winzes for purpose of proving mineral value (ore reserve) and exploiting it.

Devensian The last glacial stage in the Pleistocene, probable equivalent to the Würm glaciation of the Alps. See **Quaternary**.

deviation Departure of a drill hole from the vertical either by design or accidentally.

devitrification Deferred crystallization which, in the case of glassy igneous rocks, converts obsidian and pitchstone into dull cryptocrystalline rocks (usually termed **felsites**) consisting of minute grains of quartz and feldspar. Such devitrified glasses give evidence of their originally vitreous nature by traces of perlitic and spherulitic textures.

Devonian The oldest period of the Upper Palaeozoic, covering a time span between approx. 400 and 360 million years. The corresponding system of rocks. See **Palaeozoic**.

de-watering Partial or complete drainage by thickening, sedimentation, filtering or on screen as a process aid, to facilitate shipment or drying of product.

de-waxing In the manufacture of lubricating oils, the solvent removal of wax to reduce the **pour point**.

dextral fault A **tear fault** in which the rocks on one side of the fault appear to have moved to the right when viewed from across the fault. The opposite of a **sinistral fault**.

dextrorotatory Said of an optically active substance which rotates the plane of polarization in a clockwise direction when looking against the incoming light.

diabantite A variety of chlorite relatively rich in iron and poor in aluminium.

diabase The US etc. term for **dolerite**; also used in UK for altered dolerite.

diachronism The transgression across time planes by a geological formation. A bed of sand, when traced over a wide area, may be found to contain fossils of slightly different ages in different places, as, when deposited during a long-continued marine transgression, the bed becomes younger in the direction in which the sea was advancing. Adj. *diachronous*.

diadochy The replacement of one element by another in a crystal structure.

diagenesis Changes which take place in a sedimentary rock at low temperatures and pressures after its deposition. They include compaction, cementation, recrystallization etc. Cf. **metamorphism**.

dial A large compass mounted on a tripod, used for surveying or mapping workings in coalmines.

diallage An ill-defined, altered monoclinic pyroxene, in composition comparable with augite or diopside, with a lamellar structure; occurs typically in basic igneous rocks such as gabbro.

dialling The process of running an underground traverse with a mining **dial**.

dialogite See **rhodochrosite**.

diamagnetism Phenomenon in some materials in which the susceptibility is negative, i.e. the magnetization opposes the magnetizing force, the permeability being less than unity. It arises from the precession of spinning charges in a magnetic field. The susceptibility is generally one or two orders of magnitude weaker than typical paramagnetic susceptibility.

diaminoethanetetraacetic acid See **EDTA**.

diamond One of the crystalline forms of carbon; it crystallizes in the cubic system, rarely in cubes, commonly in forms resembling an octahedron, and less commonly in the tetrahedron. Curved faces are characteristic. It is the hardest mineral (10 in Mohs' scale); hence valuable as an abrasive, for arming rock-boring tools etc. Its high dispersion and birefringence makes it valuable as a gemstone. Occurs in blue ground, in river gravels and in shore sands. See **black diamond, bort, industrial diamond, kimberlite**.

diamond bit Drilling bit in which industrial diamonds are set in the cutting portions of the bit. Cuts hard rock and wears longer, so increasing the time between replacing bits involving a **round trip**. See **drilling rig**.

diamond dust The hardest abrasive agent, used as loose powder or paste in polishing, or embedded in metal tool parts, e.g. *rock saws* and *drill bits*.

diapir An intrusion of relatively light material into pre-existing rocks, doming the overlying cover. Applied esp. to salt and igneous intrusions. See **petroleum reservoirs**.

diaspore Alumina monohydrate, occurring in bauxite. A dimorph of **boehmite**.

diastrophism Large-scale deformation of the Earth's crust.

diatom A microscopic single-celled plant of the class Bacillariophyceae growing in both seawater and freshwater, and in soils. It secretes resistant siliceous cell walls called *frustrule*s in a large number of forms. Diatoms range from the Cretaceous to the present day but there are diatom-like organisms in the Cambrian and later rocks. See **diatomite**.

diatomite, diatomaceous earth A siliceous deposit occurring as a whitish powder consisting essentially of the frustules of diatoms. It is resistant to heat and chemical action, and is used in fireproof cements, insulating materials, as an absorbent in the manufacture of explosives and as a filter. Also *infusorial earth, kieselguhr, tripolite*.

diatom ooze A deep-sea deposit consisting essentially of the frustules of diatoms; widely distributed in high latitudes.

dichroic Said of (1) materials, such as a solution of chlorophyll, which exhibit one colour by reflected light and another colour by transmitted light; (2) minerals that display **dichroism**.

dichroism The property possessed by some crystals (e.g. tourmaline) of absorbing the ordinary and extraordinary ray to different extents. This has the effect of giving to the crystal different colours according to the direction of the incident light.

dichroite See **cordierite**.

dickite A clay mineral of the kaolin (or kandite) group, a hydrated aluminosilicate. It generally occurs in hydrothermal deposits.

dielectric Substance, solid, liquid or gas, which can sustain a steady electric field, and hence an insulator. It can be used for cables, terminals, capacitors etc.

differential grinding Comminution so controlled as to develop differences in grindability of constituents of ore.

differential thermal analysis The detection and measurement of changes of state and heats of reaction, esp. in solids and melts. The sample under investigation together with one of thermally inert material are simultaneously heated side-by-side and the difference in temperature between them noted. This becomes very marked when one of the two samples passes through a transition temperature with evolution or absorption of heat but the other does not. The technique is particularly used for clay samples. Abbrev. *DTA*.

differential titration Potentiometric titration in which the e.m.f. is noted between additions of small increments of titrant, the end point being where the e.m.f. changes most sharply.

differentiation The process of forming two or more rock types from a common magma.

diffraction The phenomenon, observed

diffraction analysis

when waves are obstructed by obstacles or apertures, of the disturbance spreading beyond the limits of the geometrical shadow of the object. The effect is marked when the size of the object is of the same order as the wavelength of the waves and accounts for the alternately light and dark bands, diffraction fringes, seen at the edge of the shadow when a point source of light is used. It is one factor that determines the propagation of radio waves over the curved surface of the Earth and it also accounts for the audibility of sound around corners.

diffraction analysis Analysis of the internal structure of crystals by utilizing the diffraction of X-rays caused by the regular atomic or ionic lattice of the crystal.

diffraction angle Angle between the direction of an incident beam of light, sound or electrons, and the direction of any resulting diffracted beam.

diffraction grating One of the most useful optical devices for producing spectra. In one of its forms, the diffraction grating consists of a flat glass plate with equidistant parallel straight lines ruled in its surface by a diamond. There may be as many as 1000 per millimetre. If a narrow source of light is viewed through a grating it is seen to be accompanied on each side by one or more spectra. These are produced by diffraction effects from the lines acting as a very large number of equally spaced parallel slits.

diffraction pattern Pattern formed by equal intensity contours as a result of diffraction effects, e.g. in optics or radio transmission.

diffractometer An instrument used in the analysis of the atomic structure of matter by the diffraction of X-rays, neutrons or electrons by crystalline materials. A monochromatic beam of radiation is incident on a crystal mounted on a goniometer. The diffracted beams are detected and their intensities measured by a counting device. The orientation of the crystal and the position of the detector are usually computer-controlled.

diffusion General transport of matter whereby molecules or ions mix through normal thermal agitation. Migration of ions may be directed and accelerated by electric fields, as in dialysis.

diffusion coefficient In the diffusion equation (*Fick's law of diffusion*), the coefficient of proportionality between molecular flux and concentration gradient. Symbol D. Units, $M^2 s^{-1}$, as for thermal diffusivity.

diffusion current In electrolysis, the maximum current at which a given bulk concentration of ionic species can be discharged, being limited by the rate of migration of the ions through the diffusion layer.

diopside

diffusion law See **Fick's law of diffusion**.

diffusion layer In electrolysis, the layer of solution adjacent to the electrode, in which the concentration gradient of electrolyte occurs.

diffusivity A measure of the rate at which heat is diffused through a material. It is equal to the thermal conductivity divided by the product of the specific heat at constant pressure and the density. Units $m^2 s^{-1}$.

digenite A cubic sulphide of copper, usually massive and associated with other copper ores. Composition probably near Cu_2S_5.

digestion In chemical extraction of values from ore or concentrate, period during which material is exposed under stated conditions to attacking chemicals.

dilatancy Property shown by some colloidal systems, of thickening or solidifying under pressure or impact. Cf. **thixotropy**.

dilatometer An apparatus for the determination of transition points of solids. It consists of a large bulb joined to a graduated capillary tube, and is filled with an inert liquid. The powdered solid is introduced, and the temperature at which there is a considerable change in volume on slow heating or cooling may be noted; alternatively, the temperature at which the volume shows no tendency to change with time may be found.

dilatometry The determination of transition points by the observation of volume changes.

dilution (1) Decrease of concentration. (2) The volume of a solution which contains unit quantity of dissolved substance. The reciprocal of *concentration*.

dilution law See **Ostwald's dilution law**.

diluvium An obsolete term for those accumulations of sand, gravel etc. which, it was thought, could not be accounted for by normal stream and marine action. In this sense, the deposits resulting from the Deluge of Noah would be *diluvial*.

dimorphism Crystallization into two distinct forms of an element or compound, e.g. carbon as diamond and graphite; FeS_2 as pyrite and *marcasite*.

Dinantian The lower Carboniferous rocks of N.W. Europe, comprising Tournaisian and Viséan Stages. See **Palaeozoic**.

dinoflagellate See **acritarchs**.

dinosaurs See panel on p. 60.

diopside A monoclinic pyroxene, typically occurring in metamorphosed limestones and

dinosaurs

The 'terrible lizards', members of the Reptile Subclass, Archosauria, were land dwellers that came to dominate life on Earth for over 140 million years from the Trias until they died out at the end of the Cretaceous. They existed perhaps 40 times as long as Man has lived on Earth. The smallest were about the size of a dog; the largest were huge. Two main orders evolved, the *saurischian* 'lizard hipped', and the *ornithischian* 'bird hipped' dinosaurs.

Megalosaurus, 'great lizard' (Fig. 1) was one of the largest flesh-eating *carnosaurs*, reaching 30 ft (9 m) in length and almost 1 tonne weight. *Tyrannosaurus* (Fig. 2) was even larger, its massive skull having teeth up to 7 in (18 cm) long. The *sauropods*, 'lizard feet', which have five toes, were plant-eaters but included the largest terrestrial animals (perhaps 130 tonnes) requiring the use of all their four legs.

Apatosaurus (Fig. 3), commonly known as *Brontosaurus*, 'thunder lizard' is perhaps the best-known dinosaur. It laid eggs about 8 in (20 cm) across in clutches of perhaps 15–20, from which the young of the size of a human baby hatched, eventually to grow to 70 ft (21 m) long and 30 tonnes in weight. Not all dinosaurs were the monsters described above. The smallest known dinosaur *Compsognathus*, '*pretty jaws*' (Fig. 4) may have reached 2 ft (20 cm) in length. Man (Fig. 9) is to scale.

Ornithischians, plant eaters, included the two-legged *ornithopods*, 'bird feet', and four-legged armoured dinosaurs, some with horns. One Family, Pachycephalosauridae, had bony heads (Fig. 5) with domes on the top of their skulls up to about 10 in (25 cm) thick, probably for head butting.

Iguanodon (Fig. 6) is one of the better known dinosaurs; *Triceratops* (Fig. 7), also well known, were plant eaters that had three massive horns. The *Pterosaurs* (Fig. 8) were reptiles highly adapted to flight by elongation of the fourth digit of the hand into a wing.

figures on next page

dolomites, and composed of calcium magnesium silicate, $CaMgSi_2O_6$, usually with a little Fe.

dioptase A rare emerald green hydrated copper silicate.

dioptre The unit of power of a lens. A convergent lens of 1 metre focal length is said to have a power of +1 dioptre. Generally, the power of a lens is the reciprocal of its focal length in metres, the power of a divergent lens being given a negative sign.

diorite A coarse-grained deep-seated (plutonic) igneous rock of intermediate composition, consisting essentially of plagioclase feldspar (typically near andesine in composition) and hornblende, with or without biotite in addition. Differs from granodiorite in the absence of quartz. See **plutonic rocks, tonalite**.

dip (1) The angle that a plane makes with the horizontal. The dip is perpendicular to the **strike** of the structure. See **drainage patterns**. (2) The angle measured in a vertical plane between the Earth's magnetic field at any point and the horizontal. Also *inclination*.

dip circle An instrument consisting of a magnetic needle or dip needle pivoted on a horizontal axis; accurate measurements of magnetic dip, or *angle of dip*, can be obtained when it is set in the magnetic north-south plane..

dip fault A fault parallel to the direction of dip. See **strike fault**.

dipping refractometer A type of refractometer which is dipped into the liquid under examination.

dip slope A land form developed in regions of gently inclined strata, particularly where hard and soft strata are interbedded. A long gentle sloping surface which coincides with the inclination of the strata below ground. See **cuesta**.

dipyre A member of the **scapolite** series, containing 20% to 50% of the meionite molecule.

directional drilling The use of special downhole drilling assemblies to turn a drill hole in the desired direction.

dinosaurs (contd)

Typical dinosaurs: Figs. 1 – 4 are Saurischia; Figs. 5 –7, Ornithischia; Fig. 8, Pterosauria; Fig. 9, Man for scale.

direction of younging   See **younging, direction of**.

direct rope haulage   Engine plane. An ascending truck is partly balanced by a descending one, motive power being applied to the drum round which the haulage rope passes.

dirt   Broken valueless mineral. Also *gangue*. US *muck*.

discharge head   Vertical distance between intake and delivery of pump, *plus* allowance for mechanical friction and other retarding resistances requiring provision of extra power.

disconformity   A break in the rock sequence in which there is no angular discordance of dip between the two sets of strata involved. Cf. **unconformity**.

discordant intrusion   An igneous intrusion that cuts across the bedding or foliation of the country rock it intrudes. See **dykes and sills**.

discovery well   The first well to reveal oil or gas in a new field or at a new level.

disintegrating mill   A mill for reducing lump material to a granular product. It consists of fixed and rotating bars in close proximity, crushing being partly by direct impact and partly by interparticulate attrition. See **beater mill**.

disk filter *American filter.* Continuous heavy-duty vacuum filter in which separating membranes are disks, each revolving slowly through its separate compartment.

dislocation A lattice imperfection in a crystal structure, classified according to type, e.g. edge dislocation, screw dislocation.

dispersed phase A substance in the colloidal state.

dispersion The dependence of wave velocity on the frequency of wave motion; a property of the medium in which the wave is propagated. In the visible region of the electromagnetic spectrum, dispersion manifests itself as the variation of refractive index of a substance with wavelength (or colour) of the light. Dispersion enables a prism to form a spectrum. See **gems and gemstones, Hartmann dispersion formula, Cauchy's dispersion formula, anomalous dispersion.**

dispersion curve A plot of frequency against wavelength for a wave in a dispersive medium. See **phonon dispersion curve.**

dispersion forces Weak intermolecular forces, corresponding to the term *a* in *van der Waal's equation.*

dispersion medium A substance in which another is colloidally dispersed.

dispersive medium Medium in which the phase velocity of a wave is a function of frequency.

dispersive power The ratio of the difference in the refractive indices (n) of a medium for the red (R) and violet (V) to the mean refractive index diminished by unity. This may be written

$$v = \frac{n_V - n_R}{n - 1}.$$

dispersivity quotient The variation of refractive index n with wavelength λ, $dn/d\lambda$.

displaced terrane Internally consistent rock masses within an orogenic area which are abruptly discontinuous with their surroundings. Sometimes *suspect terrane.* See **terrane.**

displacement pump Pump with pulsing action, produced by steam, compressed air or a plunger, causing non-return valves to prevent return flow of displaced liquid during the retracting phase of the pump cycle.

displacement series See **electrochemical series.**

disproportionation A reaction in which a single compound is simultaneously oxidized and reduced, e.g. the spontaneous reaction in water of soluble copper (I) salts to form equal amounts of copper (0) and copper (II).

disseminated values Mode of occurrence in which small specks of concentrate are scattered evenly through the gangue mineral.

dissociation The reversible or temporary breaking-down of a molecule into simpler molecules or atoms. See **Arrhenius theory of dissociation.**

dissociation constant The equilibrium constant for a process considered to be a dissociation. Commonly it is applied to the dissociation of acids in water.

disthene A less commonly used name for the mineral **kyanite.**

distribution law The total energy in a given assembly of molecules is not distributed equally, but the number of molecules having an energy different from the median decreases as the energy difference increases, according to a statistical law.

district An underground section of a coal-mine serviced by its own roads and ventilation ways: a section of a coalmine.

diurnal variation A variation of the Earth's magnetic field as observed at a fixed station, which has a period of approximately 24 hours.

divalent Capable of combining with two atoms of hydrogen or their equivalent. Also having an oxidation state of two.

diver Small plummet adjusted to a desired relative density, so that it indicates the density of the fluid in which it is immersed by its up-and-down motion. Also *cartesian diver.*

divergent junction A zone where plates move apart and new crust and lithosphere are formed. Characterized by volcanism and earthquakes, e.g. mid-Atlantic ridge.

dl- Containing equimolecular amounts of the dextrorotatory and the laevorotatory forms of a compound; racemic. Now usually written ± .

doctor test A test for sulphur in petroleum using a sodium plumbate (II) solution.

dodecyl benzene An important starting material for anionic detergents derived from petroleum. Based on the tetramer of propene.

Dogger The middle epoch of the Jurassic period. See **Mesozoic.**

doggers Flattened ovoid concretions, often of very large size, in some cases calcareous, in others ferruginous, occurring in sands or clays. They may be a metre or more in diameter.

dog-tooth spar A form of calcite in which the scalenohedron is dominant but combined with prism, giving a sharply pointed crystal like a canine tooth.

dolerite The general name for basic igneous

dolly tub

rocks of medium grain size, occurring as minor intrusions or in the central parts of thick lava flows; much quarried for road metal. Typical dolerite consists of plagioclase near labradorite in composition, pyroxene, usually augite, and iron ore, usually ilmenite, together with their alteration products.

dolly tub A large wooden tub used for the final upgrading of valuable minerals separated by water concentration in ore dressing. Also *kieve*. See **tossing**.

dolomite The double carbonate of calcium and magnesium, crystallizing in the rhombohedral class of the trigonal system, occurring as cream-coloured crystals or masses with a distinctive pearly lustre, hence the synonym *pearl spar*. A common mineral of sedimentary rocks and an important gangue mineral.

dolomitic limestone A calcareous sedimentary rock containing calcite or aragonite in addition to the mineral dolomite. Cf. **dolostone**.

dolomitization The process of replacement of calcium carbonate in a limestone or ooze by the double carbonate of calcium and magnesium (dolomite).

dolostone A rock composed entirely of the mineral dolomite. The term is synonymous with *dolomite rock*, but avoids confusion between dolomite rock and the mineral dolomite.

domain In ferroelectric or ferromagnetic material, region where there is saturated polarization, depending only on temperature. The transition layer between adjacent domains is the *Bloch wall*, and the average size of the domain depends on the constituents of the material and its heat treatment. Domains can be seen under the microscope when orientated by strong electric or magnetic fields.

dome A crystal form consisting of two similar inclined faces meeting in a horizontal edge, thus resembling the roof of a house. The term is frequently incorrectly applied to a four-faced form which is really a prism lying on an edge.

dome (1) An igneous intrusion with a domelike roof. (2) An anticlinal fold with the rock dipping outwards in all directions.

dominance See **habit**.

donor (1) The reactant in an induced reaction which reacts rapidly with the inductor. (2) That which 'gives', as in *proton-donor* or *electron-pair donor*.

Dorn effect The production of a potential difference when particles suspended in a liquid migrate under the influence of mechanical forces, e.g. gravity; the

converse of **electrophoresis**.

double bond Covalent bond involving the sharing of two pairs of electrons.

double decomposition A reaction between two substances in which the atoms are rearranged to give two other substances. In general it may be written AB + CD = AC + BD.

double-image micrometer A microscope attachment for the rapid and precise measurement of small objects, like cells or particles. The principle is that two images of the object are formed by means which allow the separation of the images to be varied and its magnitude read on a scale. By setting the two images edge-to-edge (a very precise setting) and then interchanging them, the difference of the scale readings is a measure of the diameter of the object. The double image may be formed by birefringent crystals, prisms, an interferometer-like system or a vibrating mirror.

double oblique crystals See **triclinic system**.

double-pipe exchanger Heat exchanger formed from concentric pipes, the inner frequently having fins on the outside to increase the area. The essential flow pattern is that the two streams are always parallel.

double refraction Division of an electromagnetic wave in an anisotropic medium into two components propagated with different velocities, depending on the direction of propagation. In uniaxial crystals the components are called the *ordinary* ray where the wavefronts are spherical, and the *extraordinary* ray where the wavefronts are ellipsoidal. In **biaxial** crystals both wavefronts are ellipsoidal.

doubles See **coal sizes**.

double salts Compounds having two normal salts crystallizing together in definite molar ratios.

downcast A contraction for *downcast shaft*, i.e. the shaft down which fresh air enters a mine. The fresh air may be sucked or forced down the shaft.

downhole Describes the drills, measuring instruments and equipment used down the borehole.

downthrow In a **fault**, the vertical displacement of the fractured strata. Indicated on geological maps by a tick attached to the fault line, with (where known) a figure alongside indicating the amount of downthrow.

Downtonian The lowest stage of the Old Red Sandstone facies of the Devonian System, named from its typical develop-

drainage patterns

Drainage is the removal of surplus meteoric waters by rivers and streams. The complicated network of rivers is related to the geological structure of a district, being determined by the existing rocks and superficial deposits in the case of youthful drainage systems; but in those that are mature, the courses may have been determined by strata subsequently removed. Also the natural or artificial methods of effecting this removal.

In a new land surface, the streams that develop as the result of the initial slopes, and run down dip, are *consequent* streams, lettered (c) in Fig. 1. The later, *subsequent* streams (s) are controlled by differences in resistance to erosion of geological or topological features. A stream or river that flows against the dip of strata is said to be *obsequent* (o). If the original direction of flow has been reversed, e.g. by glacial action, landslides, river capture or gradual tilting of the surface, the stream and drainage is called *reversed*.

River capture is the natural diversion of a stream or river by a neighbouring stream with greater power of erosion that erodes headwords and taps the flow of the *beheaded* stream into its own channel.

Stream order (Fig. 2) is a classification of streams according to their hierarchical position in the drainage network. In the most widely used system, first order streams, 1, are the outermost tributaries, which join together to form a second order stream, 2. Only streams of the same order, e.g. 2, 2, can combine to produce a stream of higher order. Streams 2, 2, would produce a stream of order 3 below their confluence, but streams 2, 3, would produce a stream that was still third order, 3.

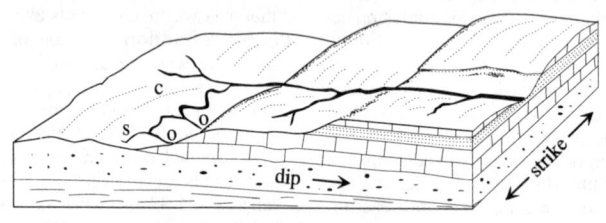

Fig. 1

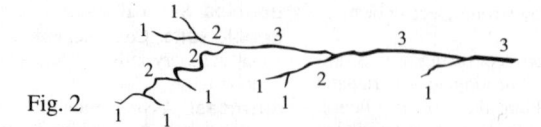

Fig. 2

Drainage pattern (Fig. 1) and stream order (Fig. 2)
c, consequent; s, subsequent; o, obsequent. The numbers refer to the order of a stream.

ment around Downton Castle in the Welsh Borderlands.

drag classifier Endless belt with transverse rakes, which moves upward through an inclined trough so as to drag settled material up and out while slow-settling sands overflow below as pulp arrives for continuous sorting into coarse and fine sands.

drainage patterns See panel on this page.

drain holes Draining of oil from strata by boring holes out from the main bore by the use of special **downhole** machinery such as **mud motor** driven drills.

dravite Brown tourmaline, sometimes used as a gemstone.

draw (1) To allow ore to run from working places, stopes, through a chute into trucks. (2) To withdraw timber props or sprags from overhanging coal, so that it falls down ready for collection. (3) To collect

drawdown

broken coal in trucks.

drawdown Fall of water level in natural reservoir, e.g. an **aquifer**, when rate of extraction exceeds rate of replenishment.

draw works Surface gear of a drilling rig.

dredge Barge or twin pontoons carrying chain of digging buckets, or suction pump, with over-gear such as jigs, sluices, trommels, tailing stackers, manoeuvring anchorlines, and power producer. Used to work alluvial deposits of cassiterite, gold, gemstones etc.

dreikanter A wind-faceted pebble typically having three curved faces (from Ger. *drei*, three). Common in desert deposits. See **ventifact, zweikanter**.

dressing Grinding of worn crushing rolls, to restore cylindrical shape. Rock crushing and screening to required sizes. Preparation of amalgamation plates with liquid mercury for gold recovery.

drift (1) A general name for the superficial, as distinct from the solid, formations of the Earth's crust. It includes typically the Glacial Drift, comprising all the varied deposits of boulder clay, outwash gravel, and sand of Quaternary age. Much of the drift is of fluvio-glacial origin. (2) A level or tunnel pushed forward underground in a metal mine, for purposes of exploration or exploitation. The inner end of the drift is called a *dead end*. (3) A heading driven obliquely through a coal seam. (4) A heading in a coalmine for exploration or ventilation. (5) An inclined haulage road to the surface. (6) Deviation of borehole from planned course. See **mining**.

drifter A cradle-mounted compressed-air rock drill, used when excavating tunnels (*drifts* or *cross-cuts*).

drifting Tunnelling along the strike of a lode, horizontally or at a slight angle.

drill (1) A hand drill or *auger*. (2) A compressed-air operated rock drill, *jack-hammer*, *pneumatic drill*. (3) Generally, the more elaborate equipment required in *power drilling*.

drill bit The actual cutting or boring tool in a *drill*. In rotary drilling it may be of the *drag* variety, with two or more cutting edges (usually hard-tipped against wear), *roller* (with rotating hard-toothed rollers), or **diamond**, with cutting face (annular if core samples are required) containing suitably embedded borts. The assembly is attached to the bottom of the **drill pipe** and rotated with it. **Drilling mud** pumped down the drill pipe maintains the pressure, cools, lubricates and carries away the debris.

drowned valleys

drill collar Heavy weight sections of pipe attached between the bit and the drill pipe. In deep bores they can weigh 100 tons and are used to damp-out torsional stresses and stabilize the drill string.

drilling (1) The operation of boring a hole in the ground, usually for exploration or for the extraction of oil, gas, water or geothermal energy. See **drilling rig**. (2) The operation of tunnelling or stoping. (3) The operation of making short holes for blasting, or deep holes with a diamond drill for prospecting or exploration. Drilling may be percussive (repeated blows on the drilling tool) or rotary (circular grinding).

drilling mud Usually a mixture of clays, water, density increasing agents like **barite** and sometimes thixotropic agents like *bentonite* pumped down through the drilling pipe and used to cool, lubricate and flush debris from the drilling assembly. It also helps to seal the bored rock and most importantly provides the hydrostatic pressure to contain the oil and gas when this is reached. Oil-based and brine muds are increasingly used. **Mud motors** powered by the mud flow can be used for drilling, particularly in directions away from the main bore axis.

drill(ing) pipe The tubes, joined by screwing together, which connect the **kelly** to the drill bit and impart rotary motion to the latter.

drilling platform Offshore platform which may be floating or fixed to the seabed and from which a number of wells may be bored radially and at various angles into the bearing strata.

drilling rig See panel on p. 66.

drill string The pipe and *bottom-hole assembly* in a borehole. See **drilling rig**.

drive Tunnel or level driven along or near a lode, vein or massive ore deposit as distinct from country rock. *Driving* is the process.

drop The vertical displacement in a downthrow fault: the amount by which the seam is lower on the other side of the fault. See **throw**.

dropping mercury electrode Polarimeter with a half element consisting of mercury dropping in a fine stream through a solution. Used in polarography, a continuously renewed mercury surface being formed at the tip of a glass capillary, the accumulating impurities being swept away with the detaching drops of mercury.

dross Small coal, inferior or worthless.

drowned Flooded, e.g. *drowned workings*.

drowned valleys Literally, river valleys which have become drowned by a rise of sea level relative to the land. This may be

65

drilling rig

The drilling rig is the derrick, surface equipment and related structures used in oilfield exploration. Offshore, in shallow waters, *jack-up* rigs, with legs based on the sea bed, are used; in deeper waters *semi-submersible* rigs (which float at all times) are employed. In very deep waters drilling ships are used. A *production platform* is the offshore platform from which the flow of oil and gas is controlled and usually stored before onward transmission to a refinery.

The drilling rig itself is shown in Figs. 1 and 2, and consists of a *derrick* raised above the *drill floor* where most of the equipment is situated. The derrick supports the weight of the *drill string*, and the lengths of *casing* when they are being installed. The drill string is made of lengths of drill pipe screwed together.

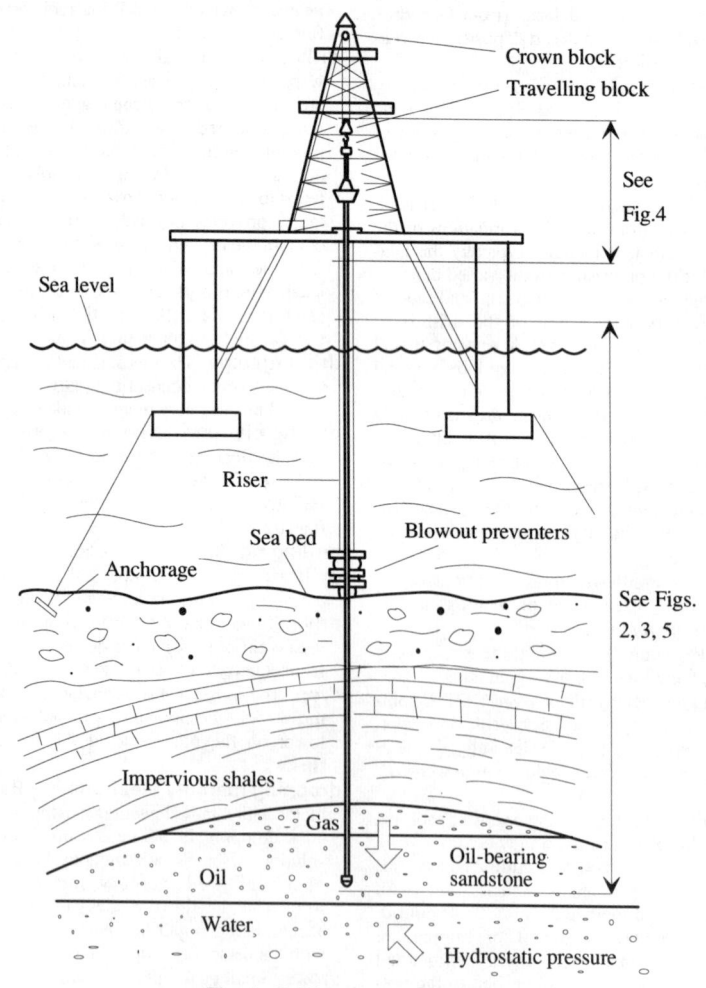

Fig. 1. Drilling rig (*not to scale*)

continued on next page

drilling rig (contd)

The rotary motion of the drill string and *drill bit* is imparted by the power-driven *rotary table* (Fig. 4). The rotary table is coupled to the *kelly*, a pipe of external hexagonal or square section at the top of the drill string by means of the *kelly bushing* which has a hole of similar section at its centre. As the drilling proceeds successive lengths of pipe are inserted between the kelly and the top of the drill string. *Drilling mud* is pumped down the drill pipe to contain the well pressure, cool and lubricate the bit, and bring up the cuttings in the space between the drill pipe and the casing or walls of the hole (Fig. 3). Drilling is usually with *tricone bits* to grind up the rock (Fig. 5); diamond-tipped hollow cylindrical bits are used to obtain sample cores.

Blowout preventers (bop) are used to control the pressure in the bore when drilling or *working over*, i.e. performing remedial operations on a producing well to restore or increase production.

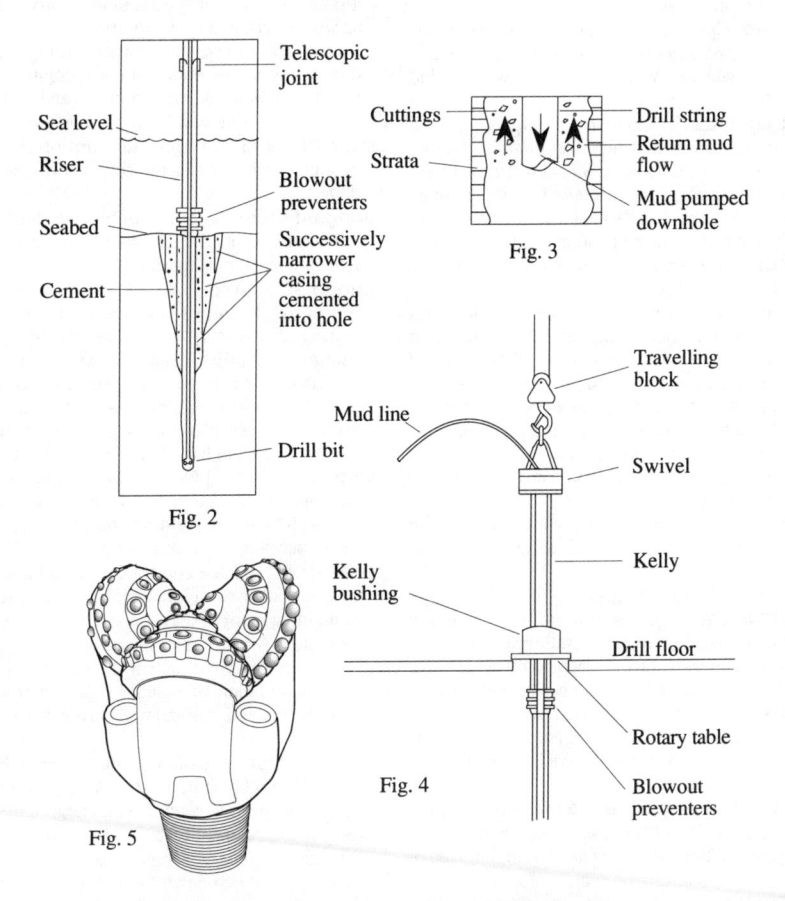

Fig. 2. drill string and casing. Fig. 3. circulation of mud. Fig. 4. drilling rig floor. Fig. 5. tricone drill bit.

due to actual depression of the land, sea level remaining stationary; or to a eustatic rise in sea level, as during the interglacial periods in the Pleistocene, when melting of the icecaps took place. See **ria, fiords**.

Drude law A law relating the specific rotation for polarized light of an optically active material to the wavelength of the incident light.

$$\alpha = \frac{k}{\lambda^2 - \lambda_0^2},$$

where a = specific rotation, k = rotation constant for material, λ_0^2 = dispersion constant for material, λ = wavelength of incident light. The law does not apply near absorption bands.

drum Cylinder or cone, or compound of these, on and off which the winding rope is paid when moving cages or skips in a mine shaft.

drum filter Thickened ore pulp is fed to a trough through which a cylindrical hollow drum rotates slowly. Vacuum draws liquid into pipes mounted internally, while solids (filter cake) are arrested on a permeable membrane wrapped round drum circumference, and are removed continuously before re-submergence.

drumlin An Irish term, meaning 'a little hill', applied to accumulations of glacial drift moulded by the ice into small hog-backed hills, oval in plan, with the longer axes lying parallel to the direction of ice movement. Drumlins often occur in groups, giving the 'basket of eggs' topography which is seen in many parts of Britain and dates from the last glaciation.

drusy Containing cavities often lined with crystals; said of mineralized lodes or veins. See **geode**.

drusy cavity See **geode**.

dry assay The determination of a given constituent in ores, metallurgical residues and alloys, by methods which do not involve liquid means of separation. See **wet assay**.

dry blowing Manual or mechanical winnowing of finely divided sands to separate heavy from light particles, practised in arid regions.

dry bone, dry bone ore See **smithsonite**.

dry gas Petroleum gas in which the lower boiling point fractions have been removed or are not naturally present.

dry ice Solid (frozen) carbon dioxide, used in refrigeration (storage) and engineering. At ordinary atmospheric pressures it sublimes slowly.

dry valley Valley produced at some former period by running water, though at present streamless. This may be due to a fall of the water table, to river capture, or to climatic change.

duff Fine coal too low in calorific value for direct sale.

dumb buddle A buddle without revolving arms or sweeps, for concentrating tin ores.

dump The heap of accumulated waste material from a metal mine, or of treated tailings from a mill or ore-dressing plant. Also *tip*.

dune An accumulation of sand formed in an area with a prevailing wind. The principal types are: *barchans*, crescent-shaped dunes which migrate in the direction of the point of the crescent; *seifs*, elongated ridges of sand aligned in the wind direction; *transverse dunes*, at right angles to the wind; and *whaleback dunes*, very large elongated dunes. Fossil sand dunes can be recognized, and they indicate desert conditions and wind conditions during past geological periods.

dune bedding A large-scale form of cross bedding characteristic of wind blown sand dunes.

dungannonite A corundum-bearing diorite containing nepheline, originally described from Dungannon, Ontario.

dunite A coarse-grained, deep-seated igneous rock, almost monomineralic, consisting essentially of olivine (> 90%), though chromite is an almost constant accessory. In several parts of the world (e.g. Bushveld Complex, South Africa) it contains native platinum and related metals. Named from Mt. Dun, New Zealand.

Duperry's lines Lines on a magnetic map showing the direction in which a compass needle points, i.e. the direction of the magnetic meridian.

durain A separable constituent of dull coal; of firm, rather granular structure, sometimes containing many spores.

duricrust A hard crust formed in or on soil in a semi-arid environment. It is formed by the precipitation of soluble minerals from mineral waters, particularly during the dry season.

dust counter Apparatus, usually portable, for collecting dust in mines, for display and check on working conditions underground.

Dwight Lloyd machine A type of continuous *sintering* machine characterized by having air drawn down through the burning bed on a travelling grate into 'wind boxes' and used in roasting pyritic ore so that by segregating and recycling the gas streams, concentrations of sulphur dioxide sufficiently high for conversion to sulphuric acid

dykes and sills

Dykes (or dikes) are discordant igneous intrusions of a tabular nature, usually nearly vertical, cutting the bedding or foliation of the country rock. Characteristically they are a metre to a few metres in width but may extend laterally for hundreds of kilometres. **Sills** are minor intrusions of igneous rocks concordantly injected between and more or less parallel to the bedding planes of the country rocks as in the figure. They occasionally break across the bedding, for example as the Great Whin Sill which may be traced across much of northern England. *Whin* and *whinstone* are old and quarrying names for basalt, dolerite etc. used e.g. as roadstone. Dykes and sills and some related intrusions, such as small laccoliths, are often grouped together as *minor intrusions* or *hypabyssal rocks*. By far the commonest rocks forming dykes and sills are basic types (e.g. dolerites and basalts). ***Cedar-tree laccoliths*** have a number of laccolithic intrusions one above another in the shape of a cedar tree. See figure below.

Dykes may die out upwards or may have been in fissures that were the feeders of lava flows, e.g. the plateau basalts. Sometimes they occur in *swarms*. See **dyke swarms**. Dykes and sills may be *composite*, when they show more than one phase of emplacement of materials of different chemical composition or mineralogical constitution, e.g. a basaltic intrusion intruded by a less basic one. The component parts do not chill against each other. If they do, the intrusion is *multiple*, not *composite*. A **ring-dyke** is a thick intrusion of plutonic dimensions and usually coarsely crystalline, of arcuate outcrop, sometimes forming a ring and usually dipping outwards. ***Cone-sheets*** are thin sheets of medium- to fine-grained igneous rock also of arcuate plan, dipping inwardly and downwards with a common focus and axis.

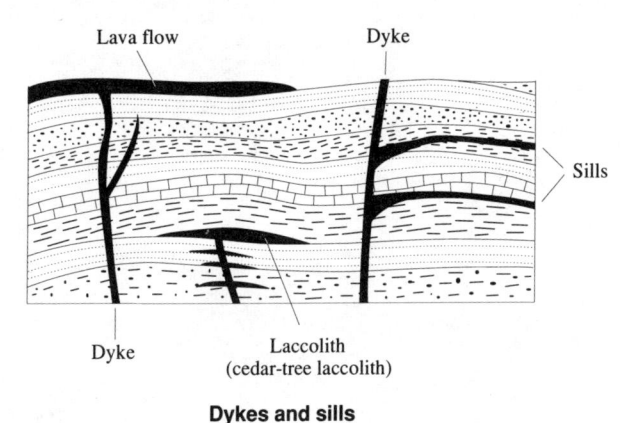

Dykes and sills

are readily obtained.

dying shift Night (*graveyard*) shift.

dykes and sills See panel on this page.

dyke phase That episode in a volcanic cycle characterized by the injection of minor intrusions, esp. dykes. The dyke phase usually comes after the major intrusions, and is the last event in the cycle. See **dykes and sills**.

dyke swarm A series of dykes of the same age, usually trending in a constant direction over a wide area. Occasionally, dykes may radiate outwards from a volcanic centre, as the Tertiary dyke swarm in Rum, Scotland; but usually they are parallel, e.g. the O.R.S. dyke swarm of S.W. Scotland, of which the trend is north-east to south-west. See **dykes and sills**.

dynamical theory of X-ray and electron diffraction A theoretical approach that takes into account the dynamical equilibrium between the incident and diffracted beams in a crystal, e.g. the effect of interference between the incident beam and multiply diffracted beams.

dynamite A mixture of nitroglycerine with kieselguhr, wood pulp, starch flour etc., making the nitroglycerine safe to handle until detonated. The most common industrial high explosive. Also *giant powder*.

dynamothermal metamorphism A regional metamorphism involving both heat and pressure.

dyne The unit of force in the CGS system of units. A force of one dyne acting on a mass of 1 g, imparts to it an acceleration of 1 cm/s^2. Approximately 981 dynes are equal to 1 g weight. 10^5 dynes = 1 newton.

dyscrasite Silver ore consisting mainly of a silver antimonide, Ag_3Sb.

dysprosium A metallic element, a member of the rare-earth group. Symbol Dy, at.no. 66, r.a.m. 162.5.

dystectic mixture A mixture with a maximum melting point. Cf. **eutectic**

E

e Symbol for the **electron** (e^-) or positron (e^+).

e Symbol for the elementary charge; 1.602×10^{-19} coulomb.

ε Symbol for molar extinction coefficient.

ε Symbol for (1) **emissivity**; (2) linear strain; (3) permittivity.

ε– Symbol for (1) substituted on the fifth carbon atom; (2) *epi-*, i.e. containing a condensed double aromatic nucleus substituted in the 1,6 positions; (3) *epi-*, i.e. containing an intramolecular bridge.

η Symbol for (1) **coefficient of viscosity**; (2) electrolytic polarization; (3) overvoltage.

E Symbol for (1) **eddy diffusivity**; (2) potential difference, esp. electromotive force of voltaic cells; (3) (with subscript) single electrode potential, thus E_H, on the hydrogen scale; E_0, standard electrode potential.

E Symbol for (1) electromotive force; (2) electric field strength; (3) **energy**; (4) **illumination**; (5) irradiance.

Eagle mounting A compact mounting of a concave diffraction grating based on the principle of the Rowland circle. The mounting suffers from less astigmatism than either the Rowland or Abney mountings, and is useful for studying higher order spectra.

Earth See panel on p. 72.

earth-pillars Occur where sediments consisting of relatively large and preferably flat stones, embedded in a soft, finer-grained matrix, are undergoing erosion, esp. in regions of heavy rainfall. As the ground is progressively lowered the flat stones protect the softer material beneath them and are thus left standing on tall, acutely conical pillars.

earthquakes See panels on pp. 74–5.

earth science A term frequently used as a synonym for geology. Has also been used to include sciences which fall outside the scope of geology, e.g. meteorology.

earthy cobalt A variety of wad containing up to about 40% of cobalt oxide. Also *asbolite*.

echelon grating A form of interferometer resembling a flight of glass steps, light travelling through the instrument in a direction parallel to the treads of the steps. The number of interfering beams is therefore equal to the number of steps. Owing to the large path difference, $t(\mu-1)$, where t is the thickness of a step and μ is the **refractive index**, the order of interference and therefore the resolving power are high, making the instrument suitable for studying the fine structure of spectral lines.

echinoderms See panels on pp. 76–7.

Echinoidea Fossil echinoids are found in strata ranging from the Lower Palaeozoic to the present. They are particularly important in the Jurassic (Clypeus Grit etc.) and Cretaceous, where, in the Chalk, they have proved useful indices of horizon, esp. the various species of *Micraster* and *Holaster*.

eckermannite A rare monoclinic alkali amphibole; a hydrous sodium lithium aluminium magnesium silicate.

eclogite A coarse-grained deep-seated metamorphic rock, consisting essentially of pink garnet, green pyroxene (some of which is often chrome-diopside) and (rarely) kyanite.

economic geology The study of geological bodies and materials that can be used profitably by man.

eddy diffusion The migration and interchange of portions of a fluid as a result of their turbulent motion. Cf. **diffusion**.

eddy diffusivity Exactly analogous, for eddy diffusion, to the **diffusivity** for molecular diffusion. Symbol *E*.

edelopal A variety of *opal* with an exceptionally brilliant play of colours.

edenite An end-member compositional variety in the hornblende group of monoclinic amphiboles: hydrated magnesium, calcium and aluminium silicate.

edge coal Highly inclined coal seams.

edge filter A type of filter in which a large number of disks are clamped on a perforated hollow shaft joining a cylinder. This is contained in another cylindrical vessel into which liquid is pumped and flows through the narrow spaces between the disks, the solids being trapped on the disk edges. Filtrate leaves via the perforations in the hollow shaft.

edge water Water pressing inward upon the gas or oil in a natural reservoir.

EDTA *Ethylene Diamine Tetra-Acetic (ethanoic) acid, diaminoethanetetracetic acid*. $CH_2N(CH_2COOH)_2$. A *chelating agent* frequently used to protect enzymes from inhibition by traces of metal ions and as an inhibitor of metal dependent proteases because of its ability to combine with metals. It is also used in special soaps to remove metallic contamination. Also *complexone*.

Edwards' roaster Long horizontal furnace through which sulphide minerals are rabbled counterwise to hot air, to remove part or all of the sulphur by ignition.

effective half-life The time required for the activity of a radioactive nuclide in the body

Earth

The third planet in order from the Sun with a mean equatorial diameter of 6378.17 km, mass 5.977×10^{24} kg and mean density 5.517. From the astronomical perspective, Earth belongs to the group of terrestrial planets, which also includes **Mercury**, **Venus** and **Mars**. It is with this group, and also the **Moon**, that its origin, structure and evolution are often compared. Earth has an atmosphere intermediate in thickness between those of Venus and Mars. It is unique in possessing vast oceans of liquid water. The complex interaction between the oceans, the atmosphere and the continental surfaces determines the energy balance, the temperature regime and hence the climate. Cloud cover is typically 50% and heat trapped within the atmosphere by the **greenhouse effect** raises the average temperature by more than 30°C above that expected for the Earth's distance from the Sun.

The present composition of the atmosphere is 77% molecular nitrogen, 21% molecular oxygen, 1% water vapour and 0.9% argon, with the balance in trace components. The high concentration of oxygen, which dates from 2000 million years ago, is a direct result of the presence of plants. The presence of oxygen allowed the formation of the high-level ozone layer, which shields the surface from solar ultraviolet radiation damaging to higher forms of life.

Earth is the most geologically active of the major planets. Its large-scale features have all been determined by the creation, destruction, relative movement and interaction of a dozen or so crustal plates – the *lithosphere* – which slide over the less rigid *asthenosphere* below. Collisions between the plates produce folded mountains, and zones of seismic activity are concentrated along the plate boundaries.

Seismic waves, such as those generated during earthquakes, reveal the internal structure of the Earth. See **earthquakes**, **seismology**. The crust lying above the Mohorovičić discontinuity ('the Moho'), is about 6 – 10 km thick beneath the oceans, but 35 km or more beneath the continents as shown in the figure.

The *lithosphere*, of which the crust is the outer part, is composed of huge slabs that are relatively rigid, floating on the weaker, more plastic region, the *asthenosphere*. See **plate tectonics**. The mantle, between the crust and the core, is thought to have a magnesium- and iron-rich silicate composition. Below it, the core may be composed of iron and nickel.

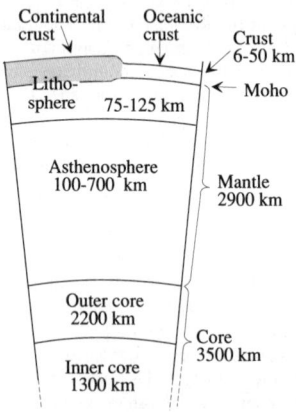

Structure of the Earth (*not to scale*)

Continental crust / Oceanic crust / Crust 6-50 km
Lithosphere 75-125 km — Moho
Asthenosphere 100-700 km
Mantle 2900 km
Outer core 2200 km
Inner core 1300 km
Core 3500 km

In planetary terms, the surface rocks of the Earth are very young. The basaltic rocks forming the ocean floors are among the youngest. The Precambrian shields, which occupy about 10% of the surface are the oldest and the nearest approximation to the cratered terrain that forms a large part of other planetary surfaces. Weathering and erosion processes mean that few traces of whatever impact craters there were now remain.

The molten metallic core gives rise to the Earth's magnetic field and *magnetosphere*. A layer of electrically charged particles (from the Sun) at a height of 200–300 kilometres forms the *ionosphere*. The funnelling of charged particles by the magnetic field to regions between latitudes of 60° and 75° create the phenomenon of the aurora. Satellite measurements have shown that the Earth is also an intense source of radio waves at kilometric wavelengths.

effervescence

to fall to half its original value as a result of both biological elimination and radioactive decay. Its value is given by

$$\frac{\tau(b_{1/2}) \times \tau(r_{1/2})}{\tau(b_{1/2}) + \tau(r_{1/2})},$$

where $\tau(b_{1/2})$ and $\tau(r_{1/2})$ are the biological and radioactive half-lives respectively.

effervescence The vigorous escape of small gas bubbles from a liquid, esp. as a result of chemical action.

efficiency of screening, numerical index of A quantity used to assess the efficiency of industrial screening procedures, i.e. sieving procedures. It is defined by the equation $E = F(D/B - C/A)$, where $E =$ the numerical index of efficiency of screening, $F =$ the percentage of the powder supply passed by the screen, $A =$ the percentage of difficult oversize in the powder supply, $B =$ the percentage of difficult undersize in the powder supply, $C =$ the percentage of the difficult oversize passed by the screen, $D =$ the percentage of the difficult undersize passed by the screen.

efflorescence (1) The loss of water from a crystalline hydrate on exposure to air, shown by the formation of a powder on the crystal surface. (2) A fine-grained crystalline deposit on the surface of a mineral or rock.

effusive Extrusive, volcanic.

Egyptian blue An artificial mineral, calcium copper silicate, $CaCuSi_4O_{12}$ thought to be the pigment of ancient Egypt.

Egyptian jasper A variety of jasper occurring in rounded pieces scattered over the surface of the desert, chiefly between Cairo and the Red Sea; used as a broochstone and for other ornamental purposes. Typically shows colour zoning.

Eh Symbol for **oxidation potential**.

Eifelian A stage in the Middle Devonian. See **Palaeozoic**.

ejecta Solid material thrown from a volcano or an impact crater. Also *ejectamenta*.

elaeolite A massive form of the mineral nepheline, greenish-grey or (when weathered) red in colour.

elastic bitumen See **elaterite**.

elasticity The tendency of a body to return to its original size or shape, after having been stretched, compressed, or deformed. The ratio of the stress called into play in the body by the action of the deforming forces to the strain or change in dimensions or shape is called the *coefficient (modulus) of elas-*

electric resistance welded tube

ticity. See the following definitions and **Hooke's law**.

elasticity of bulk The elasticity for changes in the volume of a body caused by changes in the pressure acting on it. The bulk modulus is the ratio of the change in pressure to the fractional change in volume. See **elasticity**.

elasticity of elongation The stress in this case is the stretching force per unit area of cross-section and the strain is the elongation per unit length. The modulus of elasticity of elongation is known as **Young's modulus**. See **elasticity**.

elasticity of shear, rigidity The elasticity of a body which has been pulled out of shape by a shearing force. The stress is equal to the tangential shearing force per unit area, and the strain is equal to the angle of shear, that is, the angle turned through by a straight line originally at right angles to the direction of the shearing force. See **elasticity**.

elastic limit The limiting value of the deforming force beyond which a body does not return to its original shape or dimensions when the force is removed.

elastomer A material, usually synthetic, having elastic properties akin to those of rubber.

elaterite A solid bitumen resembling dark-brown rubber. Elastic when fresh. Sometimes *mineral caoutchouc*.

elbaite One of the three chief compositional varieties of tourmaline; a complex hydrated borosilicate of lithium, aluminium and sodium. Most of the gem varieties of tourmaline are elbaites.

electrical conductivity Ratio of current density to applied electric field. Expressed in siemens per metre (S/m) or ohm^{-1} m^{-1} in S1 units. Conductivity of metals at high temperatures varies as T^{-1}, where T is absolute temperature. At very low temperatures, variation is complicated but it increases rapidly (at one stage proportional to T^{-5}), until it is finally limited by material defects of structure.

electrical double layer The layer of adsorbed ions at the surface of a dispersed phase which gives rise to the **electrokinetic effects**.

electrical prospecting Form of geophysical **prospecting** which identifies anomalous electrical effects associated with buried ore bodies. Most important techniques utilize resistivity or inverse conductivity and the inductive properties of ore minerals.

electric calamine See **hemimorphite**.

electric resistance welded tube Much used in heat exchangers, it is made continuously by forming an accurately rolled strip

earthquakes

A shaking of the Earth's crust caused, in most cases, by displacement along a fault. The place of maximum displacement is the *focus*, as in Fig. 1. The *epicentre* is the point on the surface of the Earth lying immediately above the focus of an earthquake (or nuclear explosion). On the Earth's surface, lines of equal intensity are *isoseismal lines*. Although the amount of displacement may be small, a matter of inches only, the destruction wrought at the surface may be very great, due in part to secondary causes, e.g. the severance of gas and water mains, as in the great San Francisco earthquake in 1906.

Most earthquakes take place at the boundaries of the major tectonic plates of the Earth's crust (See **plate tectonics**). They are measured around the world by *seismographs* which record the types of wave received (Fig. 2). The earthquake releases energy as *shear waves (S-wave, Secondary)* and *compressional waves (P-waves, Compression, Primary)*, as well as *surface waves (L-waves)*, e.g. *Rayleigh waves*.

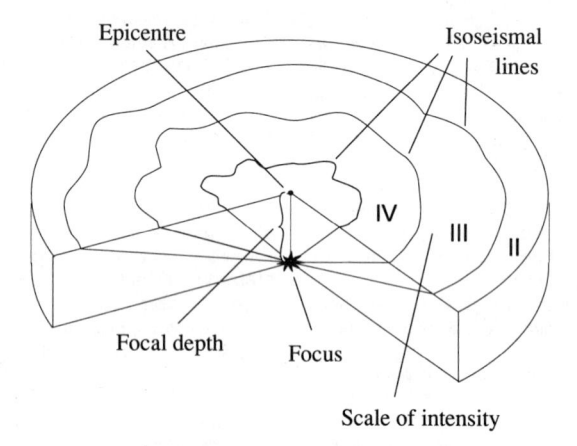

Fig. 1. Slice from Earth's crust

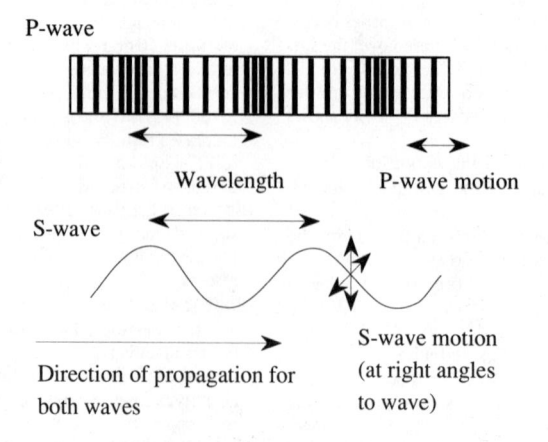

Fig. 2. P- and S-wave forms

continued on next page

earthquakes (contd)

The typical seismic wave pattern is shown in Fig. 3. P-waves can travel through any material including liquids and are faster than the other waves. S-waves will only pass through solids. The Surface or L-waves are the slowest and are confined to the Earth's crustal layers. The worldwide seismographic network enables the location, strength and depth of focus of earthquakes to be determined and sheds much light on the structure of the Earth as in Fig. 4 (see **Earth**). Since S-waves cannot traverse the core there is a shadow zone on the far side of the globe from the epicentre.

Earthquake *intensities* are expressed on a scale known as the Modified Mercalli Scale which expresses the degree of shaking as in the table. This scale is empirical and in 1935 C. F. Richter devised a scale of *magnitude* which is logarithmic, i.e. the difference between magnitude 4 and 5 is one tenth of that between 5 and 6. One of the strongest earthquakes of this century in Alaska in 1964 probably had a magnitude of 8.6. Submarine earthquakes may give rise to a **tsunami**, a wave produced in the oceans. It can be very destructive and travels great distances at very high speeds, up to 950 km/hour.

Modified Mercalli scale

I Instrumental: only detected on instruments.

II Feeble: felt by sensitive people.

III Slight: felt indoors, like a passing lorry.

IV Moderate: loose objects rock.

V Rather strong: felt generally.

VI Strong: slight damage.

VII Very strong: damage to poorly built structures.

VIII Destructive: walls fissured, chimneys fall.

IX Ruinous: damage considerable, pipes break.

X Disastrous: many buildings destroyed, rails bent.

XI Very disastrous: few masonry structures remain, all underground services destroyed.

XII Catastrophic: total destruction, waves seen on ground surface.

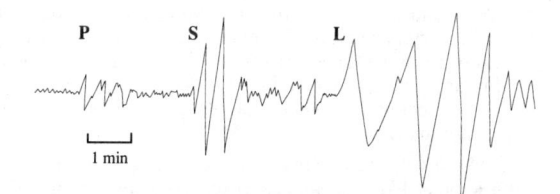

Fig. 3. Seismogram, showing arrival of P-, S- and L-waves from an epicentre 1400 km away.

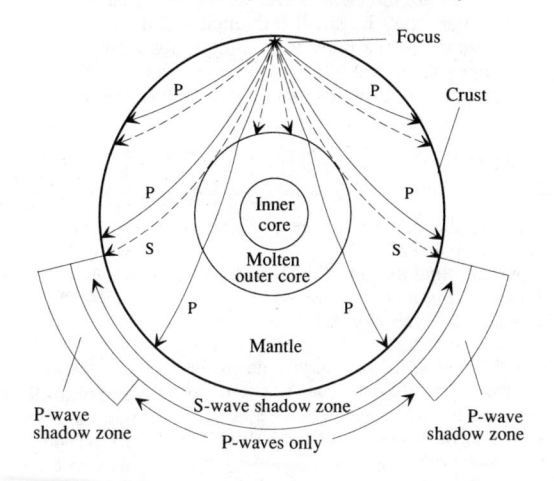

Fig. 4. Section through the centre of the Earth with paths of seismic waves. Solid lines are P-waves, dashed lines, S-waves.

echinoderms

The echinoderms are solitary marine, usually bottom-dwelling, animals belonging to the phylum Echinodermata. They have an exoskeleton of plates or ossicles of calcite and many possess a five-fold radial symmetry. Included in the phylum are echinoids and crinoids, both of considerable importance as fossils, as well as starfishes (Asteroidea), brittle-stars (Ophiuroidea), sea cucumbers (Holothuroidea) and extinct blastoids, cystoids and edrioasteroids, all of less importance.

The *echinoids*, sea urchins, are members of the class of free-swimming Echinoidea. They have a globular, heart-shaped or discoidal *test* (exoskeleton) covered in spines. The calcareous plates that comprise the test are generally fused together. The Echinoidea is divided into two subclasses, Regularia and Irregularia. In the regular echinoderms, Fig. 1. opposite, the test is nearly circular in outline. The mouth is in the centre of the underside and the anus above it on the dorsal surface. Around the anus is a ring of plates from which five narrow rays diverge, each typically consisting of two columns of plates, known as *ambulacral areas*. The regular echinoids range from the Ordovician to the present day, but are rare in the Palaeozoic. In the irregular echinoids (Fig. 2) there is a distinct bilateral symmetry to an oval or heart-chaped test. The mouth is central or in front of centre and the anus is on the central line towards the rear. Irregular echinoids developed much later than the regular subclass, and range from the Lower Jurassic to the present day, where they are dominant. Echinoids became abundant in the Jurassic limestones and even more abundant in the Cretaceous where they have been used as zonal fossils, e.g. species of *Micraster* and *Holaster*.

The *crinoids*, sea lilies, are members of the class Crinoidea which superficially have a plant-like appearance (Fig. 3). They are mainly sedentary, marine animals having a stem of columnal plates held by root-like processes and supporting a *calyx* or cup, bearing five *arms* which may be branched. The arms and calyx are the *crown*. The stem may reach several metres in length; on the arms there may be small unbranched *pinnules*. The crinoids range from the Ordovician to the present day, attaining their maximum development in the Silurian. Some Palaeozoic limestones have such large numbers of disaggregated plates of crinoids that they are referred to as *crinoidal limestone*.

figure on next page

over a mandrel and welding the edges electrically. Abbrev *ERW tube*.

electrochemical series The classification of redox halfreactions, written as reductions, in order of decreasing reducing strength. Thus the combination of any half reaction with the reverse of one further down the series will give a spontaneous reaction. For reference, the half reaction $2H^+ + e^- = H_2$ is taken as having an energy of zero. Also *electrode potential series*.

electrochemistry That branch of chemistry which deals with the electronic and electrical aspects of processes, usually, but not always, in a liquid phase.

electrodialysis Removal of electrolytes from a colloidal solution by an electric field between electrodes in pure water outside the two dialysing membranes between which is contained the colloidal solution.

electro-endosmosis Movement of liquid, under an applied electric field, through a fine tube or membrane. Also *electro-osmosis, electrosmosis*.

electrokinetic effects Phenomena due to the interaction of the relative motion with the potential between the two phases in a dispersed system. There are four: *electro-endosmosis, electrophoresis, streaming potential* and *sedimentation potential*.

electrokinetic potential Potential difference between surface of a solid particle immersed in aqueous or conducting liquid and the fully dissociated ionic concentration in

Echinoderms

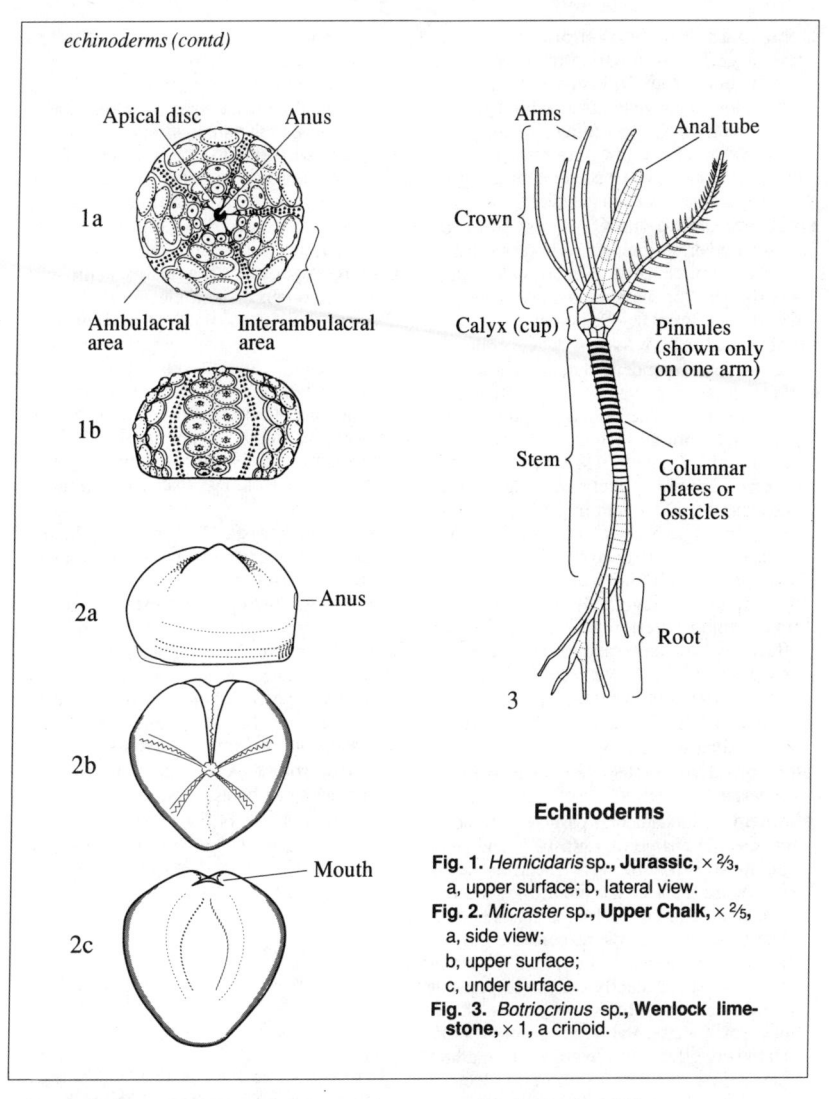

Fig. 1. *Hemicidaris* sp., **Jurassic,** × 2/3,
a, upper surface; b, lateral view.
Fig. 2. *Micraster* sp., **Upper Chalk,** × 2/5,
a, side view;
b, upper surface;
c, under surface.
Fig. 3. *Botriocrinus* sp., **Wenlock limestone,** × 1, a crinoid.

the body of the liquid. Concept important in froth flotation, electrophoresis etc. Also *zeta potential*.

electrolysis Chemical change, generally decomposition, effected by a flow of current through a solution of the chemical, or its molten state, based on ionization.

electrolyte Chemical, or its solution in water, which conducts current through ionization.

electrolyte strength Extent towards complete ionization in a dilute solution. When concentrated, the ions join in groups, as indicated by lowered mobility.

electrolytic cell An electrochemical cell in which an externally applied voltage causes a non-spontaneous change to occur, such as the breakdown of water into hydrogen and oxygen. Opposite of **galvanic cell**.

electrolytic dissociation The splitting-up (which is reversible) of substances into oppositely-charged ions.

electromagnetic prospecting Method in which distortions produced in electric waves are observed and lead to pinpointing of geological anomalies.

electromagnetic separation

electromagnetic separation Removal of ferromagnetic objects from town refuse, or 'tramp iron' from bulk materials, as they travel along a conveyor, over a drum, or into a revolving screen, by setting up a magnetic field which diverts the ferromagnetic material from the rest. Concentration of ferromagnetic minerals from gangue.

electromagnetic units Any system of units based on assigning an arbitrary value to μ_0, the permeability of free space. $\mu_0 = 4\pi\lambda 10^{-7}$ Hm^{-1} in the SI system; μ_0 is unity in the CGS electromagnetic system.

electromagnetic wave A wave comprising two interdependent mutually perpendicular transverse waves of electric and magnetic fields. The velocity of propagation in free space for all such waves is that of the velocity of light, 2.9979245×10^8 ms^{-1}. The electromagnetic spectrum ranges from wavelengths of 10^{-15} m to 10^3 m, i.e. from γ-rays through X-rays, ultraviolet, visible light, infrared, microwave, short-, medium- and long-wave radio waves. Electromagnetic waves undergo reflection and refraction, exhibit interference and diffraction effects, and can be polarized. The waves can be channelled, e.g. by waveguides for microwaves or fibre optics for light.

electrometric titration See **potentiometric titration**.

electromotive series See **electrochemical series**.

electron A fundamental particle with negative electric charge of 1.60×10^{-19} coulombs and mass 9.10×10^{-31} kg. Electrons are a basic constituent of the atom; they are distributed around the nucleus in *shells* and the electronic structure is responsible for the chemical properties of the atom. Electrons also exist independently and are responsible for many electric effects in materials. Due to their small mass, the wave properties and relativistic effects of electrons are marked. The *positron*, the antiparticle of the electron, is an equivalent particle but with a positive charge. Either electrons or positrons may be emitted in β-decay. Electrons, muons and neutrinos form a group of fundamental particles called **leptons**.

electron affinity The energy required to remove an electron from a negatively charged ion to form a neutral atom.

electron charge/mass ratio The ratio

$$e/m = 1.75 \times 10^{11} \text{ C kg}^{-1}.$$

A fundamental physical constant, the mass being the rest mass of the electron.

electron density The number of electrons per gram of a material. Approx. $\times 10^{23}$ for most light elements. In an ionized gas the equivalent electron density is the product of the ionic density and the ratio of the mass of an electron to that of a gas ion.

electron diffraction Investigation of crystal structure by the patterns obtained on a screen from electrons diffracted from the surface of crystals or as a result of transmission through thin metal films.

electronegativity The relative ability of an atom to retain or gain electrons. There are several definitions, and the term is not quantitative. It is, however, useful in predicting the strengths and the polarities of bonds, both of which are greater when there is a significant electronegativity difference between the atoms forming the bond. On the commonly used scale of Pauling, the range of values is from less than 1 (alkali metals) to 4 (fluorine).

electronic charge The unit in which all nuclear charges are expressed. It is equal to 1.60×10^{-19} coulombs.

electronic theory of valency Valency forces arise from the transfer or sharing of the electrons in the outer shells of the atoms in a molecule. The two extremes are complete transfer (*ionic bond*) and close sharing (*covalent bond*), but there are intermediate degrees of bond strength and distance.

electron mass A result of relativity theory, that mass can be ascribed to kinetic energy, is that the effective mass (m) of the electron should vary with its velocity according to the experimentally confirmed expression:

$$m = \frac{m_0}{\sqrt{1 - \left(\dfrac{v}{c}\right)^2}},$$

where m_0 is the mass for small velocities, c is the velocity of light, and v that of the electron.

electron pair Two valence electrons shared by adjacent nuclei, so forming a bond.

electron paramagnetic resonance See **electron spin resonance**.

electron probe analysis A beam of electrons is focused on to a point on the surface of the sample, the elements being detected both qualitatively and quantitatively by their resultant X-ray spectra. An accurate (1%) non-destructive method needing only small quantities (micron size) of sample.

electron spin resonance The branch of

electron-volt — **enargite**

microwave spectroscopy in which there is resonant absorption of radiation by a paramagnetic substance, possessing unpaired electrons, when the energy levels are split by the application of a strong magnetic field. The difference in energy levels is modified by the environment of the atoms. Information on impurity centres in crystals, the nature of the chemical bond and the effect of radiation damage can be found. Also *electron paramagnetic resonance*. Abbrev. **ESR**.

electron-volt General unit of energy of moving particles, equal to the kinetic energy acquired by an electron losing one volt of potential, equal to 1.602×10^{-19} J. Abbrev. *eV*.

electro-osmosis, electrosmosis See **electro-endosmosis**.

electrophoresis Motion of charged particles under an electric field in a fluid, positive groups to the cathode and negative groups to the anode.

electrostatic bonding See **electrovalence**.

electrovalence Chemical bond in which an electron is transferred from one atom to another, the resulting ions being held together by electrostatic attraction.

electroviscosity Minor change of viscosity when an electric field is applied to certain polar liquids.

element Simple substance which cannot be resolved into simpler substances by normal chemical means. Because of the existence of **isotopes** of elements, an element cannot be regarded as a substance which has identical atoms, but as one which has atoms of the same atomic number.

elementary particle Particle believed to be incapable of subdivision; the term *fundamental particle* is now more generally used.

elevation of boiling point The raising of the boiling point of a liquid by substances in solution. May be used to determine molecular weights of solutes.

ELFO Abbrev. for **Extra Light Fuel Oil**.

elution Washing of loaded ion-exchange resins to remove captured uranium ions or other seized elements using washing liquor. This liquor is the *eluent* and the enriched solution it becomes is the *eluate*. The resin (or **zeolite**) is regenerated (like rinsing water softener with brine).

eluvial, eluvium gravels Those gravels formed by the disintegration *in situ* of the rocks which contributed to their formation. Antonym *alluvial deposits*.

elvan A term applied by Cornish miners to the dyke rocks associated with the Armorican granites of that county. Elvans are actually quartz-porphyries, microgranites, and other medium- to fine-textured dyke rocks of granitic composition.

e/m The ratio of the electric charge to mass of an elementary particle. For slow moving electrons $e/m = 1.75 \times 10^{11}$ C kg^{-1}. This value decreases with increasing velocity because of the relativistic increase in mass. Also *specific charge*.

emerald The brilliant green gemstone, a form of beryl; silicate of beryllium and aluminium, crystallizing in hexagonal prismatic forms, occurring in mica-schists, calcite veins and rarely in pegmatites. See **gems and gemstones**.

emerald copper See **dioptase**.

emergence The uplift of land relative to the sea.

emery A finely granular intimate admixture of corundum and either magnetite or haematite, occurring naturally in Greece and localities in Asia Minor etc.; used extensively as an abrasive.

emissivity The ratio of emissive power of a surface at a given temperature to that of a black body at the same temperature and with the same surroundings. Values range from 1.0 for lampblack down to 0.02 for polished silver. See **Stefan-Boltzmann law**.

empirical formula Formula deduced from the results of analysis which is merely the simplest expression of the ratio of the atoms in a substance. In molecular materials it may, or may not, show how many atoms of each element the molecule contains: e.g. methanal, CH_2O, ethanoic acid, $C_2H_4O_2$, and lactic acid, $C_3H_6O_3$, have the same percentage composition, and consequently, on analysis, they would all be found to have the same empirical formula.

Emsian A stage in the Lower Devonian. See **Palaeozoic**.

emulsifier An apparatus with a rotating, stirring or other device used for making emulsions.

emulsifying agents Substances whose presence in small quantities stabilizes an emulsion, e.g. ammonium linoleate, certain benzene-sulphonic acids etc.

emulsion A colloidal suspension of one liquid in another, e.g. milk.

enantiomerism See **enantiomorphism**.

enantiomorphism Mirror image isomerism. A classical example is that of the crystals of sodium ammonium tartrates which Pasteur showed to exist in mirror image forms. Adj. *enantiomorphous*.

enargite Sulpharsenide of copper, often containing a little antimony.

encrinal limestone A crinoidal limestone. Also *encrinital limestone*.

end Solid rock at end of underground passage.

endellite US for **halloysite**.

enderbite A member of the charnockitic group of rocks, consisting of quartz, antiperthite, hypersthene and magnetite; the equivalent of a hypersthene tonalite.

endless rope haulage A method of hauling trucks underground by attachment to a long loop of rope, guided by many pulleys along the roads or haulage ways, and actuated by a power-driven drum.

endogenetic Describes (1) the processes and resultant products that originate within the Earth, e.g. volcanism, extrusive rocks; (2) the processes leading to *endomorphism*, modification of igneous rocks by assimilation of country rocks. Also *endogenous*. Cf. **exogenetic**.

endothermic Accompanied by the absorption of heat. (ΔH positive.)

energy The capacity of a body for doing work. Mechanical energy may be of two kinds: *potential energy*, by virtue of the position of the body, and *kinetic energy*, by virtue of its motion. Energy can take a wide variety of forms. Both *mechanical* and *electrical* energy can be converted into *heat* which is itself a form of energy. Electrical energy can be stored in a capacitor to be recovered at discharge of the capacitor. *Elastic potential energy* is stored when a body is deformed or changes its configuration, e.g. in a compressed spring. All forms of wave motion have energy; in electromagnetic waves it is stored in the electric and magnetic fields. In any closed system, the total energy is constant – the *conservation of energy*. Units of energy: SI unit is the **joule** (symbol J) and is the work done by a force of 1 newton moving through a distance of 1 metre in the direction of the force. The CGS unit, the *erg* is equal to 10^{-7} joules and is the work done by a force of 1 dyne moving through 1 cm in the direction of the force. The foot-pound force (ft-lb f) of the British system equals 1.356 J.

energy balance A *heat balance*.

engineering geology Geology applied to engineering practice, esp. in mining and civil engineering.

engine plane In a coalmine, a roadway on which tubs, trucks, or trains are hauled by means of a rope worked by an engine.

englacial Contained within the ice, e.g. *englacial stream*.

Engler distillation The determination of the boiling range of petroleum distillates, carried out in a definite prescribed manner by distilling 100 cm^3 of the substance and taking the temperature after every 5 or 10 cm^3 of distillate has collected. The initial and final boiling points are also measured.

Engler flask A 100 cm^3 distillation flask of definite prescribed proportions used for carrying out an **Engler distillation**.

enriched uranium Uranium in which the content of the isotope ^{235}U has been increased above its natural value of 0.7%.

enrichment (1) Effect of superficial leaching of a lode, whereby part of value is dissolved and redeposited in a lower enriched zone. See **secondary enrichment**. (2) Raising the content of the isotope ^{235}U above its natural value of 0.7%.

enstatite An orthorhombic pyroxene, chemically magnesium silicate, MgSiO$_3$; it occurs as a rock-forming mineral and in meteorites.

enstatitite A variety of orthopyroxenite composed almost entirely of **enstatite**.

enthalpy Thermodynamic property of a working substance defined as $H = U + PV$ where U is the internal energy, P the pressure and V the volume of a system. Associated with the study of heat of reaction, heat capacity and flow processes. SI unit is the joule.

entropy In thermal processes, a quantity which measures the extent to which the energy of a system is available for conversion to work. If a system undergoing an infinitesimal reversible change takes in a quantity of heat dQ at absolute temperature T, its entropy is increased by $dS = dQ/T$. The area under the absolute temperature-entropy graph for a reversible process represents the heat transferred in the process. For an *adiabatic process*, there is no heat transfer and the temperature-entropy graph is a straight line, the entropy remaining constant during the process. When a thermodynamic system is considered on the microscopic scale, equilibrium is associated with the distribution of molecules that has the greatest probability of occurring, i.e. the state with the greatest degree of disorder. *Statistical mechanics* interprets the increase in entropy in a closed system to a maximum at equilibrium as the consequence of the trend from a less probable to a more probable state. Any process in which no change in entropy occurs is said to be *isentropic*.

environmental geology The application of geological knowledge to problems created by man in the physical environment and by

Eocambrian the use of physical resources. The term includes problems of a global scale.

Eocambrian A poorly fossiliferous sequence of late Precambrian rocks, approx. the equivalent of Riphean. See **Precambrian**.

Eocene The oldest division of **Tertiary** rocks, now regarded as an epoch within the Palaeogene system.

eolian deposits See **aeolian deposits**.

eolith Literally 'dawn stone'; a term applied to the oldest-known stone implements used by early Man which occur in the Stone Bed at the base of the Crag in East Anglia, and in high-level gravel deposits elsewhere. The workmanship is crude, and some authorities question their human origin, thinking it likely that the chipping has been produced by natural causes.

Eolithic Period The time of the primitive men who manufactured and used *eoliths*: the dawn of the Stone Age. Cf. **Palaeolithic Period**, **Neolithic Period**.

eon A large part of geological time consisting of a number of **eras**, e.g. *Phanerozoic eon* which includes the **Palaeozoic**, **Mesozoic** and **Cenozoic** eras. See **geological column**.

Eötvös balance Torsion balance sensitive to minute gravitational differences in land masses.

Eötvös equation The molecular surface energy of a substance decreases linearly with temperature, becoming zero about 60°C below the critical point.

Eozoic A term suggested for the Precambrian System, but little used. It means the 'dawn of life', and is comparable with *Palaeozoic*, *Mesozoic* and *Cenozoic*.

eozoon A banded structure found originally in certain Canadian Precambrian rocks and thought to be of organic origin; now known to be inorganic and a product of dedolomitization, consisting of alternating bands of calcite and serpentine replacing forsterite.

epeirogenic earth movements Continent-building movements, as distinct from mountain-building movements, involving the coastal plain and just-submerged 'continental platform' of the great land areas. Such movements include gentle uplift or depression, with gentle folding and the development of normal tensional faults.

epicentre That point on the surface of the Earth lying immediately above the focus of an earthquake or nuclear explosion. See **earthquakes**.

epicontinental Within the limits of a continental mass, including the continental shelf.

epicontinental sea Sea covering the continental shelf or part of a continent.

epidiorite A metamorphosed gabbro or dolerite in which the original pyroxene has been replaced by fibrous amphibole. Other mineral changes have also taken step in the conversion, by dynamothermal metamorphism, of a basic igneous rock into a green schist.

epidosites Metamorphic rocks composed of epidote and quartz. See **epidotization**.

epidote A hydrated aluminium iron silicate, occurring in many metamorphic rocks in lustrous yellow-green crystals, and as an alteration product in igneous rocks. Also *pistacite*.

epidotization A process of alteration, esp. of basic igneous rocks in which the feldspar is albitized with the separation of epidote and zoisite representing the anorthite molecule of the original plagioclase.

epigenetic Ore deposits formed later than the rocks enclosing them. See **syngenetic**.

epimorph A natural cast of a crystal.

epistilbite A white or colourless zeolite; hydrated calcium aluminium silicate, crystallizing in the monoclinic system.

epitaxy Unified crystal growth or deposition of one crystal layer on another.

epoch A subdivision of a geological period, the time equivalent of a **series**, e.g. the **Wenlock** epoch of the **Silurian** period.

epsomite, Epsom salts Hydrated magnesium sulphate, $MgSO_4.7H_2O$. Occurs as incrustations in mines, in colourless acicular to prismatic crystals. Many chemical and other uses.

equal falling particles Particles possessing equal *terminal velocities*; the underflow, oversize product of a classifier.

equation of state An equation relating the volume, pressure and temperature of a given system, e.g. van der Waals' equation.

equilibrium The state reached in a reversible reaction when the reaction velocities in the two opposing directions are equal, so that the system has no further tendency to change.

equilibrium constant The ratio, at equilibrium, of the product of the active masses of the molecules on the right side of the equation representing a reversible reaction to that of the active masses of the molecules on the left side.

equilibrium of floating bodies For a body which floats, partly immersed in liquid, the weight of the body is equal to the weight of the fluid it displaces. The ratio of the

equivalent weight

volume of the body to the volume immersed is the ratio of the density of the fluid to that of the body. See **Archimedes' principle**.

equivalent weight That quantity of one substance which reacts chemically with a given amount of a standard. In particular, the equivalent weight of an acid will react with one mole of hydroxide ions, while the equivalent weight of an oxidizing agent will react with one mole of electrons. Also *equivalent*.

Er Symbol for **erbium**.

era A geological time unit within an eon, the formal chronostratigraphic unit above a **system**, e.g. *Mesozoic* era. See **geological column**.

erbium A metallic element, a member of the rare earth group. Symbol Er, at.no. 68, r.a.m. 167.26. Found in the same minerals as dysprosium (gadolinite, fergusonite, xenotime), and in euxenite.

erg Unit of work or energy in CGS system. 1 erg of work is done when a force of 1 dyne moves its point of application 1 cm in the direction of the force. See **energy, fundamental dynamical units**.

erionite One of the less common zeolites; a hydrated aluminium silicate of sodium, potassium and calcium.

erosion The removal of the land surface by weathering, corrasion, corrosion and transportation, under the influence of gravity, wind and running water. Also, the eating away of the coastline by the action of the sea. Soil erosion may result from factors such as bad agricultural methods, excessive deforestation, overgrazing.

erratics Stones, ranging in size from pebbles to large boulders, which were transported by ice, which, on melting, left them stranded far from their original source. They furnish valuable evidence of the former extent and movements of ice sheets.

erubescite See **bornite**.

eruptive rocks A term sometimes used for all igneous rocks; but usually applied to **volcanic** or **extrusive rocks**.

erythrite Hydrated cobalt arsenate, occurring as pale reddish crystals or incrustations. Also *cobalt bloom*.

escape boom Means of escape from an offshore platform. Pivoted chutes with a bouyant outer end are normally secured inboard. In emergency they are released and rotate outwards until the outer end floats and personnel can slide to safety.

escarpment A long cliff-like ridge developed by denudation where hard and soft inclined strata are interbedded, the outcrop of each hard rock forming an *escarpment*, such

ethene

as those of the Chalk (Chiltern Hills, North and South Downs) and the Jurassic limestones (Cotswold Hills). Generally an escarpment consists of a short steep rise (the *scarp face*) and a long gentle slope (the *dip slope*). Cf. **cuesta**.

Eschka's reagent Mixture of MgO and Na_2CO_3 (2 : 1); used for estimation of the sulphur content of fuels.

esker A long, narrow, steep-sided, sinuous ridge of poorly stratified sand and gravel deposited by a subglacial or englacial stream. Found in glaciated areas.

ESR Abbrev. for **Electron Spin Resonance**.

essential minerals Components present, by definition, in a rock, the absence of which would automatically change the name and classification of the rock. Cf. **accessory minerals**.

essexite A coarse-grained deep-seated igneous rock, a nepheline monzogabbro or nepheline monzodiorite containing labradorite, alkali-feldspar, titanaugite, kaersutite and biotite. Named from Essex Co., Mass.

essonite Original spelling of **hessonite**.

estuarine deposition Sedimentation in the environment of an estuary. The deposits differ from those which form in a deltaic environment, chiefly in their closer relationship to the strata of the adjacent land, and are usually of finer grain and of more uniform composition. Both are characterized by brackish water and sediments which contain land-derived animal and plant remains.

estuarine muds So-called *estuarine muds* are, in many cases, silts admixed with sufficient true clay to give them some degree of plasticity; they are characterized by a high content of decomposed organic matter.

estuary An inlet of the sea at the mouth of a river; developed esp. in areas which have recently been submerged by the sea, the lower end of the valley having been thus drowned. See **fiords**. Cf. **delta**.

etched figures, etch-figures Small pits or depressions of geometrical design in the faces of crystals, due to the action of some solvent. The actual form of the figure depends upon the symmetry of the face concerned, and hence they provide invaluable evidence of the true symmetry in distorted crystals.

ethanol The IUPAC name for *ethyl alcohol*. C_2H_5OH, mp $-114°C$, bp $78.4°C$; a colourless liquid, of vinous odour, miscible with water and most organic solvents, rel.d. 0.789; formed by the hydrolysis of ethyl chloride or of ethyl hydrogen sulphate.

ethene The IUPAC name for *ethylene*.

$H_2C = CH_2$, mp $-169°C$, bp $-103°C$, a gas of the alkene series, contained in illuminating gas and in gases obtained from the cracking of petroleum. Used for synthetic purposes, e.g. polythene and ethylene oxide, and for maturing fruit in storage.

ether *Alkoxyalkane.* (1) Any compound of the type R—O—R', containing two identical, or different, alkyl groups united to an oxygen atom; they form a homologous series $C_nH_{2n+2}O$. (2) Specifically, diethyl ether.

ethyl alcohol See **ethanol**.

ethylene See **ethene**.

ethylene diamine tetra-acetic acid See **EDTA**.

Eu Symbol for **europium**.

euclase A monoclinic mineral, occurring as prismatic, usually colourless, crystals; hydrated beryllium aluminium silicate.

eucrite A coarse-grained, usually ophitic, deep-seated basic gabbro containing plagioclase near bytownite in composition, both ortho- and clino-pyroxenes, together with olivine. Eucrite is an important rock type in the Tertiary complexes of Scotland.

eucryptite A hexagonal lithium aluminium silicate, commonly found as an alteration product of spodumene.

eudialyte, eudialite A pinkish-red complex hydrated sodium calcium iron zirconosilicate. It crystallizes in the rhombohedral system and occurs in some alkaline igneous rocks.

eugeosyncline A geosyncline, characterized by intermittent volcanic activity.

euhedral crystals See **idiomorphic crystals**.

europium Metallic element of the rare earth group. Symbol Eu, at.no. 63, r.a.m. 151.96, valency 2 and 3. Contained in black monazite, gadolinite, samarskite, xenotime.

Eurypterida An order of Crustaceans ranging from Ordovician to Permian and represented by such types as *Eurypterus* and *Stylonurus*, the latter reaching almost 2 m in length with a scorpion-like appearance. They are entirely aquatic.

eustatic movements Changes of sea level, constant over wide areas, due probably to alterations in the volumes of the seas. These may be due to variations in the extent of the polar icecaps, large-scale crustal movements in ocean basins, or submarine volcanic activity.

eutaxitic Descriptive of the streaky banded structure of certain pyroclastic rocks, as contrasted with the smooth layered structure of flow-banded lavas.

eutectic Relating to mixture of two or more substances having a minimum melting point. Such a mixture behaves in certain respects like a pure compound. The solid phases co-exist at an isobarically invariant point which is the minimum melting temperature for the assemblage of solids.

eutectic point The point in the binary or ternary constitutional diagram indicating the composition of the eutectic alloy, or mixture of minerals, and the temperature at which it solidifies.

euxenite Niobate, tantalate and titanate of yttrium, erbium, cerium, thorium and uranium, and valuable on this account. Commonly massive and brownish-black in colour.

evaporation The conversion of a liquid into vapour, at temperatures below the boiling point. The rate of evaporation increases with rise of temperature, since it depends on the saturated vapour pressure of the liquid, which rises until it is equal to the atmospheric pressure at the boiling point. Evaporation is used to concentrate a solution.

evaporator A still designed to evaporate moisture or solvents to obtain the concentrate.

evaporite Sedimentary deposit of material previously in aqueous solution and concentrated by the evaporation of the solvent. Normally found as the result of evaporation in lagoons or shallow enclosed seas, e.g. salt and anhydrite.

evolutionary operation Introduction of sectionally controlled variants into a commercial process, during transfer of laboratory research into better production.

exfoliation The splitting off of thin folia or sheets of rock from surfaces exposed to the atmosphere, particularly in regions of wide temperature variation. It is one of the processes involved in spheroidal weathering.

exinite A hydrogen-rich **maceral** which is found in coal.

exogenetic Describes (1) processes originating at or near the surface of the Earth, e.g. weathering, denudation, and the rocks, (esp. sedimentary rocks), ore deposits and land forms to which they give rise; (2) the processes leading to *exomorphism*, modification of the country rocks by igneous rocks intruding them. Also *exogenous*. Cf. **endogenetic**.

exothermic Accompanied by the evolution of heat.

expansion of gases All gases have very nearly the same coefficient of expansion, namely 0.003 66 per kelvin when kept at constant pressure. See **absolute temperature**.

experimental petrology

experimental petrology A branch of petrology concerned with the laboratory study of rocks and minerals under different physical and chemical conditions.

exploitation well Well drilled in a proved deposit.

exploration well Well drilled in unproven ground to test for oil after a positive seismic or other survey.

explosion A rapid increase of pressure in a confined space. Explosions are generally caused by the occurrence of exothermic chemical reactions in which gases are produced in relatively large amounts.

explosive There are two main classes, 'permitted' and 'non-permitted', i.e. those which are safe for use in coalmines and those which are not. Ammonium nitrate mixtures are mostly used in coalmines; nitroglycerine derivatives in metal mines. ANFO (*Ammonium Nitrate and Fuel Oil*) is now widely used in hard rock mining. Explosives are used as propellants (*low explosives*) and for blasting (*high explosives*), in both civil and military applications.

explosive fracturing The use of an explosive charge to crack strata and increase oil flow round a borehole. Invented by Col. Roberts in Titusville, Pennsylvania as the *Roberts torpedo*.

explosive limits The upper and lower limits of the ratio of an inflammable gas to air within which the mixture can explode. Petrol vapour will only explode between 7.6% and 1.4 % vapour to air by volume.

extension ore In assessment of reserves, ore which has not been measured and sampled, but is inferred as existing from geological reasoning supported by proved facts regarding adjacent ore.

extinction coefficient   See **molar absorbance**.

extraction A process for dissolving certain constituents of a mixture by means of a liquid with solvent properties for one of the components only. Substances can be extracted from solids, e.g. grease from fabrics with petrol; or from liquids, e.g. extraction of an aqueous solution with ethoxyethane, the efficiency depending on the partition coefficient of the particular substance between the two solvents.

extraction metallurgy First stage or stages of ore treatment, in which gangue minerals are discarded, and valuable ones separated and prepared for working up into finished metals, rare earths or other saleable products. Characteristically, the methods used do not change the physical structure of these products save by comminution.

extractive distillation A technique for improving, or achieving in cases impossible without it, distillation separation processes by the introduction of an additional substance which changes the system equilibrium.

extra-lateral rights   See apex law.

extra light fuel oil A heating oil with a Redwood viscosity of 32″. Abbrev. *ELFO*.

extraneous ash In raw coal, the so-called 'free dirt' or associated shale and enclosing beds.

extrinsic Said of electrical conduction properties arising from impurities in the crystal.

extrusive rocks Rocks formed by the consolidation of magma on the surface of the ground as distinct from *intrusive rocks* which consolidate below ground. Commonly referred to as *lava flows*; normally of fine grain or even glassy.

eyepiece graticule Grid incorporated in the eyepiece for measuring objects under the microscope. Special type used in particle-size analysis consists of a rectangular grid for selecting the particles and a series of graded circles for use in sizing the particles. Also *micrometer eyepiece, ocular micrometer* (US).

eyot (Pronounced *ait*). A small island.

F

f Symbol for (1) activity coefficient, for molar concentration; (2) partition function.

F Symbol for **fluorine**.

F Symbol used, following a temperature (e.g. 41°F) to indicate the **Fahrenheit scale**.

F Symbol for faraday.

[F] A Fraunhofer line in the blue of the solar spectrum of wavelength 486.1527 nm. It is the second line in the Balmer hydrogen series, also Hβ.

fabric The sum of all the textural and structural features of a rock.

face (1) See **crystal**. (2) The exposed surface of coal or other mineral deposit in the working place where mining, winning or getting is proceeding.

face-airing The operation of directing a ventilating current along the face of a working place. See **flushing**.

face-centred crystal lattice See **unit cell**.

facet A flat polished face on a gemstone.

facies The appearance or aspect of any rock; the sum total of its characteristics. Used of igneous, metamorphic and esp. sedimentary rocks.

facies, stratigraphic The sum of the rock and fossil features of a sedimentary rock. They include *lithofacies*: mineral composition, grain size, texture, colour, cross bedding and other sedimentary features; and *biofacies*: the fossil plant and animal characteristics of the rock.

faecal pellets Animal excrement, often in the form of rods or ovoid pellets, found in sedimentary rocks. See **coprolite**.

Fahrenheit scale The method of graduating a thermometer in which freezing point of water is marked 32° and boiling point 212°, the fundamental interval being therefore 180°. Fahrenheit has been largely replaced by the Celsius (Centigrade) and Kelvin scales. To convert °F to °C subtract 32 and multiply by 5/9. For the *Rankine* equivalent add 459.67 to °F; this total multiplied by 5/9 gives the *Kelvin* equivalent.

fairfieldite A hydrated phosphate of calcium and manganese, crystallizing in the triclinic system as prismatic crystals or fibrous aggregates.

Fajans-Soddy law of radioactive displacement The atomic number of an element decreases by two upon emission of an α-particle, and increases by one upon emission of a β-particle.

fall (1) The collapse of the roof of a level or tunnel, or of a flat working place or stall; the collapse of the hanging wall of an inclined working place or stope. (2) A mass of stone which has fallen from the roof or sides of an underground roadway, or from the roof of a working place.

false amethyst An incorrect name given to a purple gemstone. Applied (wrongly) to purple fluorite, and sometimes to purple corundum. See **Oriental amethyst**.

false bedding See **cross bedding**.

false diamond Several natural minerals are sometimes completely colourless and, when cut and polished, make brilliant gems. These include zircon, white sapphire and white topaz. All three, however, are birefringent and can be easily distinguished from true diamond by optical and other tests. Many artificial diamond *simulants* are now made, some difficult to detect. See **cubic zirconia**, **YAG**.

false ruby Some species of garnet (*Cape ruby*) and some species of spinel (*balas ruby* and *ruby spinel*) possess the colour of ruby, but have neither the chemical composition nor the physical attributes of true ruby.

false topaz A name wrongly applied to yellow quartz. See **citrine**.

Famennian The highest stage in the Devonian. See **Palaeozoic**.

family The group of radioactive nuclides which form a decay series.

fan (1) A detrital cone found at the foot of mountains and also in the deep sea (submarine fan). (2) Fan cleavage; an axial-plane cleavage in which the cleavage planes fan out.

fan drift Ventilating passage along which air is moved by means of a fan.

fanglomerate A conglomerate formed by lithification of a fan.

fan shaft Mine shaft or pit at the top of which a ventilating fan is placed.

fassaite A monoclinic pyroxene rich in aluminium, calcium and magnesium, and poor in sodium; found in metamorphosed limestones and dolomites.

fast (1) A heading or working place which is driven in the solid coal, in advance of the open places, said to be in the *fast*. (2) A hole in coal which has had insufficient explosive used in it, or which has required undercutting. (3) In shaft sinking, a hard stratum under poorly consolidated ground, on which a **wedging crib** can be laid.

fat coals Coals which contain plenty of volatile matter (gas-forming constituents).

fathom Six feet. In general mining, the volume of a 6 ft cube; in gold mining, often a volume 6 ft by 6 ft by the thickness of the reef; in lead mining, sometimes a volume 6 ft by 6 ft by 2 ft. It is the unit of performance of a rock drill - 'fathoms per shift'. Abbrev. *Fm*.

faujasite One of the less common zeolites; a hydrated silicate of sodium, calcium and aluminium. It exhibits a wider range of molecular absorption than any other zeolite.

fault A fracture in rocks along which some displacement (the *throw* of the fault) has taken place. The displacement may vary from a few millimetres to thousands of metres. Movement along faults is the common cause of earthquakes.

fault breccia A fragmental rock of breccia type resulting from shattering during the development of a fault.

fayalite A silicate of iron, Fe_2SiO_4, crystallizing in the orthorhombic system; a common constituent of slags but occurring also in igneous rocks, chiefly of acid composition, including pitchstone, obsidian, quartz-porphyry, rhyolite, and also in ferrogabbro.

FCC Abbrev. for *Face-Centred Cubic*.

F-centred lattice Face-centred crystal lattice. See **unit cell**.

F-diagram The cumulative residence time distribution in a continuous flow system, plotted on dimensionless co-ordinates. Important in assessing the performance of chemical reactors, kilns etc.

Fe Symbol for **iron**.

feather ore A plumose or acicular form of the sulphide of lead and antimony. Also *jamesonite*. The name is also used for *stibnite*.

feed Forward motion of drill or cutter.

feeder A mechanical appliance for supplying broken rock or crushed ore, at a predetermined rate, to some form of crusher or concentrator.

feldspar, felspar A most important group of rock-forming silicates of aluminium, together with sodium, potassium, calcium, or (rarely) barium, crystallizing in closely similar forms in the monoclinic and triclinic systems. The chief members are *orthoclase* and *microcline* (potassium feldspar); *albite* (sodium feldspar); and the *plagioclases* (sodium-calcium feldspar). The form *felspar*, though still commonly used, perpetuates a false derivation from the German *Fels* (rock); actually it is from the Swedish *feldt* (field).

feldspathic sandstone See **arkose**.

feldspathoids *Foids*. A group of rock-forming minerals chemically related to the feldspars, but undersaturated with regard to silica content, and therefore incapable of free existence in the presence of magmatic silica. The chief members of the group are **haüyne**, **leucite**, **nepheline**, **nosean** and **sodalite**.

felsic Mnemonic for fe̲ldspars, fe̲lspathoids and quartz (si̲lica) actually present as mineral constituents in a rock. Also applied to rocks largely composed of feldspars and quartz. See **mafic**.

felsite Fine-grained igneous rocks of acid composition, occurring as lavas or minor intrusions, and characterized by felsitic texture; a fine patchy mosaic of quartz and feldspar, resulting from the devitrification of an originally glassy matrix.

felspar See **feldspar**.

femic constituents Those minerals which are contrasted with the salic constituents in determining the systematic position of a rock in the *C.I.P.W.* scheme of classification. Note that these are the *calculated* components of the 'norm'; the corresponding *actual* minerals in the 'mode' are said to be *mafic*, i.e. rich in magnesium and iron.

fenite A metasomatic leucocratic alkali-syenite usually associated with cabonatites.

fenitization The process of conversion of granite rocks into fenite by alkali metasomatism.

ferberite Iron tungstate, the end-member of the wolframite group of minerals, the series from $FeWO_4$ to $MnWO_4$.

fergusite An alkaline **syenite** containing large crystals of pseudoleucite in a matrix of aegirine-augite, olivine, apatite, sanidine and iron oxides.

fergusonite A rare mineral occurring in pegmatites; it is a niobate and tantalate of yttrium, with small amounts of other elements.

ferri-, ferro-. Prefixes denoting trivalent and divalent iron respectively.

ferrimolybdite Hydrated molybdate of iron. Most so-called molybdite is ferrimolybdite. It occurs as a yellowish alteration product of molybdenite.

ferro-actinolite An end-member compositional variety in the monoclinic amphiboles; essentially $Ca_2Fe_5Si_8O_{22}(OH,F)_2$, but the name is applied to a member of the actinolite series with more than 80% of this molecule.

ferro-edenite An end-member compositional variety in the monoclinic amphiboles; a hydrous sodium, iron, calcium and aluminium silicate.

ferroelectric materials Dielectric materials (usually ceramics) with domain structure, which exhibit spontaneous electric polarization. Analogous to ferromagnetic materials

ferrogedrite (see **ferromagnetism**). They have relative permittivities of up to 10^5, and show dielectric hysteresis. Rochelle salt was first to be discovered. Others include barium titanite and potassium dihydrogen phosphate.

ferrogedrite See **gedrite**.

ferrohastingsite A compositional variety in the hornblende group of monoclinic amphiboles.

ferromagnesian Containing a relatively large proportion of iron and magnesium, as in minerals, e.g. hypersthene, or rocks, e.g. peridotite, theralite.

ferromagnetism Phenomenon in some magnetically ordered materials in which there is a bulk magnetic moment and the magnetization is large. The electron spins of the atoms in microscopic regions, *domains*, are aligned. In the presence of an external magnetic field the domains oriented favourably with respect to the field grow at the expense of the others and the magnetization of the domains tends to align with the field. Above the *Curie temperature*, the thermal motion is sufficient to off-set the aligning force and the material becomes *paramagnetic*. Certain elements (iron, nickel, cobalt), and alloys with other elements (titanium, aluminium) exhibit permeabilities up to 10^4 (*ferromagnetic materials*). Some show marked hysteresis and are used for permanent magnets, magnetic amplifiers etc.

ferrous [iron(II)] oxide FeO. Black oxide of iron.

ferrous [iron(II)] sulphate $FeSO_4.7H_2O$. Also the mineral **melanterite**.

ferruginous deposits Sedimentary rocks containing sufficient iron to justify exploitation as iron ore. The iron is present, in different cases, in silicate, carbonate, or oxide form, occurring as the minerals chamosite, thuringite, siderite, haematite, limonite etc. The ferruginous material may have formed contemporaneously with the accompanying sediment, if any, or may have been introduced later.

fibre camera Instrument for measuring the X-ray diffraction pattern of fibrous materials.

fibrolite A variety of the aluminium silicate **sillimanite** occurring as felted aggregates of exceedingly thin fibrous crystals; also used when the mineral is cut as a gemstone.

Fick's law of diffusion The rate of diffusion in a given direction is proportional to the negative of the concentration gradient, i.e.

$$\text{molar flux} = -D \frac{\partial C}{\partial x},$$

where D is the diffusion coefficient.

field emission Emission that arises, at normal temperature, due to a high-voltage gradient causing an intense electric field at a metallic surface and stripping electrons from the surface atoms.

field-emission microscope Microscope in which the positions of the atoms in a surface are made visible by means of the electric field emitted on making the surface the positive electrode in a high-voltage discharge tube containing argon at very low pressure. When an argon atom passes over a charged surface atom, it is stripped of an electron, and thus is drawn toward the negative electrode, where it hits a fluorescent screen in a position corresponding to that of the surface atom.

field lens Lens placed in or near the plane of an image to ensure that the light to the outer parts of the image is directed into the subsequent lenses of the system, and thus uniform illumination over the field of view is ensured.

field of view The area over which the image is visible in the eyepiece of an optical instrument. It is usually limited by a circular stop in the focal plane of the eye lens (*eyepiece*).

fiery mine A mine in which there is a possibility of explosion from gas or coal dust.

filar micrometer Modified eyepiece graticule with movable scale or movable crosshair.

filling (1) The loading of tubs or trucks with coal, ore, or waste. (2) The filling-up of worked-out areas in a metal mine.

film sizing Concentration of finely divided heavy mineral by gently sloped surfaces which may be plane, riffled, or vibrated. See **buddle, Wilfley table**.

filter-press action A differentiation process involving the mechanical separation of the still liquid portion of a magma from the crystal mesh. The effective agent is pressure operating during crystallization.

filtrate The liquid freed from solid matter after having passed through a filter.

filtration The separation of solids from liquids by passing the mixture through a suitable medium, e.g. filter paper, cloth, glass wool, which retains the solid matter on its surface and allows the liquid to pass through.

fining-upwards cycle A sedimentary cycle in which coarse-grained material grades up into finer-grained material, e.g. turbidite deposits.

fiords, fjords Narrow winding inlets of the sea bounded by mountain slopes; formed by the drowning of steep-sided valleys, deeply excavated by glacial action; in many cases a rock bar partially blocks the entrance and impedes navigation.

'fire' Flashes of different spectral colours seen in diamond and other gemstones as a result of *dispersion*. See **gems and gemstones**.

fire bank A slack or rubbish heap or dump, at surface on a colliery, which becomes fired by spontaneous combustion.

fireclay Clay consisting of minerals predominantly of SiO_2 and Al_2O_3, low in Fe_2O_3, CaO, MgO etc. Those clays which soften only at high temperatures are used widely as refractories in metallurgical and other furnaces. Fireclays occur abundantly in the Carboniferous System, as 'seat earths' underneath the coal seams.

fire damp The combustible gas contained naturally in coal; chiefly a mixture of methane and other hydrocarbons; forms explosive mixtures with air.

fire-damp cap Blue flame which forms over the flame of a safety-lamp when sufficient fire damp is present in colliery workings.

fireman (1) In a metal mine, a miner whose duty it is to explode the charges of explosive used in headings and working places. (2) In a coalmine, an official responsible for safety conditions underground.

fire opal A variety of opal (cryptocrystalline silica) characterized by a brilliant orange-flame colour. Particularly good specimens, prized as gemstones, are of Mexican origin.

fire stink The smell given off underground when a fire is imminent, e.g. in the **gob**; also, smell of sulphuretted hydrogen from decomposing pyrite.

firestone A stone or rock capable of withstanding a considerable amount of heat without injury. The term has been used with reference to certain Cretaceous and Jurassic sandstones employed in the manufacture of glass furnaces.

firn See **névé**.

first-order reaction One in which the rate of reaction is proportional to the concentration of a single reactant, i.e.

$$\frac{dc}{dt} = -kc,$$

where c is the concentration of the reagent.

first weight The first indications of roof pressure which occur after the removal of coal from a seam.

Fischer-Tropsch process Method of obtaining fuel oil from coal, natural gas etc. Cf. **hydrogenation**. The 'synthesis gas', hydrogen and carbon monoxide in proportional volumes 2:1, is passed, at atmospheric or slightly higher pressure and temperature up to 200°C, through contact ovens containing circulating water with an iron or cobalt catalyst. The gases are washed out and the resultant oil contains alkanes and alkenes, from the lower members up to solid waxes; fractionation yields petrols, diesel oils etc.

fishing tool A tool attached to a drill string and designed to catch and retrieve components lost downhole. Also *overshot tool*.

fissile Capable of nuclear fission, i.e. breakdown into lighter elements of certain heavy isotopes (^{232}U, ^{235}U, ^{239}Pu) when these capture neutrons of suitable energy. Also *fissionable*.

fission-track analysis Examination of defects caused in solids, including minerals, by the spontaneous fission of the heavy nuclide ^{238}U. Fission tracks are principally found in micas, sphene, zircon and apatite, and are mainly used for age determination.

fissure A cleft in rock determined in the first instance by a fracture, a joint plane, or fault, subsequently widened by solution or erosion; may be open, or filled in with superficial deposits, often minerals of pneumatolytic and hydatogenetic provenance. See **grike**.

fissure eruption Throwing-out of lava and (rarely) volcanic 'ashes' from a fissure, which may be many kilometres in length. Typically there is no explosive violence, but a quiet welling-out of very fluid lava. Recent examples are known from Iceland.

fixed points Temperature which can be accurately reproduced and used to define a temperature scale and for the calibration of thermometers. The temperature of pure melting ice and that of steam from pure boiling water at one atmosphere pressure define the Celsius and Fahrenheit scales. The *International Practical Temperature Scale* defined ten fixed points ranging from the triple point of hydrogen (13.81 K) to the freezing point of gold (1337.58 K). See **gold** melting point, **Kelvin thermodynamic scale of temperature**.

fjords See **fiords**.

flag Natural flagstones are sedimentary rocks of any composition which can be readily separated, on account of their distinct stratification, into large slabs. They are often fine-grained sandstones interbedded with shaly or micaceous partings along which they can be split.

flame ionization gauge Gauge used in gas chromatography, as a very sensitive detector for the separate fractions in the effluent carrier gas, by burning it in a hydrogen flame, the electrical conductivity of which is measured by inserting two electrodes in the

flame speed   **fluidized bed**

flame with an applied voltage of several hundred volts.

flame speed The rate at which a flame front will travel through an inflammable mixture of gas and air. Highly dependent on the fuel and the physical conditions.

flame test The detection of the presence of an element in a substance by the colouration imparted to a Bunsen flame.

Flandrian The final post-glacial, temperate stage of the Quaternary, by definition from 10 000 years ago to the present. Equivalent to the Holocene. See **Quaternary**.

flare gas See **flare stack**.

flare stack Vertical pipe for the safe dispersal of hydrocarbon vapours from an oil rig or refinery. Steam is injected into the gas flow to ensure complete combustion.

flaser structure A streaky, patchy structure in a dynamically metamorphosed rock. Flaser gabbro, for example, shows an apparent crude flow structure round unaltered granular lenses.

flash photolysis Photolysis induced by light flashes of short duration but high intensity, e.g. from a laser.

flash point The temperature at which a liquid, heated in a Cleveland cup (open test) or in a Pensky-Martens apparatus (closed test), gives off sufficient vapour to flash momentarily on the application of a small flame. The *fire point* is ascertained by continuing the test.

flatback stope A stope, in overhand stoping, worked upwards into a lode more or less parallel to level. See **overhand stopes**.

flèches d'amour Acicular, hair-like crystals of rutile, a crystalline form of the oxide of titanium, TiO_2, embedded in quartz. Used as a semi-precious gemstone. Also *love arrows* (the literal translation), *cupid's darts*, or *Venus' hair stone*.

flint Concretions of silica, sometimes tabular, but usually irregular in form, particularly abundant on the bedding planes of the Upper Chalk. See **chert**, **paramudras**, **silica**.

flint gravel A deposit of gravel in which the component pebbles are dominantly of flint, e.g. the Tertiary and fluvioglacial gravels in S.E. England.

float (1) Values so fine that they float on the surface of the water when crushed or washed, e.g. *float gold*. (2) Surficial deposit of rock or mineral detached from the main dyke or vein. (3) Term for blocks of bedrock within soil or superficial deposits encountered during prospecting or drilling, e.g. **erratics**.

floating roof A tightly fitting but free cover which floats on an otherwise open tank of fuel. There is no space for the build up of inflammable vapour.

float stone A coarse, porous, friable variety of impure silica, which on account of its porosity, floats on water until saturated.

flocculation The coalescence of a finely divided precipitate into larger particles.

flocculation Coagulation of ore particles by use of reagents which promote formation of flocs, as a preliminary to settlement and removal of excess water by thickening and/or filtration.

flocculent Existing in the form of cloud-like tufts or flocs.

flood basalts Widespread plateau basalts originating from fissure eruptions.

flood plain A plain of stratified alluvium bordering a stream and covered when the stream floods.

floor The upper surface of the stratum underlying a coal seam.

flopgate Diverting gate which directs moving material into alternative routes. Can be worked by remote control.

flos ferri A stalactite form of the orthorhombic carbonate of calcium, *aragonite*, some of the masses resembling delicate coralline growths; deposited from hot springs.

flotation See **froth flotation**.

flow improver Chemicals added to oil passing through a pipeline which reduce frictional losses.

flow sheet A diagram showing the sequence of operations used in a process of production with a given plan, e.g. the extraction and refining of metals.

flow-structure A banding, often contorted, resulting from flow movements in a viscous magma, adjacent bands differing in colour and/or degree of crystallization. It is also shown by the alignment of phenocrysts or of minute crystals and crystallites, in the groundmass of lavas and, more rarely, minor intrusions. Also loosely used of metamorphic rocks.

fluid inclusion See **inclusion**.

fluidization A technique whereby gas or vapour is passed through solids so that the mixture behaves as a liquid, and is of special significance when the solid is a catalyst to induce reactions in the fluidizing medium.

fluidized bed If a fluid is passed upward through a bed of solids with a velocity high enough for the particles to separate from one another and become freely supported in the fluid, the bed is said to be fluidized. Then the total fluid frictional force on the particles is equal to the effective weight of the bed. Fluidized beds are used in chemical industry because of the intimate contact between solid and gas, the high rates of heat transfer and the uniform temperatures within the bed, and the high heat transfer coefficients from the bed to the walls of the containing vessel.

flume A flat-bottomed timber trough, or other open channel, generally nowadays formed in concrete, for the conveyance of water, e.g. to ore-washing plant, or as a bypass.

fluorapatite The commonest form of apatite, $Ca_5(PO_4)_3F$.

fluorescence Emission of radiation, generally light, from a material during illumination by radiation of usually higher frequency, or from the impact of electrons.

fluorimetry Analytical method in which fluorescence induced by ultraviolet light or X-rays is measured. Also *fluoremetry*.

fluorine Pale greenish-yellow gas, the most electronegative (non-metallic) of the elements and the first of the halogens. Chemically highly corrosive and never found free. Symbol F, at.no. 9, r.a.m. 18.9984, valency 1, mp −223°C, bp −187°C. Its abundance in the Earth's crust is 544 ppm and in seawater 1.3 ppm. Its ionic size (1.33Å) is almost identical with those of OH^- and O^{2-}, and it enters into many silicate minerals having OH groups in their structures. There are many independent fluorine minerals of which the most important is fluorite, CaF_2, and it is also present in apatite. Fluorine has an essential biological role, toxic in excess. Its industrial uses include the separation of isotopes of uranium.

fluorite, fluorspar Calcium fluoride, CaF_2, crystallizing in the cubic system, commonly in simple cubes. Occasionally colourless; yellow, green, but typically purple; the coloured varieties may fluoresce strongly in ultraviolet light. See **blue john**.

fluorocarbons Hydrocarbons in which some or all of the hydrogen atoms have been replaced by fluorine. The fluorinated derivatives of methane are widely used as refrigerating agents and propellants for aerosols.

fluorochrome A molecule or chemical group which fluoresces on irradiation. Can be made to bind to a specific site and thus localize it.

fluorophore A group of atoms which give a molecule fluorescent properties.

fluorspar See **fluorite**.

flushing The operation of clearing off accumulation of fire damp or noxious gases underground by means of air currents. See face-airing.

flute cast A hollow eroded by turbulent flow and subsequently filled by sediment. More properly called a *mould*, it is common in **turbidite** deposits, and can be used to determine the direction of flow of the depositing currents.

fluvial, fluviatile Relating to the flow of a river or stream.

fluviatile deposits Sand and gravel deposited in the bed of a river.

fluvioglacial Relating to the meltwater streams flowing from a glacier; the deposits and landforms produced by such streams, e.g. **kames**.

flysch Sediments derived from the Alpine orogeny. More generally applied to almost any **turbidite** deposits, derived from developing, large-scale, fold structures.

foam See **froth**.

foam plug Mass of foam generated and blown into underground workings to seal off a fire or keep out oxygen, where a fire risk exists.

foam separation Removal of solutes or ions from a liquid by bubbling air through in the presence of surface active agents which tend to be adsorbed on to the bubbles. Cf. **froth flotation**, for larger particles.

focal length of a lens The distance measured along the principal axis, between the principal focus and the second principal point. In a thin lens both principal points may be taken to coincide with the centre of the lens.

focus (1) See **earthquakes**. (2) Point to which rays converge after having passed through an optical system, or a point from which such rays appear to diverge. In the first case the focus is said to be *real*; in the second case, *virtual*. The *principal focus* is the focus for a beam of light rays parallel to the principal axis of a lens or spherical mirror.

foid A term meaning **feldspathoid** used by international agreement on rock classification.

foidite Internationally ('**IUGS**') agreed usage for volcanic rocks containing more than 60% feldspathoids ('foid') by volume among light-coloured constituents. The most abundant feldspathoid name should be used if possible, e.g. nephelinite, leucitite etc. See **volcanic rocks**.

foidolite Internationally ('**IUGS**') agreed usage for plutonic rocks containing more than 60% feldspathoids ('foid') by volume among light-coloured constituents. The most abundant feldspathoidal name should be used if possible, e.g. nephelinolite, leucitolite etc. See **plutonic rocks**.

folding See panel on p. 91.

foliation The arrangement of minerals normally possessing a platy habit (e.g. the micas, chlorites and talc) in folia and leaves,

folding

Folding (*bending*) of strata is usually the result of compression that causes the formation of the geological structures known as anticlines, synclines, monoclines, isoclines etc. The amplitude (i.e. vertical distance from crest to trough) of a fold ranges from a centimetre to thousands of metres. The *dip* and *strike* of a bed are shown in the figure for **drainage patterns** on p. 64.

The fold is usually of bedded rocks but may be applied to **foliation** or **cleavage**, and as a descriptive term can include primary structures. In an *anticline* the beds dip outwardly, from the central line to form an arch, with the stratigraphically older rocks at the *core*, as in the figure. In a *syncline* they dip inwardly with the stratigraphically younger rocks in the centre. The *axial plane* (or *axial surface*) of a fold connects the hinge lines of successive strata and divides the fold into equal parts. The intersection of the axial plane with any bed of the fold is the *axis*. A *crest line* is the line through the highest points on the same bed in a vertical section on an anticlinal fold; a *trough line* is the line through the lowest points on a bed in a synclinal fold. If the beds in both *limbs* have not been inverted from their original position the fold is *normal*; it is *symmetrical* if the two limbs are equally inclined (in opposite directions), and *overturned* (or *overfolded*) if one limb is inverted. If the two limbs are inverted and more or less parallel the fold is *isoclinal* and, if the two limbs are approximately flat, the fold is *recumbent*. The *pitch* is the inclination of the fold axis. A fold *plunges* if the axis is not horizontal, the amount of plunge being the angle between the axis and the horizontal in a vertical plane.

When the order of the stratigraphical succession is unknown, so that the fold is treated purely geometrically, the shapes are said to be *antiforms*, closing upwards, or *synforms*, closing downwards. In large composite folds, of regional extent, with subordinate smaller folds, the main arching upwards is an *anticlinorium (pl. anticlinoria)*; if it is downwards in a general trough shape, a *synclinorium* (pl. *synclinoria*). A *monocline* is a steepened, somewhat step-like part of a gently inclined or horizontal bed.

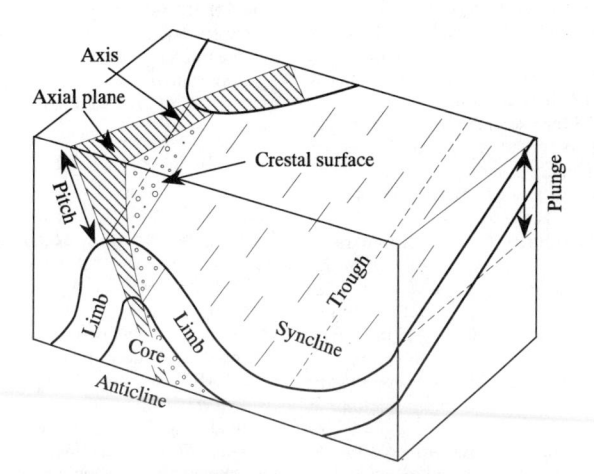

Position of axial plane or surface, crestal surface, plunge and pitch in folded beds

lying with their principal faces and cleavages in parallel planes; such arrangement is due to development under great pressure during regional metamorphism.

following dirt Following *stone*. A thin bed of loose shale above coal; a parting between the top of a coal seam and the roof. See **pug**.

fool's gold See **pyrite**.

footwall The lower wall of country rock in contact with a vein or lode. The upper wall is the *hanging wall*.

footway A colliery shaft in which ladders are used for descending and ascending.

Foraminifera See **Protozoa**.

forearc basin A sedimentary basin developed in the gap between a volcanic arc and its **subduction zone**.

forepoling A mining method for progressing through loosely consolidated ground by driving roof-support poles forward over frames.

foreset beds Gently inclined cross-bedded units progressively covering **bottomset beds** and covered in turn by **topset beds**. Foreset beds form the major bulk of a delta.

form A complete assemblage of crystal faces similar in all respects as determined by the symmetry of a particular class of crystal structure. Thus the *cube*, consisting of 6 similar square faces, and the *octahedron*, consisting of 8 faces, each an equilateral triangle, are crystal *forms*. The number of faces in a form ranges from 1 (the pedion) to 48 (the hexakis-octahedron). A natural crystal may consist of one form or many.

formation A lithostratigraphical rock unit used as a basis for rock mapping. A formation has a recognizable lithological identity and is divisible into *members* or combined into *groups*. It has been casually used in Britain but more precisely in North America.

formula The representation of the types and relative numbers of atoms in a compound, or the actual number of atoms in a molecule of a compound. It uses chemical symbols and subscripts, e.g. H_2SO_4 or C_6H_6.

fornacite, furnacite A basic chromarsenate of copper and lead, crystallizing in the monoclinic system.

forsterite An end-member of the olivine group of minerals, crystallizing in the orthorhombic system. Chemically forsterite is a silicate of magnesium, Mg_2SiO_4. It typically occurs in metamorphosed impure dolomitic limestones.

forsterite-marble, ophicalcite A characteristic product of the contact metamorphism of magnesian (dolomitic) limestones containing silica of organic or inorganic origin.

fossil The relic or trace of some plant or animal which has been preserved by natural processes in rocks of the past.

fossil zone The stratigraphical horizon characterized by a *zone fossil*. See **zone**.

fouls The cutting-out of portions of the coalseam by 'wash outs'; barren ground.

foul solution In cyanide process, one contaminated by salts or metals other than gold and silver, and which is not fit to be recirculated in the process.

Fourier analysis The determination of the harmonic components of a complex waveform (i.e. the terms of a Fourier series that represents the waveform) either mathematically or by a wave-analyser device.

Fourier optics The application of Fourier analysis and the use of Fourier transforms to problems in optics, in particular to image formation. The *Fraunhofer diffraction* pattern is the Fourier transform of the distribution of amplitude of light across the diffracting object. The distribution of amplitude in the Fraunhofer pattern is modified by the optical system and the image formed is the transform of this modified distribution. The same principle is used in X-ray crystal structure analysis where an 'image' of the atomic arrangement is constructed mathematically from the X-ray diffraction pattern.

Fourier principle The principle that all repeating waveforms can be resolved into sine wave components consisting of a fundamental and a series of harmonics at multiples of this frequency. It can be extended to prove that non-repeating waveforms occupy a continuous frequency spectrum.

Fourier transform A mathematical relation between the energy in a transient and that in a continuous energy spectrum of adjacent component frequencies. The Fourier transform $F(u)$ of the function $f(x)$ is defined by

$$F(u) = \int_{-\infty}^{+\infty} e^{-iut} f(t)\, dt .$$

Some writers use e^{+iut} instead of e^{-iut}.

Fourier transform spectroscopy The production of a spectrum by taking the **Fourier transform** of a two-beam interference pattern.

foyaite A widely distributed variety of nepheline-syenite, described originally from the Foya Hills in Portugal. Typically it contains about equal amounts of nepheline and potassium feldspar, associated with a subordinate amount of coloured mineral such as aegirine.

fp Abbrev. for **freezing point**.

Fr Symbol for **francium**.

fracking, fracturing Forcing liquid out into the strata round a well bottom to to

fractional crystallization increase the permeability of the petroleum formation. The liquid contains sand or other material which remains in the fissures and prevents them closing. See **proppants, tertiary production, hot dry rock**.

fractional crystallization In chemistry, the separation of substances by the repeated partial crystallization of a solution. In geology, the separation of a cooling magma into parts by crystallization of different minerals at successively lower temperatures.

fractional distillation The process of distillation used for the separation of the various components of liquid mixtures. An effective separation can only be achieved by the use of **fractionating columns** attached to the still.

fractional distribution In an assembly of particles having different values of some common property such as size or energy, the fraction of particles in each range of values is called the fractional distribution in that range.

fractionating column A vertical tube or column attached to a still and usually filled with packing, e.g. Raschig rings, or intersected with bubble plates. An internal reflux takes place, resulting in a gradual separation between the high-boiling and the low-boiling fractions inside the column, whereby the fractions with the lowest boiling point distil over. The efficiency of the column depends on its length or on the number of bubble plates used, and also on the ability of the packing to promote contact between the vapour and liquid phases.

fractionation See **fractional distillation**.

fracture The broken surface of a mineral as distinct from its cleavage. The fracture is described, in different cases, as conchoidal (shell-like), platy or flat, smooth, hackly (like that of cast-iron) or earthy. Thus calcite has a perfect rhombohedral cleavage, but conchoidal fracture.

fracture cleavage A set of closely-spaced, parallel joints. Common in shallowly deformed metamorphic rocks.

fragmental Applied to rocks that are formed of fragments of pre-existing rocks. Also *clastic*.

francium The heaviest alkali metal. Symbol Fr, at.no. 87. No stable isotopes exist. ^{223}Fr of half-life 22 min is the most important.

franklinite Iron-zinc spinel, occurring rarely, as at the type locality, Franklin, New Jersey.

Frasch process Method of mining elemental sulphur, by drilling into deposit and flushing it out by hot compressed air as a foam of low density. The sulphur is melted by superheated water.

Frasnian The lower stage of the Upper Devonian rocks of Europe. See **Palaeozoic**.

free-air anomaly A gravity anomaly which had been corrected for the height of the measured station above datum but without allowance for the attractive effect of topography. Cf. **Bouguer anomaly**.

free atom Unattached atom assumed to exist during reactions. See **free radicals**.

free energy The capacity of a system to perform work, a change in free energy being measured by the maximum work obtainable from a given process.

free milling Descriptive term for gold or silver ore, which contains its metal in amalgamable state.

free radicals Groups of atoms in particular combinations capable of free existence under special conditions, usually for only very short periods (sometimes only microseconds). Because they contain unpaired electrons, they are paramagnetic, and this fact has been used in determining the degree of dissociation of compounds into free radicals. The existence of free radicals such as methyl (CH_3.) and ethyl (C_2H_5.) has been known for many years.

free-settling In classification of finely ground mineral into equal-settling fractions, use of conditions in which particles can fall freely through the environmental fluid, as opposed to the packed conditions of mineral settling.

freeze sinking Shaft sinking, or penetration of water-logged strata, in which refrigerated brine is circulated to freeze the ground and make establishment of an imperviously lined shaft possible.

freezing The conversion of a liquid into the solid form. This process occurs at a definite temperature for each substance, this temperature being known as the *freezing point*. The freezing of a liquid invariably involves the extraction of heat from it, known as *latent heat of fusion*. See **latent heat, depression of freezing point**.

freezing point The temperature at which a liquid solidifies; the same as that at which the resulting solid melts (the *melting point*). The freezing point of water is used as the lower fixed point in graduating a thermometer. Its temperature is defined as 0°C (273.15 K). Abbrev. *fp*. See **water, depression of freezing point**.

French chalk The mineral, talc, ground into a state of fine subdivision, its softness and its perfect cleavage contributing to its special properties when used as a filler or

fundamental dynamical units

The basic equations of dynamics are such as to be the same for any system of fundamental units. Unit force acting on unit mass produces unit acceleration; unit force moved through unit distance does unit work; unit work done in unit time is unit power. Four systems are, or have been, in general use, the SI system now being the only one employed in scientific work. See **SI units**.

	System				
	ft lb sec	gravitational	CGS	SI	dimensions
length	foot (ft)	foot	centimeter (cm)	metre (m)	L
mass	pound (lb)	slug	gram (g)	kilogram (kg)	M
time	second (s)	second	second	second	T
velocity	ft/s	ft/s	cm/s	m/s	LT^{-1}
acceleration	ft/s^2	ft/s^2	cm/s^2	m/s^2	LT^{-2}
force	poundal (pdl)	pound force (lbf)	dyne	newton (N)	MLT^{-2}
work	ft pdl	ft lbf	erg	joule (J)	ML^2T^{-2}
power	ft pdl/s	ft lbf/s	erg/s	watt (W)	ML^2T^{-3}

Notes: (1) There is no name for the unit of power except in the SI system. It is possible to express power in the ft lb sec system and in the gravitational system by the horse-power (550 ft lbf s^{-1}) and in the CGS system by the watt (10^7 erg s^{-1}). (2) The unit of force (lbf) in the gravitational system is also known as the pound weight (lb wt).

dry lubricant.

Frenkel defect Disorder in the crystal lattice, due to some of the ions (usually the cations) having entered interstitial positions, leaving a corresponding number of normal lattice sites vacant. Likely to occur if one ion (in practice the cation) is much smaller than the other, e.g. in silver chloride and bromide.

freshwater sediments Sediments which are accumulating or have accumulated in freshwater, i.e. river, lake or glaciofluvial environments.

fresnel A unit of optical frequency, equal to 10^{12}Hz = 1THz (terahertz).

Fresnel-Arago laws Laws concerning the conditions for the interference of beams of polarized light: (1) two rays of light emanating from the same polarized beam, and polarized in the same plane, interfere in the same way as ordinary light; (2) two rays of light emanating from the same polarized beam and polarized at right angles to each other will interfere only if they are brought into the same plane of polarization; (3) two rays of light polarized at right angles and emanating from ordinary light will not interfere if brought into the same plane of polarization.

friable Of ore, easily fractured or crumbled during transport or comminution.

fringing reef A coral reef directly attached to or bordering the shore of an island or continent, having a rough table-like surface exposed at low tide. Cf. **barrier reef, atoll**.

froth Foam. A gas-liquid continuum in which bubbles of gas are contained in a much smaller volume of liquid, which is expanded to form bubble walls. The system is stabilized by oil, soaps or emulsifying agents which form a binding network in the bubble walls.

froth flotation Process in dominant use for concentrating values from low-grade ores. After fine grinding, chemicals are added to a pulp (ore and water) to develop differences in surface tension between the various mineral species present. The pulp is then copiously aerated, and the preferred (*aerophilic*) species clings to bubbles and floats as a mineralized froth, which is skimmed off.

frue vanner Endless rubber belt which is driven slowly upslope while finely ground ore is washed gently downslope. Belt is given a side shake to aid distribution, and wash water is so adjusted that heavy material stays on belt, while light gangue is washed down to bottom end of pulley system round which belt circulates.

fuchsite A green variety of muscovite (white mica) in which chromium replaces

fuel oils

some of the aluminium.

fuel oils Oils obtained as residues in the distillation of petroleum; used, either alone or mixed with other oils, for domestic heating and for furnace firing (particularly marine furnaces); also as fuel for internal combustion engines.

fugitive Descriptive of the dissolved volatile constituents of magma, which are commonly lost by evaporation when the magma is erupted as lava, and which are partly responsible for metasomatic alteration when magma is intruded.

fulgurites *Lightning tubes.* Tubular bodies, branching or irregularly rod-like, produced by lightning in loose unconsolidated sand; caused by the vitrification of the sand grains forming silica glass. Although of very narrow cross-section, some specimens have been found to exceed 6 m in length. Also *lechatelierite.*

Fuller's earth A non-plastic clay consisting essentially of the mineral montmorillonite, and similar in this respect to bentonite. Used originally in 'fulling', i.e. absorbing fats from wool, hence the name. The Fullers' Earth of English stratigraphy is a small division of the Jurassic System in the S. Cotswolds.

fumaroles Small vents on the flanks of a volcanic cone, or in the crater itself, from which gaseous products emanate.

fume Cloud of airborne particles, generally visible, of low volatility and less than a micrometre in size, arising from condensation of vapours or from chemical reaction.

fundamental dynamical units See panel on p. 94.

fusion drilling

fungible Oil products which are interchangeable and can therefore be mixed during transport. Makes it difficult to trace the origins of a given sample.

furfural $C_4H_3O.CHO$. A colourless liquid, bp 162°C, obtained by distilling pentoses with diluted hydrochloric acid. Used as a solvent, particularly for the selective extraction of crude rosin, also as raw material for synthetic resins. Used in petroleum refining for the selective extraction of aromatics and naphthenes and to allow their subsequent recovery. Also *fural, furfuraldehyde.*

furnacite See **fornacite**.

fusain Mineral charcoal, the soiling constituent of coal, occurring chiefly as patches or wedges. It consists of plant remains from which the volatiles have been eliminated.

fuse A thin waterproof canvas length of tube containing gunpowder arranged to burn at a given speed for setting off charges of explosive.

fusion (1) The process of forming new atomic nuclei by the fusion of lighter ones; principally the formation of helium nuclei by the fusion of hydrogen and its isotopes. The energy released in the process is referred to as *nuclear energy* or *fusion energy.* (2) The conversion of a solid into a liquid state; the reverse of *freezing.* Fusion of a substance takes place at a definite temperature, the melting point, and is accompanied by the absorption of latent heat of fusion.

fusion drilling Method of hard-rock boring with a paraffin-oxygen jet which melts the rock, the slag being decrepitated and flushed out by a water spray.

G

g Abbrev. for **gram(me)**.
g Symbol for **osmotic coefficient**.
g Symbol for **acceleration due to gravity**. 981.274 cm/s^2.
γ Symbol for (1) substituted on the carbon atom of a chain next but two to the functional group; (2) substituted on one of the central carbon atoms of an anthracene nucleus; (3) substituted on the carbon atom next but two to the hetero-atom in a heterocyclic compound; (4) a stereoisomer of a sugar; (5) ratio of specific heats of a gas; (6) surface tension; (7) propagation coefficient; (8) Gruneisen constant; (9) molar activity coefficient; (10) coefficient of cubic thermal expansion; (11) the greatest refractive index in a biaxial crystal; (12) **electrical conductivity**; (13) magnetic field intensity in the CGS system, commonly used in magnetic methods of surveying; equal to 10^{-5} gauss. Synonym *nanotesla*.
G Symbol for (1) thermodynamic potential; (2) Gibbs function; (3) **free energy**.
G Symbol for (1) the constant of **gravitation**; (2) shear modulus, rigidity; (3) conductance.
Γ Symbol for surface concentration excess.
Ga Symbol for **gallium**.
gabbro A coarse-grained plutonic rock, consisting essentially of plagioclase, near labradorite in composition, and clinopyroxene, with or without olivine in addition. The *gabbro clan* includes also norite, eucrite, troctolite, kentallenite etc. See **plutonic rocks**.
gadolinite Silicate of beryllium, iron and yttrium, often with cerium; occurs in pegmatite.
gadolinium A rare metallic element; trivalent; a member of the rare earth group. Symbol Gd, at.no. 64, r.a.m. 157.25. Only known in combination; obtained from the same sources as europium.
gadolinium gallium garnet See **GGG**.
gahnite A mineral belonging to the spinel group; occurs as grey octahedral cubic crystals. Also *zinc spinel* (see **spinel**), the composition being zinc aluminate, $ZnAl_2O_4$.
gal Unit of acceleration used in gravity measurements. 1 cm/s^2. In honour of Galileo. Often used as *milligal*.
galatin dynamite High explosive containing nitroglycerine, sodium nitrate, collodion cotton, and such inert fillers as wood meal and sodium carbonate.
galaxite A rare manganese aluminium spinel, $MnAl_2O_4$.
galena Lead sulphide, PbS; commonest ore of lead, occurring as grey cubic crystals, often associated with zinc blende, in mineralized veins. Also *lead glance*.
gallery A tunnel or passage in a mine.
gallium A metallic element in the third group of the periodic system. Symbol Ga, at.no. 31, r.a.m. 69.72, rel.d. 5.9, oxidation state +3, mp 30.15°C. Used in fusible alloys and high temperature thermometry. Gallium arsenide is an important semi-conductor.
galvanic cell An electrochemical cell from which energy is drawn. Cf. **electrolytic cell**.
gamma-radiation Electromagnetic radiation of high quantum energy emitted after nuclear reactions or by radioactive atoms when nucleus is left in excited state after emission of α- or β-particle.
gamma-ray energy Energy of a gamma-ray photon given by $h\nu$ where ν is the frequency and h is Planck's constant. The energy may be determined by diffraction by a crystal or by the maximum energy of photoelectrons ejected by the γ-rays. The depth of penetration into a material is determined by the energy.
gang A train or *journey* of tubs or trucks.
gangue Valueless rock or mineral aggregates in an ore.
gangway Main haulage road, or level underground.
gannister, ganister A particularly pure and even-grained siliceous grit or loosely cemented quartzite, occurring in the Upper Carboniferous of northern England, and used in the manufacture of silica-bricks.
gape Aperture below which a crushing machine can receive and work on entering ore.
garnet A group of cubic minerals which are silicates of di- and tri-valent metals and occur typically in metamorphic rocks, e.g. garnetiferous schists. Some species are of value as gems, rivalling ruby in colour. See **andradite, grossular, melanite, pyrope, spessartine, uvarovite**.
garnierite A bright green nickeliferous serpentine, hydrated nickel magnesium silicate. It occurs in serpentinite as a decomposition product of olivine, and in other deposits, and is an important ore of nickel.
gas Explosive mixture of combustible gases with air, particularly *methane* and *carbon monoxide*. Also used for accumulations of combustion products, e.g. *carbon dioxide*. See **afterdamp, black damp, choke damp, fire damp, natural gas, white damp**.
gas cap The free gas phase overlying liquid hydrocarbon in a reservoir.
gas chromatography. See **gas-liquid**

gas drain

chromatography.
gas drain A tunnel or borehole for conducting gas away from old workings.
gas drilling See **air drilling**.
gas generator Chemical plant for producing gas from coal, e.g. water gas, by alternating combustion of coal and reduction of steam.
gas lift Method of pumping oil from the bottom of a well by releasing compressed liquid gas there. On vaporization it lifts and entrains the oil.
gas-liquid chromatography A form of partition or adsorption chromatography in which the mobile phase is a gas and the stationary phase a liquid. Solid and liquid samples are vaporized before introduction on to the column. The use of very sensitive detectors has enabled this form of chromatography to be applied to submicrogram amounts of material. Abbrev. *GLC*.
gas tar Coal tar condensed from coal gas, consisting mainly of hydrocarbons. Distillation of tar provides many substances, e.g. ammoniacal liquor, 'benzole', naphtha and creosote oils, with a residue of pitch. Dehydrated, it is known as 'road tar', and used as a binder in road-making.
gastropods See panel on p. 98.
gas well A deep boring, generally in an oilfield, which yields natural gas rather than oil. See **natural gas**.
gateway A road through the worked-out area (goaf) for haulage in longwall working of coal. Road connecting coal working with main haulage. Also *gate road*.
gathering line Small-bore pipes which collect oil or gas from peripheral wells and take them to central distributing station.
gathering motor Light electric loco used to move loaded coal trucks from filling points to main haulage system.
Gault A blue-grey clayey formation in the Cretaceous (Albian) of England. Also *gault clay*.
gauge door A door underground for controlling the supply of air to part of the mine.
gauss CGS electromagnetic unit of magnetic flux density; equal to 1 maxwell cm^{-2}, each unit magnetic pole terminating 4π lines. Now replaced by the SI unit of magnetic flux density, the tesla (T). $1T = 10^4$ gauss.
gay-lussite A rare grey hydrated carbonate of sodium and calcium, occurring in lacustrine deposits.
Gd Symbol for **gadolinium**.
Ge Symbol for **germanium**.
geanticline A regional upwarping of the crust of the Earth. Cf. **geosyncline**.

geobotanical surveying

Gedinnian The oldest stage in the Devonian Period. See **Palaeozoic**.
gedrite An orthorhombic amphibole, containing more aluminium and less silicon than anthophyllite. The iron-aluminium end-member has been called *ferrogedrite*. Gedrite occurrences are restricted to metamorphic and metasomatic rocks.
gegenions The simple ions, of opposite sign to the colloidal ions, produced by the dissociation of a colloidal electrolyte. Also *counterions*.
gehlenite The calcium aluminium end-member of the melilite group of minerals, $Ca_2Al_2SiO_7$.
Geiger counter An instrument for measuring ionizing radiation, with a tube carrying a high voltage wire in an atmosphere containing argon plus halogen or organic vapour at low pressure, and an electronic circuit which quenches the discharge and passes on an impulse to record the event. Also *Geiger-Müller counter*.
Geissler pump A glass vacuum pump which operates from the water supply.
gel The apparently solid, often jelly-like, material formed from a colloidal solution on standing. A gel offers little resistance to liquid diffusion and may contain as little as 0.5% of solid matter. Some gels, e.g. gelatin, may contain as much as 90% water, yet in their properties are more like solids than liquids.
gelignite Explosive used for blasting, composed of a mixture of nitroglycerine (60%), guncotton (5%), woodpulp (10%) and potassium nitrate (25%).
gem gravels Sediments of the gravel grade containing appreciable amounts of gem minerals, and formed by the disintegration and transportation of pre-existing rocks, in which the gem minerals originated. They are really placers of a special type, in which the heavy minerals are not gold or tin, but such minerals as garnets, rubies, sapphires etc. As most of the gem minerals are heavy and chemically stable, they remain near the point of origin, while the lighter constituents of the parent rocks are washed away, a natural concentration of the valuable components resulting.
gems and gemstones See panels on pp. 100–101.
geobotanical indicator See **geobotanical surveying**.
geobotanical surveying Form of geochemical prospecting. (1) Identification and systematic surveying of distribution of metallophile plant species, e.g. *calamine violet* associated with zinc anomalies in cen-

gastropods

Members of the class of Gastropoda, they are marine, freshwater or terrestrial molluscs usually with coiled calcareous shells, typically helical, and include snails, whelks, limpets and cowries. The shell is a single piece (*univalve*), as in the figure, and most are coiled *dextrally*, i.e. with the *aperture* on the right looking upwards towards the *apex*. If coiled to the left it is said to be *sinistral*. In some gastropods the *whorls* are nearly in the same plane.

The earliest gastropods were in the Lower Cambrian, but they only became important in the Ordovician. Freshwater and terrestrial forms date from the Upper Palaeozoic. Gastropods were abundant in the Tertiary and have been used for correlation, and very large numbers of species are extant today.

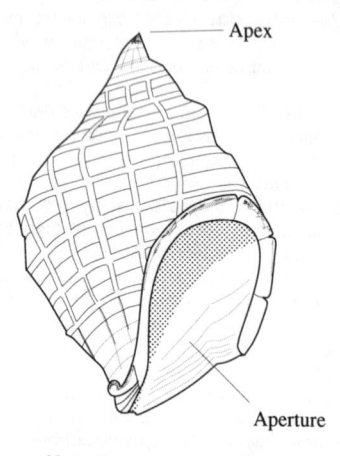

Nucella sp. × 1. Red Crag, Pleistocene

---

tral Europe (*geobotanical indicators*). (2) Identification and systematic surveying of pathological conditions in plants caused by metal toxaemia. (3) Systematic sampling of vegetation to identify anomalous concentrations of metals in plant tissues.

geochemical prospecting Application of **geochemistry** to mineral exploration by the systematic analysis of bedrock, soil, stream, river and ground water, stream gravels and vegetation for the purposes of identifying anomalous concentrations of particular elements of economic interest, or elements commonly associated with such ore bodies. See **geobotanical surveying, soil sampling, stream sampling, tracers.**

geochemistry The study of the chemical composition of the Earth's crust.

geochronology Study of time with respect to the history of the Earth, primarily through the use of *absolute*, or *isotopic*, and *relative age-dating* methods.

geode Hollow, rounded rock, mineral nodule or concretion, often lined with crystals which have grown inwards. Also *drusy cavity*.

geognosy An old term for absolute knowledge of the Earth, as distinct from geology, which includes various theoretical aspects.

geoid The figure of the Earth, considered as a smooth oblate spheroid or ellipsoid, and taken as the reference for *geodetic levelling*.

geo-isotherms Lines or surfaces of equal temperature within the Earth.

geological column See panel on p. 102.

geological time The time extending from the end of the Formative Period of Earth history to the beginning of the Historical Period. See **geological column**.

geological work of rivers See **rivers, geological work of.**

geology The study of the planet Earth. It embraces mineralogy, petrology, geophysics, geochemistry, physical geology, palaeontology and stratigraphy. It increasingly involves the use of the chemical, physical, mathematical and biological sciences. See **earth science.**

geomagnetic effect The effect of the Earth's magnetic field on cosmic rays by which positively charged particles are deflected towards the east.

geomorphology The study of landforms and their relationship to the underlying geological structure.

geophones Array of sound detectors used to collect information in seismic surveys from planned explosions.

geophysical prospecting Prospecting by using quantitative physical measurements directly or indirectly, including magnetic, gravitational, electrical, electromagnetic, seismic and radioactive methods.

geosyncline A major elongated downwarp of the Earth's crust, usually hundreds of kilometres long and filled with sediments and lavas many kilometres in thickness. The rocks are generally deformed and metamorphosed later.

geothermal gradient The rate at which the temperature of the Earth's crust increases with depth.

geothermal power Power generated by using the heat energy of crustal rocks. Active volcanic areas have traditionally been a source, but more recently deep boreholes in areas with a high geothermal gradient have shown economic potential.

germanium A metalloid element in the fourth group of the periodic system. It is greyish-white in appearance. Symbol Ge, at.no. 32, r.a.m. 72.59, rel.d. 5.47, mp about 958°C. There are only 1.5 ppm in the Earth's crust, where the element mainly occurs substituting for silicon in silicates, and in coal. The principal industrial source is zinc smelter flue dust, and it occurs in coal flue dusts. Rare germanium minerals include *argyrodite*. The main use of the metal, which has exceptional properties as a semi-conductor, is in the manufacture of solid-state rectifiers or diodes in microwave detectors, and, in a highly pure state, in transistors and microchips.

German lapis See **Swiss lapis**.

gersdorffite Metallic grey sulphide-arsenide of nickel, occurring as cubic crystals or in granular or massive forms.

get To win or mine.

geyser A volcano in miniature, from which hot water and steam are erupted periodically instead of lava and ashes, during the waning phase of volcanic activity. Named from the Great Geyser in Iceland, though the most familiar example is probably 'Old Faithful' in the Yellowstone Park, Wyoming. The eruptive force is the sudden expansion which takes place when locally heated water, raised to a temperature above boiling point, flashes into steam. Until the moment of eruption, this had been prevented by the pressure of the overlying column of water in the pipe of the geyser, which is usually terminated upwards by a sinter crater. Also *gusher*.

geyserite See **sinter**.

GGG *Gadolinium gallium garnet*. A simulant of diamond.

ghost crystal A crystal within which may be seen an early stage of growth, outlined by a thin deposit of dust or other mineral deposit.

giant *Hydraulic giant*. See **monitor**.

giant powder Dynamite.

Gibbs' adsorption theorem Solutes which lower the surface tension of a solvent tend to be concentrated at the surface, and conversely.

Gibbs-Duhem equation For binary solutions at constant pressure and temperature, the chemical potentials (μ_1, μ_2) vary with the mole fractions (x_1, x_2) of the two components as follows:

$$\frac{\partial \mu_1}{\partial \ln x_1} = \frac{\partial \mu_2}{\partial \ln x_2}.$$

gibbsite Hydroxide of aluminium, $Al(OH)_3$, occurring as minute mica-like crystals, concretional masses, or incrustations. An important constituent of bauxite. Also *hydrargillite*.

gilsonite See **uintaite**.

gin (1) A hand hoist which consists of a chain or rope barrel supported in bearings and turned by a crank. (2) A portable tripod carrying lifting tackle.

gismondine A rare zeolite; a hydrated calcium aluminium silicate which occurs in cavities in basaltic lavas.

Givetian A stage in the Middle Devonian. See **Palaeozoic**.

glacial action All processes relating to the action of glacier ice, comprising: (*a*) the grinding, scouring, plucking and polishing effected by ice, armed with rock fragments frozen into it; and (*b*) the accumulation of rock debris resulting from these processes.

glacial deposits Deposits including spreads of boulder clay, sheets of sand and gravel occurring as outwash fans, outwash deltas and kames; also deposits of special topographical form, such as drumlins and eskers.

glacial erosion The removal of rock materials by the action of glaciers and associated meltwater streams. Includes grinding, scouring, plucking, grooving and polishing by rock fragments contained in the ice.

glacial sands Sands covering extensive areas in advance of sheets of boulder clay, and together with glacial (largely fluvioglacial) gravels, represent the outwash from ice sheets.

glaciation Both the processes and products arising from the presence of ice masses on the Earth. The effects are most obvious on land but there is increasing evidence that the shallow sea floors too were affected. Glaciation, traditionally connected with the Pleistocene Period, is now known from older geological periods including the Permo-Carboniferous and the Precambrian.

gems and gemstones

These are minerals but also some rocks or other natural material such as amber, coral, pearl and shells, which when cut and polished possess the qualities of *beauty* and *durability* that make them suitable for personal adornment or as ornaments. *Rarity* is also an essential element that plays perhaps the most important part in determining the value of a gem; value is much lower for most *synthetic* or *artificial* stones.

The durability and beauty of gemstones depends on various physical and chemical properties. Most gemstones are well crystallized and derive many properties from their crystalline structure. Hardness is usually possessed in a high degree; diamond is the hardest, 10 on Mohs' scale, and ruby and sapphire, varieties of corundum, are both 9. *Specific gravity* is important in determining the identity of a stone. The optical properties powerfully influence the appearance of gemstones; stones may be transparent, translucent (e.g. opal) or opaque (e.g. turquoise). *Lustre* is largely a surface effect due to reflection. *Refraction* causes several beautiful effects. When light strikes the surface of the stone, some is reflected, and some enters the stone but is bent or refracted (Snell's law). When the light reaches the back of the stone, some is reflected (total internal reflection) enhancing the brilliance of the stone (Fig. 1). The cutting of the stone is intended to increase this effect. The different component colours of white light are refracted to differing extents, giving the play of colours which is known as *fire*, and of which the measure is the *dispersion*.

The refractive index (*indices* in doubly refracting stones) may be measured in a *refractometer*, which depends on the *critical angle* at the interface of gemstone and refractometer prism (Fig. 2). The cause of the *colour* of gemstones is complex, dependent on their chemical composition, and *dichroism* is the stone's property of having different colours when viewed in different crystallographic directions, e.g. green tourmaline, sapphire.

Gemstones are cut to enhance their beauty; some examples of cutting are shown in Fig. 3. *Precious stones* are generally taken to be diamond, ruby, sapphire, emerald and opal (and somewhat differently, pearl). *Semi-precious stones* are arbitrarily and less clearly defined but usually include beryls (aquamarine, morganite, heliodor), chrysoberyl (including alexandrite), cordeirite, garnets (pyrope, almandine, spessartine, grossular, demantoid), olivine (peridot), sphene, spinels, topaz, tourmaline, zircon, jade (jadeite, nephrite), lapis-lazuli and turquoise.

Artificial or imitation gemstones may in themselves be of great beauty. In recent years more synthetic stones and simulants have been developed for diamond and other stones.

figures on next page

glacier A large mass of ice. Three kinds can be recognized: (1) valley glacier; (2) Piedmont glaciers which overflow from valleys and coalesce on the lower ground; and (3) large continental ice sheets (e.g. Greenland) and smaller icecaps (e.g. Iceland).

glacier lake See **lake**.

glance Opaque mineral with a resinous or shining lustre.

glauberite Monoclinic sulphate of sodium and calcium, occurring with rock salt, anhydrite etc., in saline deposits.

Glauber salt Properly termed *mirabilite* (hydrated sodium sulphate, $Na_2SO_4.10H_2O$). A monoclinic mineral formed in salt lakes, deposited by hot springs, or resulting from the action of volcanic gases on seawater.

glauconite Hydrated silicate of potassium, iron and aluminium, a green mineral occurring almost exclusively in marine sediments, particularly in greensands. It is generally found in rounded fine-grained aggregates of ill-formed platelets although it has a mica structure. The manner of its formation is somewhat uncertain.

glaucophane A blue monoclinic amphibole

gems and gemstones (contd)

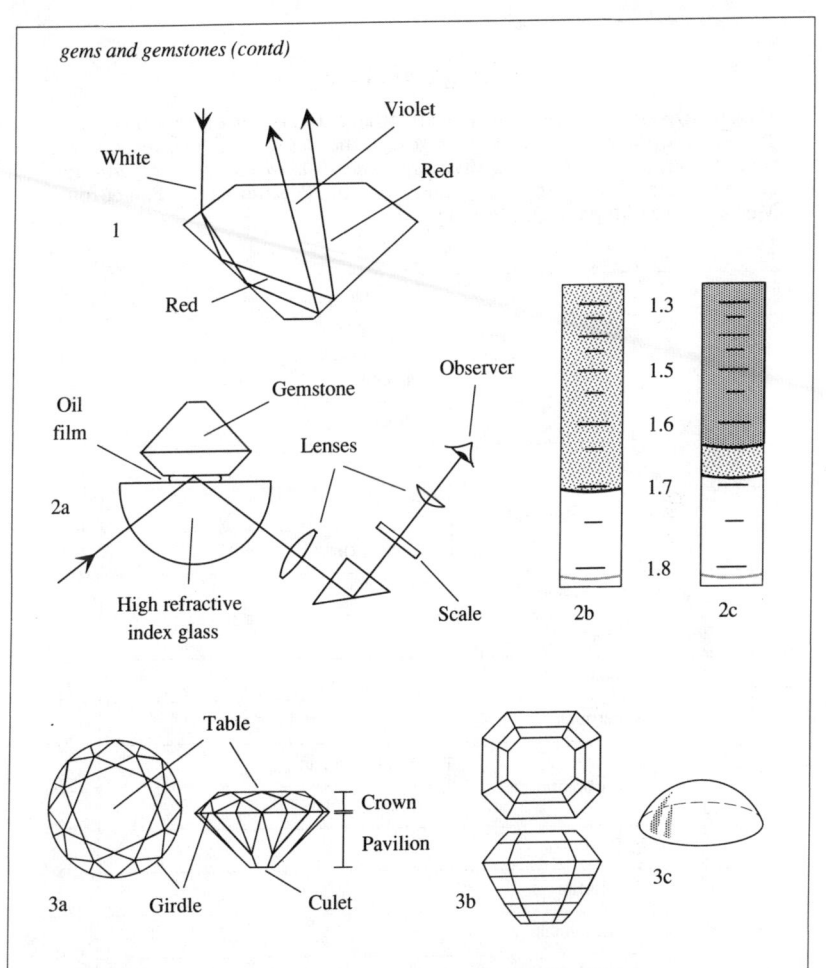

Fig. 1. Dispersion and total internal reflection in a gemstone.

Fig. 2. (a) Principle of the refractometer. The oil must be of high refractive index so that air is excluded and internal reflection occurs at the gemstone interface. **(b, c) Appearance of a refractometer scale: singly refracting (b),** showing 1.72 as in spinel; doubly refracting (c), showing 1.65, 1.69 as in peridot.

Fig. 3. (a) Brilliant cut, used for many stones, notably diamond. **(b)** Step or trap cut, used esp. for emerald. **(c)** Cabochon, used esp. for garnet.

occurring in crystalline schists formed at high pressures. A hydrated sodium iron magnesium silicate, the name is used for an end-member compositional variety of amphibole.

gliding planes In minerals, planes of molecular weakness along which movement can take place without actual fracture. Thus calcite crystals or cleavage masses can be distorted by pressure and pressed into quite thin plates without actual breakage. See **slip planes**.

glimmerite An ultrabasic rock composed mainly of mica, usually of the biotite variety.

globigerina ooze A deep-sea deposit covering a large part of the ocean floor (one quarter of the surface of the globe); it consists chiefly of the minute calcareous shells of the foraminifer, *Globigerina*.

geological column

The subdivisions of geological time are arranged in a vertical sequence from the oldest at the bottom to the youngest at the top or as the corresponding stratigraphic units of these subdivisions. The *Proterozoic* and *Archean* are considered as Eons. See the panels on the **Precambrian**, **Palaeozoic**, **Mesozoic**, **Tertiary** and **Quaternary**.

Eon	Era	Period			Age, Ma
Phanerozoic	Cenozoic	Tertiary	Quaternary		1.64
			Neogene	Pliocene	5.2
				Miocene	23
			Palaeogene	Oligocene	35
				Eocene	56
				Palaeocene	65
	Mesozoic	Cretaceous			146
		Jurassic			208
		Triassic			245
	Palaeozoic	Permian			290
		Carboniferous	Pennsylvanian		323
			Mississippian		362
		Devonian			408
		Silurian			439
		Ordovician			510
		Cambrian			570
Precambrian		Proterozoic			2500
		Archaean			4600

globulites Crystallites (i.e. incipient crystals) of minute size and spherical shape occurring in natural glasses such as pitchstones.

GLORIA Abbrev. for *Geological Long Range Inclined Asdic*. Long range sidescan sonar by which very large areas of the ocean floor have been surveyed.

gloryhole Combination of open pit mining with underground tunnel through which spoil is removed after gravitating down.

gmelinite A pseudohexagonal zeolite, white or pink in colour and rhombohedral in form, resembling chabazite. Chemically, it is hydrated silicate of aluminium, sodium and calcium.

gneiss A metamorphic rock of coarse grain size, characterized by a mineral banding, in which the light minerals (quartz and feldspar) are separated from the dark ones (mica and/or hornblende). The layers of dark minerals are foliated, while the light bands are granulitic. See **metamorphism**.

gneissose texture A rock texture in which

foliated and granulose (granulitic) bands alternate.

goaf, gob The space left by the extraction of a coal seam, into which waste is packed. Also *loose waste*.

gob fire A fire occurring in a worked-out area, due to ignition of timber or broken coal left in the gob.

gob heading, gob road A roadway driven through the gob after the filling has settled.

gob stink A smell indicating spontaneous combustion or a fire in the gob.

go-devil Cylindrical plug with brushes, scrapers and rollers on its periphery, able to move under the oil pressure through a pipeline and clean it. Also *pig, rabbit*.

goethite Orthorhombic hydrated oxide of iron with composition FeO.OH. Dimorphous with lepidocrocite.

gold A heavy, yellow, metallic element, occurring in the free state in nature. Symbol Au, at.no. 79, r.a.m. 196.967, rel.d. at 20°C 19.3, mp 1337.58 K, electrical resistivity about 0.02 microhm metres. Abundance in the Earth's crust is 0.004 ppm. Most of the metal is retained in gold reserves but some is used in jewellery, dentistry and for decorating pottery and china. In coinage and jewellery, the gold is alloyed with varying amounts of copper and silver. *White gold* is usually an alloy with nickel, but as used in dentistry this alloy contains platinum or palladium.

gold amalgam A variety of native gold containing approximately 60% of mercury.

golden beryl A clear yellow variety of the mineral beryl, prized as a gemstone. Heliodor is a variety from South West Africa. Golden beryl has been used as a name for *chrysoberyl*.

golden spike Colloq. for a marker point in a stratotype section that defines a boundary between two geological divisions.

Gondwanaland The hypothetical Palaeozoic continent of the southern hemisphere which broke up and drifted apart to form bits of the present continents of South America, Africa, India, Australia and Antarctica.

goniatite Any cephalopod belonging to the order Goniatitida. See **ammonoids**.

gonnardite A rare zeolite; a hydrated sodium, calcium, aluminium silicate.

gooseberry stone The garnet **grossular**, which was named for the resemblance of this green variety, both in form and colour, to the gooseberry which has the botanical name *Ribes grossularia*.

gorge A general term for all steep-sided, relatively narrow valleys, e.g. canyons, overflow channels etc.

goslarite Hydrated zinc sulphate, a rare mineral precipitated from water seeping through the walls of lead mines; formed by the decomposition of sphalerite. See **white vitriol**.

gossan The leached, oxidized material found in surface exposures of an ore deposit; represents the residue left after **secondary enrichment** of a mineral vein or lode. Often stained brown by iron oxides and rich in quartz, gossans are an indication of mineral deposits below the surface, although they may not be of any value themselves.

Gothlandian Obsolete name used in Europe for **Silurian**.

Gouy layer Diffuse layer of counterions surrounding charged lattices at surface of particle immersed in liquid.

graben An elongated downthrown block bounded by faults along its length. Cf. **horst**.

graded bedding Bedding which shows a sorting effect with the coarser material at the base progressively changing upwards to finer sediment at the top. Occasionally such grading may be *reversed*.

grahamite A member of the asphaltite group.

grain (1) See **rift and grain**. (2) Average size of mineral crystals composing a rock. Direction in which it tends to split. Of a metal, texture of broken surface (smooth-, rough- etc.).

grains See **coal sizes**.

grain size The average size of the grains or crystals in a sample of metal or rock. See **particle size**.

grammatite Synonym for **tremolite**.

gram(me) The unit of *mass* in the CGS system. It was originally intended to be the mass of $1 cm^3$ of water at 4°C but was later defined as one-thousandth of the mass of the International Prototype Kilogramme, a cylinder of platinum-iridium kept at Sèvres. Abbrev. *g*.

gram(me)-atom The quantity of an element whose mass in grams is equal to its relative atomic mass. A **mole** of atoms.

gram(me)-ion Mass in grams of an ion, numerically equal to that of the molecules or atoms constituting the ion.

gram(me)-molecular volume See **molar volume**.

gram(me)-molecule See **mole**.

granite A coarse-grained igneous rock containing megascopic quartz, averaging 25%, much feldspar (orthoclase, microcline, sodic plagioclase), and mica or other coloured minerals. In the wide sense, granite includes alkali-granites, adamellites and granodiorites, while the *granite clan* includes the medium- and fine-grained equivalents of these rock types. See **plutonic rocks**.

granite-aplite See aplite.

granite-porphyry Porphyritic microgranite, a rock of granitic composition but with a groundmass of medium grain size in which larger crystals (phenocrysts) are embedded.

granite series A series relating the different types of granitic rock with respect to their time and place of formation in an orogenic belt, starting with early-formed deep-seated autochthonous granites and ending with post-tectonic high-level plutons.

granitic texture See **granitoid texture**.

granitization A metamorphic process by which rocks can be changed into granite *in situ*.

granitoid texture A rock fabric in which the minerals have no crystal form and occur in shapeless interlocking grains. Such rocks are in the coarse grain size group. Also *xenomorphic granular texture*.

granoblastic texture An arrangement of equigranular mineral grains in a rock of metamorphic origin similar to that of a normal granite but produced by recrystallization in the solid and not by crystallization from a molten condition. The grains show no preferred orientation.

granodiorite An igneous rock of coarse grain size, containing abundant quartz and at least twice the amount of plagioclase over orthoclase, in addition to coloured minerals such as hornblende and biotite. See **plutonic rocks**.

granophyre An igneous rock of medium grain size, in which quartz and feldspar are intergrown as in graphic granite.

granule A rock or mineral with a grain size between 2 and 4 mm.

granulite A granular-textured metamorphic rock, a product of regional metamorphism.

granulitic texture The texture of a granulite, sometimes referred to as *granulose* or *granoblastic*, is an arrangement of shapeless interlocking mineral grains resembling the granitic texture but developed in metamorphic rocks. Fewer than 10% of the grains have a preferred orientation.

granulitization The process in regional metamorphism of reducing the components of a solid rock to grains. If the reduction of the size of the particles goes farther, mylonite is produced.

graphic formula A formula in which every atom is represented by the appropriate symbol, valency bonds being indicated by dashes; e.g. H—O—H, the graphic formula for water.

graphic granite Granite of pegmatitic facies, in which quartz and alkali-feldspar are intergrown in such a manner that the quartz simulates runic characters. Also *runite*.

graphic texture A rock texture in which one mineral intimately intergrown with another occurs in a form simulating ancient writing, esp. runic characters; produced by simultaneous crystallization of two minerals present in eutectic proportions. See **graphic granite**.

graphite One of the two naturally occurring forms of crystalline carbon, the other being diamond. It occurs as black, soft masses and, rarely, as shiny crystals (of flaky structure and apparently hexagonal) in igneous rocks; in larger quantities in schists particularly in metamorphosed carbonaceous clays and shales, and in marbles; also in contact metamorphosed coals and in meteorites. Graphite has numerous applications in trade and industry now much overshadowing its use in 'lead' pencils. Much graphite is now produced artificially in electric furnaces using petroleum as a starting material. Also *black lead*, *plumbago*.

graphitization The transformation of amorphous carbon to graphite brought about by heat. It results in a volume change due to the alteration in atomic lattice layer spacing. It is reversible under bombardment by high-energy neutrons and other particles.

grapnel, grappel An extracting tool used in boring operations.

graptolites See panel on p. 105.

grating An arrangement of alternate reflecting and non-reflecting elements, e.g. wire screens or closely spaced lines ruled on a flat (or concave) reflecting surface, which, through diffraction of the incident radiation analyses this into its frequency spectrum. An *optical grating* can contain a thousand lines or more per cm. A *standing-wave* system of high-frequency sound waves with their alternate compressive and rarefied regions can give rise to a diffraction grating in liquids and solids. With a *criss-cross* system of waves, a three-dimensional grating is obtainable.

grating spectrum An optical spectrum produced by a **diffraction grating**.

gravel The name of the aggregate consisting dominantly of pebbles, though usually a considerable amount of sand is intercalated. The grain size variously defined as 2–20 mm. In the Stratigraphical Column, gravels of different ages and origins occur abundantly, e.g. in South East England,

graptolites

A group of extinct colonial marine organisms that existed only during the Palaeozoic Era. They are generally placed in the phylum Chordata but, sometimes, the Coelenterata. They consist of one or more branches or *stipes*, along which individuals in the colony occur in rows.

The most primitive forms are the *dendroid graptolites* (Fig. 1). They persisted from the Middle Cambrian to the Lower Carboniferous. The Graptoloidea is the class of graptolites that ranged from the Lower Ordovician to the Lower Devonian (Figs. 2–4). They occur commonly in black shales and are of great importance for stratigraphical correlation in the Ordovician (when they reached their acme) and Silurian, when they were waning.

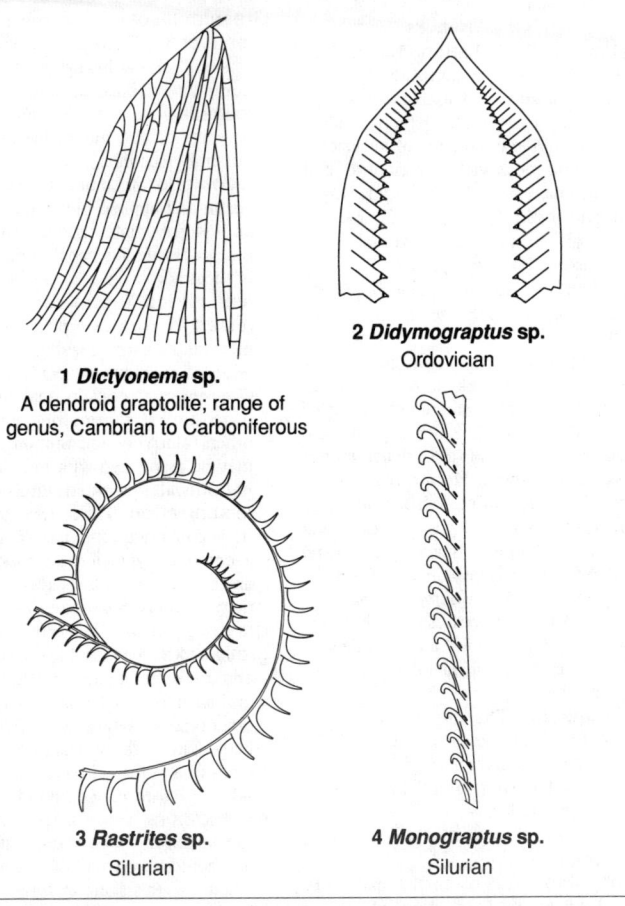

1 *Dictyonema* sp.
A dendroid graptolite; range of genus, Cambrian to Carboniferous

2 *Didymograptus* sp.
Ordovician

3 *Rastrites* sp.
Silurian

4 *Monograptus* sp.
Silurian

where they consist chiefly of well-rounded flint pebbles originally derived from the Chalk. These gravels are mainly of fluviatile and fluvioglacial origin, but marine gravels are also common in the littoral zone. The indurated equivalent of gravel is conglomerate. See **particle size**, **Wentworth scale**.

gravimetric analysis The chemical analysis of materials by the separation of the constituents and their estimation by weight.

gravitation The name given to that force of nature which manifests itself as a mutual attraction between masses, and whose mathematical expression was first given by Newton, in the law which states: 'Any two particles of matter attract one another with a force directly proportional to the product of their masses and inversely proportional to the square of the distance between them.' This may be expressed by the equation:

$$F = G \frac{m_1 m_2}{d^2},$$

where F is the force of gravitational attraction between bodies of mass m_1 and m_2, separated by a distance D. G is the constant of gravitation, 6.67×10^{-11} N m^2 kg^{-2}.

gravitational differentiation The production of igneous rocks of contrasted types by the early separation of denser crystals such as olivine, pyroxenes etc. which become concentrated in the basal parts of intrusions. The ultramafic rocks such as peridotites and picrites may have been formed in this way.

gravitational field That region of space in which at all points a gravitational force would be exerted on a test particle.

gravity plane An inclined plane on which the descending full trucks pull up the ascending empty ones.

gravity separation Use of differences between relative densities of roughly-sized grains of mineral to promote settlement of denser species while less-dense grains are washed away. See **buddle, dense-media process, frue vanner, Humphreys spiral, hydrocyclone, jig, sluice, Wilfley table,**.

gravity stamp Set of 5 heavy pestles, lifting about 10 in at 90 times a minute by a cam and allowed to fall on ore spread in mortar box, for crushing purposes. Also *Californian stamp*.

gravity tectonics Processes of rock deformation and folding which are activated by gravity applied over considerable periods of geological time.

gravity transport The movement of material under the influence of gravity. It includes downhill movement of weathering products, and movement of unweathered material in landslides.

gray See **ionizing radiation: units of measurement**.

Gray-King Test Test of coking quality of coal under prescribed conditions of heating to 600°C.

graywacke See **greywacke**.

grease table Sloping table anointed with petroleum jelly, over which diamondiferous concentrate is washed, the diamond adhering strongly while the gangue is worked away.

Great Ice Age See **Pleistocene Period**.

greenalite A septechlorite related to the chlorites chemically and to the serpentines structurally. A hydrous, iron silicate of composition $Fe_6Si_4O_{10}(OH)_3$.

green carbonate of copper See **malachite**.

greenhouse effect Phenomenon by which thermal radiation from the Sun is trapped by water vapour and carbon dioxide on a planet's surface, thus preventing its re-emission as long-wave radiation. This leads to the temperature at the planet's surface being considerably higher than would otherwise be the case. The effect is most pronounced for **Venus** and the **Earth**.

Greenland spar See **cryolite**.

greenockite Cadmium sulphide occurring as small yellow hexagonal crystals in cavities in altered basic lavas.

greensand A sand or sandstone with a greenish colour, due to the presence of the mineral glauconite.

greenstone An omnibus term lacking precision and applied indiscriminately to basic and intermediate igneous rocks in which much chlorite has been produced as a result of metamorphism.

green vitriol **Melanterite**.

greisen A rock composed essentially of mica and quartz, resulting from the alteration of a granite by percolating solutions. Greisens often contain small amounts of fluorite, topaz, tourmaline, cassiterite, and other relatively uncommon minerals, and may be associated with mineral deposits, as in Cornwall. See **pneumatolysis**.

greisenization The process by which granite is converted to greisen. Greisenization is a common type of wall rock alteration in areas where granite is traversed by hydrothermal veins. Sometimes *greisening*.

grey copper ore See **tetrahedrite**.

greywacke A sandstone containing silt, clay and rock fragments in addition to quartz grains. It is much more poorly sorted than other types of sandstone, and often occurs on beds which show gradation in grain size from fine at the top to coarse at the bottom. Also *graywacke*. See **turbidite**.

greywethers Grey-coloured rounded blocks of sandstone or quartzite left as residual boulders on the surface of the ground when less resistant material was denuded. From a distance they resemble sheep grazing. See **sarsen**.

Griffin mill Pendulum mill in which a hanging roller bears against a stationary bowl as it rotates, crushing passing ore.

grike, gryke A fissure in limestone rock caused by the solvent action of rainwater. See **karst**.

grindability Empirical assessment of response of ore to pulverizing forces, applied under specified conditions.

grinding Comminution of minerals by dry or, more usually, wet methods, mainly in rod, ball, or pebble mills.

grit (1) Siliceous sediment, loose or indurated, the component grains being angular. (2) Sometimes applied to a hard coarse-grained sandstone.

grizzly The set of parallel bars or grating used for the coarse screening of ores, rocks etc.

grossular An end-member of the garnet group, the composition being represented by $Ca_3Al_2Si_3O_{12}$; formed in the contact-metamorphism of impure limestone. Commonly contains some iron, and is greenish, brownish, or pinkish. Also *gooseberry stone*. See **hydrogrossular**.

ground The mineralized deposit and rocks in which it occurs, e.g. payground, payable reef; barren ground, rock without value.

groundmass In igneous rocks which have crystallized in two stages, the groundmass is the finer-grained portion, in which the phenocrysts are embedded. It may consist wholly of minute crystals, wholly of glass, or partly of both.

ground sluicing Bulk concentration of heavy minerals *in situ*, by causing a stream of water to flow over unconsolidated alluvial ground with just enough force to flush away the lighter, less valuable sands leaving the heavier ones to be removed for further treatment.

ground water Water occupying space in rocks. It may be *juvenile*, having arisen from a deep magmatic source, or *meteoric*, the result of rain percolating into the ground. Also *groundwater*.

group (1) A vertical column of the periodic system, containing elements of similar properties. (2) Metallic radicals which are precipitated together during the initial separation in qualitative analysis. (3) A number of atoms which occur together in several compounds.

group A stratigraphic rock unit consisting of two or more **formations**.

group reaction The reaction by which members of a **group** are precipitated.

Grüneisen's relation The coefficient of volume expansion of a solid β, is given by $β = γκX$, where κ is the compressibility, C the specific heat capacity of unit volume and γ is Grüneisen's constant which for most materials lies between 1 and 2 and is practically independent of temperature.

grunerite A monoclinic calcium-free amphibole; a hydrated silicate of iron and magnesium, differing from cummingtonite in having Fe Mg. Typically found in metamorphosed iron-rich sediments.

gryke See **grike**.

guard magnet Strong magnet, usually suspended above moving stream of lump ore, to remove steel from broken drills etc. (**tramp iron**), which might damage the crushing machine.

guides Timbers, ropes, or steel rails at sides of shaft used to steady the cage or skip.

gulching The noise which generally precedes a fall or settlement of overlying strata.

guncotton A **nitrocellulose** (cellulose hexanitrate) with a high nitrogen content. It burns readily and explodes when struck or strongly heated. Used for explosives.

gun perforation See **perforation**.

Günz Name of an early glacial stage of the Pleistocene Epoch in the Alps. See **Quaternary**.

gusher See **geyser**.

Gutenberg discontinuity The seismic-velocity discontinuity separating the mantle of the Earth from the core at a depth of approximately 2900 km.

Gutzeit test A method of determining arsenic, by adding metallic zinc and hydrochloric acid. The evolved gases darken mercury (II) salts.

guyot A flat-topped seamount, a topographic feature of the ocean floor.

G-value A constant in radiation chemistry denoting the number of molecules reacting as a result of the absorption of 100 eV radiation energy.

gypsum Crystalline hydrated calcium sulphate, $CaSO_4.2H_2O$. Occurs massive as *alabaster*, fibrous as *satin spar*, and as clear, colourless, monoclinic crystals known as *selenite*. Used in making plaster of Paris, and plaster and plasterboard used in building.

gypsum plate A thin plate of gypsum used in the determination of the sign of the birefringence of crystals in a polarizing microscope.

gyratory breaker A widely used form of rock-breaker, in which an inner cone gyrates in a larger outer hollow cone.

gyrolite A hydrated calcium silicate, formula $Ca_2Si_3O_7(OH)_2.H_2O$. Often occurs in amygdales with apophyllite.

Gzelian The youngest epoch of the Pennsylvanian period.

H

h Symbol for (1) Planck's constant. (2) specific enthalpy.

H Symbol for (1) **hydrogen**; (2) henry.

Ha, Hb, Hc etc. The lines of the Balmer series in the hydrogen spectrum. Their wavelengths are: Ha, 656.299; Hb, 486.152; Hc, 434.067; Hd 410.194 nm. The series continues into the ultraviolet, where about 20 more lines are observable.

H Symbol for **enthalpy**.

[H] One member of the strongest pair (H and K) of Fraunhofer lines in the solar spectrum, almost at the limit of visibility in the extreme violet. Their wavelengths are [H], 396.8625 nm; [K], 393.3825 nm; and the lines are due to ionized calcium.

Haber process Currently the most important process of fixing nitrogen, in which the nitrogen is made to combine with hydrogen under influence of high temperatures (400°–500°C), high pressure ($\times 10^7$ N/m^2), and catalyst of finely-divided iron from iron(III) oxides, in large continuous enclaves. Many variants operate at different pressures according to the catalyst. The product ammonia may be dissolved in water or condensed, and unreacted gases recycled.

habit A term used to cover the varying development of the crystal forms possessed by any one mineral. Thus calcite may occur as crystals showing the faces of the hexagonal prism, basal pinacoid, scalenohedron and rhombohedron. According to the relative development or *dominance* of one or other of these forms, the habit may be prismatic, tabular, scalenohedral, or rhombohedral.

hackmanite A fluorescent variety of sodalite, showing on freshly fractured surfaces a pink colour which fades on exposure to light, but which returns if kept in the dark or subjected to X-rays or ultraviolet light. See **tenebrescence**.

hade The angle of inclination of a fault plane, measured from the vertical.

haematite See **hematite**.

hafnium A metallic element in the fourth group of the periodic system. Symbol Hf, at. no. 72, r.a.m. 178.49, rel.d. 12.1, mp about 2200°C. It occurs in zirconium minerals, where its chemical similarity but relatively high neutron absorption makes it a troublesome impurity in zirconium metal for nuclear engineering. Used to prevent recrystallization of tungsten filaments.

hagatalite A variety of **zircon** which contains an appreciable quantity of the rare earth elements.

Haidinger fringes Optical interference fringes produced by transmission and reflection from two parallel, partly reflecting surfaces, e.g. a plate of optical glass. The fringes are produced by division of amplitude of the wave front and are circular fringes formed at infinity (cf. **contour fringes**). Used extensively in interferometry, e.g. Fabry and Pérot, Michelson interferometers.

Haldane apparatus An apparatus for the analysis of air; used for the analysis of mine gases.

half-life The time taken for the activity of a radionuclide to lose half its activity by radioactive decay. Also *half-value period*. The half-life T is related to the *decay constant* λ by $T = (\ln 2)/\lambda$. See **ionizing radiation**.

half-wave plate A plate of doubly refracting, uniaxial crystal cut parallel to the optic axis, of such thickness that, if light is transmitted normally through it, a phase difference of half a period is introduced between the ordinary and extraordinary waves. A half-wave plate is used in Laurent's polarimeter.

half-width A measure of sharpness on any function $y = f(x)$ which has a maximum value y_m at x_0 and also falls off steeply on either side of the maximum. The half-width is the difference between x_0 and the value of x for which $y = y_m/2$. Used particularly to measure the width of spectral lines or of a response curve.

halides Fluorides, chlorides, bromides, iodides and astatides.

halite *Common or rock salt*. The naturally occurring form of sodium chloride, crystallizing in the cubic system; forming deposits of considerable thickness in close association with anhydrite and gypsum, esp. in the Permian and Triassic rocks. The salt is pumped out as brine or mined. Salt domes, with which oil or gas may be associated, occur in many parts of the world.

halloysite One of the clay minerals, a hydrated form of kaolinite and member of the kandite group; consists of hydrated aluminium silicate.

halo Product of diffusion in rock surrounding an ore deposit of traces of mineral or element being sought, identified by geochemical tests. The halo may be the only surface indication of an ore deposit at depth. See **geochemical prospecting**.

halogen One of the seventh group of elements in the periodic table, for which there is one electron vacancy in the outer energy level, viz., F, Cl, Br, I, At. The

halotrichite

main oxidation state is −1.
halotrichite Hydrated sulphate of iron and aluminium, occurring rarely as yellowish fibrous silky colourless crystals. Also *iron alum*.
hambergite Basic beryllium borate, crystallizing in the orthorhombic system.
hammer-drill A compressed-air rock drill in which the piston is not attached to the steel or borer but moves freely.
hammer mill See **impact crusher**.
Hancock jig One in which ore is jigged up and down with some throw forward in a tank of water, the heavy mineral stratifying down and being separately removed.
hand specimen A piece of rock or mineral of a size suitable for megascopic study, further investigation or preserving in a collection.
hangfire Unexpected delay or failure of explosive charge to detonate, thus creating a dangerous situation.
hanging valley A tributary valley not graded to the main valley. It is a product of large-scale glaciation and due to the glacial overdeepening of the main valley relative to the hanging valley. There may be rapids or waterfalls from the tributary to the main valley. See **valleys**.
hanging wall Rock above the miner's head, usually the country rock above the deposit being worked.
hardening of oils The hydrogenation of oils in the presence of a catalyst, usually finely divided nickel, in which the unsaturated acids are transformed into saturated acids, with the result that the glycerides of the unsaturated acids become hard.
hard heading Sandstone or other hard rock met with in making headings or tunnels in a coalmine.
Hardinge mill Widely used grinding mill, made in three sections: a flattish cone at the feed end, a cylindrical drum centrally and a steep cone; the assembly being hung horizontally between trunnions.
hardness The resistance which a mineral offers to abrasion. The absolute hardness is measured with the aid of a sclerometer. The comparative hardness is expressed in terms of Mohs' scale, and is determined by testing against ten standard minerals: (1) talc, (2) gypsum, (3) calcite, (4) fluorite, (5) apatite, (6) orthoclase, (7) quartz, (8) topaz, (9) corundum, (10) diamond. Thus a mineral with 'hardness 5' will scratch or abrade fluorite, but will be scratched by orthoclase. Hardness varies on different faces of a crystal, and in some cases (e.g. kyanite) in different directions on any one face.

Harz jig

hard pan, hardpan (1) A hardened impervious layer of soil cemented by iron oxides and hydroxides (*iron pan*), silica, carbonates etc, sometimes clayey. (2) A layer of partly cemented gravel below the surface in a gold placer.
hard-rock geology An informal term for the geology of igneous and metamorphic rocks. Cf. **soft-rock geology**.
hard-rock mining Term used to distinguish between deposits soft enough to be detached by mechanical excavator and those which must first be loosened by blasting.
hard-rock phosphate A phosphatic deposit resulting from the leaching of calcium carbonate out of a phosphatic limestone, leaving a phosphatic residue. Applied specifically to the phosphate deposits of Florida which have this origin.
Hardy and Schulze 'law' The efficiency of an ion used as a coagulating agent is roughly proportional to its state of oxidation.
Harker diagram A variation diagram in which chemical analyses of rocks are plotted to show their relationships. The constituents are plotted as ordinates against the silica content as abscissa.
harmotome A member of the zeolite group, hydrated silicate of aluminium and barium, crystallizing in the monoclinic system, though the symmetry approaches that of the orthorhombic system. Best known by reason of the distinctive cruciform twin groups that are not uncommon. Occurs in mineralized veins.
hartite A naturally occurring hydrocarbon compound, $C_{20}H_{34}$, crystallizing in the triclinic system.
Hartmann dispersion formula An empirical expression for the variation of refractive index n of material with the wavelength of light λ;

$$n = n_0 + \frac{c}{(\lambda - \lambda_0)^a},$$

where n_0, c, λ_0 are constant for a given material. For glass a is about 1.2.
Hartman test Test for aberration of a lens, in which a diaphragm containing a number of small apertures is placed in front of the lens and the course of the rays is recorded by photographing the pencils of light in planes on either side of the focus.
harzburgite An ultrabasic igneous rock belonging to the peridotite group. It consists almost entirely of olivine and orthopyroxene, usually with a little chromite, magnetite and diopside.
Harz jig Concentrating appliance in which

hastingsite

water is pulsed through a submerged fixed screen, across which suitably sized ore moves. Heaviest particles gravitate down and through, and lighter ones overflow.

hastingsite A monoclinic amphibole, hydrated sodium calcium iron aluminosilicate. The name is used for an end-member compositional variety of amphibole.

haulage level Underground tramming road in, or parallel to, strike of the ore deposit, usually in footwall. Broken ore gravitates, or is moved, to ore chutes and drawn to trucks in this level.

hausmannite A blackish-brown crystalline form of manganese oxide, occurring with other manganese ores. Crystallizes in the tetragonal system but is often found massive.

Hauterivian A stage in the Lower Cretaceous. See **Mesozoic**.

haüyne A feldspathoid, crystallizing in the cubic system, consisting essentially of silicate of aluminium, sodium and calcium, with sodium sulphate; occurs as small blue crystals chiefly in phonolites and related rock types.

Hawaiian-type eruption An eruption characteristic of a shield volcano with large quantities of fluid lava, mainly as lava fountains from fissures, and only rare explosive phenomena.

hawaiite An olivine-bearing oligoclase or andesine trachyandesite.

hawk's eye A dark-blue form of silicified crocidolite found in Griqualand West; when cut *en cabochon*, it is used as a semi-precious gemstone. Cf. **tiger's eye**.

HCP Abbrev. for *Hexagonal Close Packing*.

head (1) A superficial deposit consisting of angular fragments of rock, originating from the breaking up of rock by alternate freezing and thawing of its contained water, followed by downhill movement. Head is found in valleys in periglacial regions, i.e. those formerly near the edge of an ice sheet. In south-eastern England, also *coombe rock*. (2) An advance main roadway driven in solid coal. (3) The difference in air pressure producing ventilation. (4) The top portion of a seam in the coalface. (5) The complete falling unit in a stamp battery, or merely the weight at the end of the stem.

headframe, headgear The steel or timber frame at the top of a shaft, which carries the sheave or pulley for the hoisting rope, and serves such other purposes as, e.g. acting as transfer station for hoisted ore, or as loading station for man and materials.

heading Passageway through solid coal.

headings *Head tin*. Concentrate settling

heavy liquids

nearest to the entry point of a concentrating device such as a **sluice** or **buddle**.

head motion Vibrator, a sturdy device which gives reciprocating movement to shaking tables, used in gravity methods of concentration.

heap leaching This, perhaps aided by heap roasting, is the dissolution of copper from oxidized ore by solvation with sulphuric acid. The resulting liquor is run over scrap iron to precipitate out its copper.

heap sampling See **quartering**.

heat Heat is energy in the process of transfer between a system and its surroundings as a result of temperature differences. However, the term is still used also to refer to the energy contained in a sample of matter. Also for **temperature**, e.g. forging or welding *heat*.

heat exchange The process of using two streams of fluid for heating or cooling one or the other, either for conservation of heat or for the purpose of adjusting process streams to correct processing temperatures.

heat-flow measurement The measurement of the amount of heat leaving the Earth.

heat flux The total flow of heat in *heat exchange*, in appropriate units of time and area.

heat of formation The net quantity of heat evolved or absorbed during the formation of one mole of a substance from its component elements in their standard states.

heat of solution The quantity of heat evolved or absorbed when one mole of a substance is dissolved in a large volume of a solvent.

heat transfer coefficient The rate of heat transfer q between two phases may be expressed as

$$q = hA(T_1 - T_2),$$

where A is the area of the phase boundary, (T_1-T_2) is their difference in temperature, and the heat transfer coefficient h depends on the physical properties and relative motions of the two phases.

heat transfer salt Molten salts used as a heating or quenching medium. Usually mixtures of sodium or potassium nitrate or nitrite, range 200°C to 600°C. Abbrev. *HTS*.

heave (1) The horizontal distance separating parts of a faulted seam, bed, vein or lode, measured normal to the fault plane. (2) Rising of the floor of a mine.

heavy ground Unstable roof rock requiring special care and support.

heavy liquids Liquids, organic or solutions of heavy salts, of relative densities adjustable in the range 1.0 to 4.1, used to separate ore constituents into relatively heavy (sink) and light (float) fractions with fair precision, and to carry out specific gravity tests on minerals including gemstones. They include carbon tetrachloride, bromoform, methylene iodide and Clerici's solution.

heavy-media separation   *Dense-media separation.* Method of upgrading ore by feeding it into liquid slurry of intermediate density, the heavier fraction sinking to one discharge arrangement and the light ore overflowing. Used to remove shale from coal (*sink-float process*) and to reject waste from ore.

heavy metal   Loose term for polluting metal ions that are persistent and potentially toxic in the environment; includes arsenic, cadmium, lead, mercury, vanadium and zinc.

heavy metal replacement   See **isomorphous replacement**.

heavy mineral   A detrital mineral from a sedimentary rock having a higher than normal specific gravity. Commonly applied to minerals which sink in bromoform (density 2.9).

heavy spar   See **barytes**.

hectorite   A rare lithium-bearing mineral in the montmorillonite group of clay minerals.

hedenbergite   An important calcium-iron pyroxene, $CaFeSi_2O_6$, occurring as black crystals, and also as a component molecule in many of the rock-forming clinopyroxenes.

held water   Water kept above the natural water table through capillary force. US *free water*.

helictic   Descriptive of the S-shaped trails of inclusions found in the minerals of some metamorphic rocks, esp. abundant in garnet and staurolite. These inclusion trails are often continuous with the mineral alignments of surrounding crystals and may help to indicate the crystallization history of the rock.

heliodor   A beautiful variety of clear yellow beryl.

heliotrope   See **bloodstone**.

helium   Inert gaseous element. Symbol He, at.no. 2, r.a.m. 4.0026, with extremely stable nucleus identical to α-particle. It liquefies at temperatures below 4K, and undergoes a phase change to a form known as *liquid helium II* at 2.2K. The latter form has many unusual properties believed to be due to a substantial proportion of the molecules existing in the lowest possible quantum energy state. Liquid helium is the standard coolant for devices working at cryogenic temperatures. The abundance of helium in the Earth's crust is 0.003 ppm and in the atmosphere 5.2 ppm (vol). Almost all of it is of radiogenic origin. It is obtained from gas wells.

Helmert's formula   An empirical formula giving the value of g, the acceleration due to terrestrial gravity, for a given latitude and altitude: $g = 9.80616 - 0.025928 \cos2\lambda + 0.000069 \cos^22\lambda - 0.000003086 H$, where λ is the latitude and H is the height in metres above sea level; g is in ms^{-2}.

Helmholtz double layer   Electrical double layer. This assumes an interphase between a relatively insoluble solid and an ambient ionized liquid, in which oppositely charged ions tend to concentrate in layers. See **electrokinetic potential**.

hematite, haematite   Oxide of iron, Fe_2O_3, crystallizing in the trigonal system. It occurs in a number of different forms: kidney iron ore massive; specular iron ore in groups of beautiful, lustrous, rhombohedral crystals as, for example, from Elba; bedded ores of sedimentary origin, as in the Precambrian throughout the world; and as a cement and pigment in sandstones.

hemicrystalline rocks   Those rocks of igneous origin which contain some interstitial glass, in addition to crystalline minerals. Cf. **holocrystalline rocks**.

hemimorphism   The development of polar symmetry in minerals, in consequence of which different forms are exhibited at the ends of bi-terminated crystals. Hemimorphite shows this character in a marked degree.

hemimorphite   An orthorhombic hydrated silicate of zinc; one of the best minerals for demonstrating polar symmetry, the two ends being distinctly dissimilar. US *calamine* or *electric calamine*.

Hercynian orogeny   The late Palaeozoic orogeny in Europe.

hercynite   See **spinel**.

hessite   Silver telluride, a metallic grey pseudo-cubic mineral occurring in silver ores in various parts of the world.

hessonite   *Cinnamon stone.* A variety of garnet containing a preponderance of the grossular molecule, and characterized by a pleasing reddish-brown colour.

Hess's law   The net heat evolved or absorbed in any chemical change depends only on the initial and final states, being independent of the stages by which the final state is reached.

heterodesmic structure   Structure which includes two or more types of crystal bonding.

heterogeneous   Of a system consisting of more than one phase.

heteromorphous rocks   Rocks of closely similar chemical composition, but which contain different mineral assemblages.

Hettangian The oldest stage in the Jurassic. See **Mesozoic**.

heulandite One of the best-known zeolites, often beautifully crystalline, occurring as coffin-shaped monoclinic crystals in cavities in basic igneous rocks. In composition similar to hydrated calcium sodium aluminium silicate.

hex- Colloquialism for *uranium* (VI) *fluoride*, the compound used in the separation of uranium isotopes by gaseous diffusion.

hexagonal packing System in which many metals crystallize, thus achieving minimum volume. Each lattice point has twelve equidistant neighbours in such a cell construction.

hexagonal system A crystal system in which three equal coplanar axes intersect at an angle of 60°, and a fourth, perpendicular to the others, is of a different length. See **Bravais lattices**.

hexane C_6H_{14}. There are five compounds with this formula: normal hexane, a colourless liquid, of ethereal odour, bp 69°C, rel.d. 0.66, is an important constituent of petrol and of solvent petroleum ether or ligroin.

Hf Symbol for **hafnium**.

Hg Symbol for **mercury**.

hiatus A break or gap in the stratigraphical record, because of either non-deposition or erosion.

hiddenite See **spodumene**.

high explosive Explosive in which the active agent is in chemical combination and is readily detonated by application of shock. Nitrated cotton, nitroglycerine and ammonium nitrate are widely used, diluted to required explosive strength by inert fillers such as kieselguhr or wood pulp. See **gelignite**.

Highgate resin A popular name for the fossil gum-resin occurring in the Tertiary London Clay at Highgate in North London. Also *copaline, copalite*.

high grading Selective mining, in which subgrade ore is abandoned unworked. Also, theft of valuable concentrates or specimens such as nuggets of gold.

high-intensity separation Dry concentration of small particles of mineral in accordance with their relative ability to retain ionic charge after passing through an ionizing field.

Highland Boundary Fault One of the most important dislocations in the British Isles, extending from Helensburgh to Stonehaven and separating the Highlands of Scotland from the Midland Valley.

high-pressure hose Armoured hose, reinforced with circumferentially embedded wire, and hence able to withstand moderately high pressure and rough usage.

high-tension separation Electrostatic separation, in which small particles of dry ore fall through a high-voltage d.c. field, and are deflected from gravitational drop or otherwise separated in accordance with the electric charge they gather and retain.

hillebrandite *Dicalcium silicate hydrate.* Occurs as white fibrous aggregates in impure thermally metamorphosed limestones and in boiler scale.

Hilt's Law An expression of the observation that the more deeply buried a coal seam, the higher is the rank of its coal.

hindered settling Hydraulic classification of sand-sized particles in accordance with their ability to gravitate through a column of similar material expanded by a rising current of water.

hinge fault A fault along which the displacement increases from zero at one end to a maximum at the other end.

histogram A graphical representation of class frequencies as rectangles against class interval, the value of frequency being proportional to the area of the corresponding rectangle.

Historical Geology That major branch of geology that is concerned with the evolution of the Earth and its environment from its origins to the present day.

history of geology The study of the history or development of geological knowledge.

hitch (1) A fault of minor importance, usually not exceeding the thickness of a seam. (2) A ledge cut in a rock face to hold mine timber in place.

Ho Symbol for **holmium**.

hogback A ridge with a sharp summit and steep slopes on both sides, usually 20°.

hoist An engine with a drum, used for winding up a load from a shaft or in an underground passage such as a winze.

hold-up In any process plant, the amount of material which must always be present in the various reactors etc. to ensure satisfactory operation.

hole control Adjusting the composition of the drilling mud and the drill pressure and rate to accommodate the changes in the rock formation which it is penetrating.

holmium A metallic element, a member of the rare earth group. Symbol Ho, at.no. 67, r.a.m. 164.930. It occurs in euxenite, samarskite, gadolinite and xenotime.

holmquistite A rare lithium-bearing calcium-free variety of orthorhombic amphibole.

holoaxial A term applied to those classes of crystals characterized by axes of symmetry only; such crystals are not symmetrical about planes of symmetry.

Holocene The younger, temperate, epoch of the Quaternary period. Its base is taken as 10 000 years before the present (*BP*). Synonym *Recent*. See **Quaternary**.

holocrystalline rocks Igneous rocks in which all the components are crystalline; glass is absent. Cf. **hemicrystalline rocks**.

hologram See **holography**.

holography A method of recording and reconstructing the wave front emanating from an illuminated object. Coherent light from a laser is split in two; one is a reference beam and the other illuminates the object. The waves scattered by the object and the reference beam are recombined to form an interference pattern on a photographic plate, the *hologram*; this records both the amplitude and phase of the scattered light. When the hologram is itself illuminated by light from a laser or other point source, two images are produced, one is virtual but the other is real and can be viewed directly. So a 3-dimensional image of the object can be produced.

holohedral Term applied when a crystal is complete, showing all possible faces and angles.

homodesmic structure Crystal form with only one type of bond (either ionic or covalent).

homogeneous Said of a system consisting of only one phase, i.e. a system in which the chemical composition and physical state of any physically small portion are the same as those of any other portion.

homopolar Having an equal distribution of charge, as in a covalent bond between like atoms.

homotaxis A term introduced by Huxley in 1862 to indicate that strata or sequences of strata in different areas sharing the same fossil characteristics are not necessarily the same age. A faunal assemblage may originate in locality A, be gradually dispersed or migrate to locality B and eventually reach locality C. The strata accumulating at these three localities are *homotaxia* although not necessarily contemporaneous.

hone, honestone, whetstone Terms applied to fine-textured even-grained indurated sedimentary rocks which may be used as oilstones for imparting a keen edge to cutting tools. Honestone has been largely replaced now by emery and silicon carbide products.

Hooke's law For an elastic material the *strain* is proportional to the applied *stress*. The value of the stress at which a material ceases to obey Hooke's law is the *limit of proportionality*. See **elasticity**.

hook-up In pilot plant testing, flexible assembly of machines into continuous flow line before final treatment.

'Hope sapphire' Synthetic stone having the composition of spinel and blue colour which turns purple in artificial light. First produced in the attempts to synthesize sapphire.

hopper A container or surge bin for broken ore, used to hold small amounts.

hopper crystal A crystal which has grown faster along its edges than in the centres of its faces, so that the faces appear to be recessed. This type of skeletal crystallization is often shown by rock salt.

horizon The surface separating two beds of rock. It has no thickness, and is more frequently used in the sense of a thin bed or time plane with a characteristic lithofacies or biofacies, persistent over a wide area.

horn A steep-sided mountain peak formed by the coalescence of three or more cirques.

hornblende Important members of the amphibole group of rock-forming minerals. They are of complex composition, essentially silicate of calcium, magnesium and iron, with smaller amounts of sodium, potassium, hydroxyl and fluorine; crystallizes in the monoclinic system; occurs as black crystals in many different types of igneous and metamorphic rocks, including hornblende-granite, syenite, diorite, andesite etc., and hornblende-schist and amphibolite.

hornblende-gneiss A coarse-grained metamorphic rock, containing hornblende as the dominant coloured constituent, together with feldspar and quartz, the texture being that typical of the gneisses. Differs from hornblende-schist in grain size and texture only.

hornblende-schist A type of green schist, formed from basic igneous rocks by regional metamorphism, and consisting essentially of sodic plagioclase, hornblende and sphene, frequently with magnetite and epidote. See **glaucophane**.

hornblendite An igneous rock composed almost entirely of hornblende.

hornfels A fine-grained rock which has been partly or completely recrystallized by contact metamorphism.

horn lead Sometimes applied to the mineral **phosgenite**.

horn silver See **cerargyrite**.

horsehead The curved part at the end of the arm of an oil well pump; it keeps the cable

horst An elongated uplifted block bounded by faults along its length. Cf. **graben**.

horst faults Two parallel normal faults heading outwards and throwing in opposite directions, the resulting structure being termed a *horst*.

host rock See **country rock**.

hot dry rock Potential source of heat energy from hot underground rocks. Water is pumped down an injection well into artificially induced fractures, and recovered, hot, in a second well drilled into the fracture system. The fracturing is initially caused by injecting water under very high pressure (*hydrofracting, hydrofracking*). Heat is extracted at the surface in a heat exchanger and the water recycled. Favourable conditions are provided by radiothermal granites.

hot-fluid injection Pumping steam, hot water or gas into the formation to increase flow of low specific gravity oil. Also *hot footing*.

Houdry process Catalytic cracking of petroleum, using activated aluminium hydrosilicate.

howieite A black triclinic hydrated silicate of sodium, manganese and iron.

Hoxnian An interglacial stage of the late Pleistocene. See **Quaternary**.

hübnerite, huebnerite Manganese tungstate, the end-member of the wolframite group of minerals, the series from $MnWO_4$ to $FeWO_4$.

Huff separator High-tension or electrostatic separator used to concentrate small particles of dry ore.

hullite See **chlorophaeite**.

humic acids Complex organic acids occurring in the soil and in bituminous substances formed by the decomposition of dead vegetable matter.

humite An orthorhombic magnesium silicate, also containing magnesium hydroxide. Found in impure marbles. The humite group also includes chondrodite, clinohumite and norbergite.

hummer screen Type used to grade smallish minerals, using a.c. to provide vibration by solenoid action.

Humphreys spiral Spiral sluice which combines separation of mineral sands by simple gravitational drag with mild centrifugal action as the pulp cycles downward.

Hungarian cat's eye An inferior greenish cat's eye obtained in the Fichtelgebirge in Bavaria. No such stone occurs in Hungary.

Huntington mill Wet-grinding mill in which 4 cylindrical mullers, hung inside a steel tub, bear outward as they rotate, thus grinding the passing ore.

Huronian A major division of the Proterozoic of the Canadian Shield, typically exposed on the northern shores of Lake Huron.

hushing, hush A washing away of the surface soil to lay bare the rock formation for prospecting.

hutch (1) A small train or wagon. (2) A basket for coal. (3) A compartment of a jig used for washing ores. The concentrate which passes through a jig screen is called the *hutchwork*.

hyacinth, jacinth The reddish-brown variety of transparent zircon, used as a gemstone. The name has also been used for a brownish grossular from Ceylon.

hyalite A colourless transparent variety of *opal*, occurring as globular concretions and crusts. Also *Müller's glass*.

hyalo- A prefix for volcanic rock names, meaning *containing glass*.

hyalophane Feldspar containing barium, with up to 30% $BaAl_2Si_2O_8$.

hyalopilitic texture A texture of andesitic volcanic rocks in which the groundmass consists of small microlites of feldspar embedded in glass.

hybrid rocks Rocks which originate by interaction between a body of magma and its wall rock or roof rock, which may be another igneous, or sedimentary, or metamorphic rock.

hydrargillite See **gibbsite**.

hydration The addition of water to anhydrous minerals, the water being of atmospheric or magmatic origin. Thus anhydrite, by hydration, is converted into gypsum, and feldspars into zeolites.

hydraulic air compressor Arrangements in which water falling to the bottom of a shaft entrains air which is released in a tunnel at depth, while the water rises to a lower discharge level.

hydraulic blasting In fiery mines, rock breaking by means of a ram-operated device acting on a **hydraulic cartridge**.

hydraulic cartridge An apparatus for splitting rock, mass concrete etc. It consists of a long cylindrical body which has numerous pistons projecting from one side and moving in a direction at right angles to the body (under hydraulic pressure from within the body), which is placed in a hole drilled to receive it.

hydraulic classifier Device in which vertically flowing column of water is used to carry up and carry on light and small mineral particles while heavy and large ones sink.

hydraulic control Use of fluid in sensing mechanism to actuate a signalling or a correcting device, in response to pressure changes.

hydraulic fracturing Method of increasing oil flow from less permeable strata by forcing liquid into it under very high pressure. See **fracking**.

hydraulic mining, hydraulicking The operation of breaking down and working a bank of gravel, alluvium, poorly consolidated or decomposed bedrock by high-pressure water jets as in mining gold, tin *placers* or china clay deposits.

hydraulic stowing The filling of worked-out portions of a mine with water-borne waste material. The water drains off and is pumped to the surface.

hydrides Compounds formed by the union of hydrogen with other elements. Those of the non-metals are generally molecular liquids or gases, certain of which dissolve in water (oxygen hydride) to form acid (e.g. hydrogen chloride) or alkaline (e.g. ammonia) solutions. The alkali and alkaline earth hydrides are crystalline, salt-like compounds, in which hydrogen behaves as the electronegative element. They contain H^--ions and, when electrolysed, give hydrogen at the anode. Transition elements give alloy or interstitial hybrids.

hydrocarbons A general term for organic compounds which contain only carbon and hydrogen. They are divided into saturated and unsaturated hydrocarbons, aliphatic (alkane or fatty) and aromatic (benzene) hydrocarbons. Crude oil is essentially a complex mixture of hydrocarbons.

hydrocarbons See **native hydrocarbons**.

hydrocerussite A colourless hydrated basic carbonate of lead occurring as an encrustation on native lead, galena, cerussite and other lead minerals.

hydrocyanic acid An aqueous solution of hydrogen cyanide. Dilute solution called *prussic acid*. Monobasic. Forms cyanides. Very poisonous.

hydrocyclone A small cyclone extractor for removing suspended matter from a flowing liquid by means of the centrifugal forces set up when the liquid is made to flow through a tight conical vortex. Used to separate solids in mineral pulp into coarse and fine fractions. Fluent stream enters tangentially to cylindrical section and coarser sands gravitate down steep-sided conical section to controlled apical discharge. Bulk of pulp, containing finer particles, overflows through a pipe inserted in the central vortex. Classification into fractions is aided by the centrifugal force with which the pulp is delivered to the appliance, which can handle large tonnages.

hydrofining Process of removing undesirable impurities, particularly sulphur and unsaturated compounds, from petroleum fractions using hydrogen at high temperature and pressure over a platinum catalyst. Also *hydrotreating*. Obsolete term *hydroforming*.

hydrofluoric acid Aqueous solution of hydrogen fluoride. Dissolves many metals, with evolution of hydrogen. Etches glass owing to combination with the silica of the glass to form silicon fluoride, hence it is stored in, e.g. polythene or gutta-percha vessels.

hydroforming See **hydrofining**.

hydrogen See panel on p. 116.

hydrogenation Chemical reactions involving addition of hydrogen, present as a gas, to a substance, in the presence of a catalyst. Important processes are: the hydrogenation of coal; the hydrogenation of fats and oils; the hydrogenation of naphthalene and other substances. See **Fischer-Tropsch process**.

hydrogeology The study of the geological aspects of the Earth's water.

hydrogrossular A hydrated variety of garnet, close to grossular in composition but with some of the silica replaced by water. A massive variety from South Africa is sometimes called 'Transvaal Jade'.

hydrohematite, turgite $Fe_2O_3.nH_2O$. Probably a mixture of the two minerals *hematite* and *goethite*, the former being in excess. It is fibrous and red in the mass, with an orange tint when powdered.

hydrolysis (1) The formation of an acid and a base from a salt by interaction with water; it is caused by the ionic dissociation of water. (2) The decomposition of organic compounds by interaction with water, either in the cold or on heating, alone or in the presence of acids or alkalis; e.g. esters form alcohols and acids; oligosaccharides and polysaccharides on boiling with dilute acids yield monosaccharides.

hydromagnesite Hydrated magnesium hydroxide and carbonate, occurring as whitish amorphous masses, or rarely as monoclinic crystals in serpentines.

hydrometallurgy Extraction of metals or their salts from crude or partly concentrated ores by means of aqueous chemical solutions; also electrochemical treatment including electrolysis or ion exchange.

hydrometer An instrument by which the relative density of a liquid may be determined by measuring the length of the stem of the hydrometer immersed, when it floats in the liquid with its stem vertical.

hydrophane A variety of opal which, when

hydrogen

The least dense element, forming diatomic molecules H_2. Symbol H, at.no. 1, r.a.m. 1.00797, valency 1. It is a colourless, odourless, diatomic gas, water being formed when it is burnt; mp $-259.14°C$, bp $-252.7°C$, density 0.08988 g/dm^3 at s.t.p.

It is cosmically the most abundant of all elements but in its elemental form is not of major importance in the Earth's crust (abundance 1520 ppm; 0.53 ppm by volume in the atmosphere). Combined in the form of water H_2O, however, hydrogen is of the greatest importance in the Earth's hydrosphere, atmosphere and biosphere, and hydrogen is part of the basis of all life.

In the lithosphere, hydrogen occurs in many minerals as the hydroxyl ion (OH^{1-}); in hydrated minerals, e.g. gypsum, $CaSO_4.2H_2O$, and as hygroscopic water ('H_2O-') which is absorbed from the atmosphere by soils and rocks. On heating, this H_2O may be driven off, which also happens to the '*water of crystallization*' of the hydrated minerals with, however, breakdown of the structure. The $(OH)^{1-}$ ion has an atomic radius very similar to that of oxygen and enters many minerals, esp. the amphiboles, micas and clay minerals, and metallic hydroxides.

There are numerous simple hydroxide minerals as well as basic salts. In the igneous rocks, amphiboles and micas are common; in sedimentary rocks, clay minerals include kaolinite, clay micas, montmorillonites and chlorites. Among the metamorphic rocks, the micas, talc and serpentine may be common. Although the total hydroxyl content of these minerals is not high, the volumes of rock are enormous, so the total quantity is significant.

Hydrogen is released in volcanic emanations in relatively small quantities, but the amounts of water vapour from this source are huge. Other volcanic gases include hydrogen sulphide, hydrochloric acid, methane, hydrofluoric acid and ammonium salts. Petroleum and natural gas, both of biological origin from decayed plants and bacteria, are important in sediments. In natural gas the most important compound is methane, CH_4.

---

dry, is almost opaque, with a pearly lustre, but becomes transparent when soaked with water, as implied in the name.

hydrophilic colloid A colloid which readily forms a solution in water.

hydrophobic colloid A colloid which forms a solution in water only with difficulty.

hydrothermal Relating to the action of hot water, and mineral deposits formed by such processes. The term is sometimes restricted to water of magmatic origin.

hydrothermal metamorphism A change in the mineral composition and texture of a rock which was effected by heated water.

hydrothermal mineralizarion Processes involving hot water, usually of magmatic origin, by which mineral deposits may be formed.

hydroxyapatite The hydroxyl-bearing variety of the phosphate apatite.

hydroxyl -OH. A monovalent group consisting of a hydrogen atom and an oxygen atom linked together.

hydrozincite, zinc bloom A monoclinic hydroxide and carbonate of zinc. It is an uncommon ore, occurring with smithsonite as an alteration product of sphalerite in the oxide zone of some lodes.

hygrometer Instrument for measuring or giving output signal proportional to atmospheric humidity. Electrical hygrometers make use of *hygristors*.

hypabyssal rocks Literally, igneous rocks that are not quite abyssal (i.e. *deep-seated, plutonic*), occurring as minor intrusions. The three main divisions, based on mode of occurrence, are *plutonic, hypabyssal* and *volcanic*. See **dykes and sills**.

hypersthene An important rock-forming silicate of magnesium and iron, $(Mg,Fe)SiO_3$, crystallizing in the orthorhombic system; an essential constituent of norite, hypersthene-pyroxenite, hypersthenite, hypersthene-andesite and charnockite. Strictly, an ortho-pyroxene containing 50%–70% of the enstatite molecule.

hypersthenite A coarse-grained igneous rock, consisting essentially of only one

hypidiomorphic

component, hypersthene, together with small quantities of accessory minerals.

hypidiomorphic *Subhedral.* Term referring to the texture of igneous rocks in which some of the component minerals show crystal faces, the others occurring in irregular grains. Cf. **idiomorphic crystals**.

hypogene Said of rocks formed, or agencies at work, under the Earth's surface.

hypsochrome A radical which shifts the absorption spectrum of a compound towards the violet end of the spectrum. A *batho-chrome* shifts it the other way.

hypsoflore A radical which tends to shift the fluorescent spectrum of a compound toward shorter wavelengths. A *bathoflore* shifts it the other way.

hypsometer An instrument used for determining the boiling point of water, either with a view to ascertaining altitude, by calculating the pressure, or for correcting the upper fixed point of the thermometer used.

Hz Symbol for hertz, the SI unit of frequency.

I

i Symbol for van't Hoff's factor.
i- Abbrev. for (1) *optically inactive*; (2) *iso-*, i.e. containing a branched hydrocarbon chain.
I Symbol for **iodine**.
I Symbol for ionic strength.
I Symbol for (1) electric current; (2) luminous intensity.
Iapetus Ocean An ocean which is thought to have existed from late Precambrian to Lower Palaeozoic times in the general position of the present Atlantic Ocean. The *Iapetus suture* along which the ocean opened, separating northern and southern faunal provinces, and eventually closed, is thought to have traversed Ireland and the Solway Firth. The name is from *Iapetus*, father of *Atlas*, after whom the Atlantic Ocean was named.
IAS Abbrev. for **Image Analysis System**.
ice action The work and effects of ice on the Earth's surface. See **glacial erosion, glaciation, glacier**.
ice age A period when glacial ice spread over regions which were normally ice-free. *The Ice Age* is a synonym of the Pleistocene epoch.
ice contact slope The steep slope of material originally deposited at an ice front and in contact with it.
Iceland agate A name quite erroneously applied to the natural glass *obsidian*.
Iceland spar A very pure transparent and crystalline form of calcite, first brought from Iceland. It has perfect cleavage, is noted for its double refraction, and hence is used in construction of the Nicol prism.
I-centred lattice *Body-centred crystal lattice. See* **unit cell**.
ichor The name applied by Sederholm to highly penetrating granitic liquids, charged with magmatic vapours (emanations), which he believed to operate in palingenesis.
icositetrahedron A solid figure having 24 trapezoidal faces, and belonging to the cubic system. Exemplified by some garnets.
iddingsite An alteration product of olivine consisting of goethite, quartz, montmorillonite group clay materials, and chlorite.
ideal crystal Crystal in which there are no imperfections or alien atoms.
idioblast A crystal which grew in metamorphic rock and is bounded by its own crystal faces. Cf. **idiomorphic crystals**. Adj. *idioblastic*. See **porphyroblastic**.
idiomorphic crystals Igneous rock minerals which are bounded by the crystal faces peculiar to the species. Cf. **allotriomorphic** (anhedral) and **hypidiomorphic** (subhedral).
idocrase See **vesuvianite**.
igneous complex A group of rocks, occurring within a comparatively small area, which differ in type but are related by similar chemical or mineralogical peculiarities. This indicates derivation from a common source.
igneous intrusion A mass of igneous rock which crystallized before the magma reached the Earth's surface; including **dykes and sills, stocks, bosses** and **batholiths**.
igneous cycle, magmatic cycle The sequence of events usually followed in igneous activity; it consists of an eruptive phase, a plutonic phase, and a phase of minor intrusion.
igneous rocks Rock masses generally accepted as being formed by the solidification of magma injected into the Earth's crust, or extruded on its surface.
ignimbrite A pyroclastic rock consisting originally of lava droplets and glass fragments which were so hot at the time of deposition that they were welded together. Also *welded tuff*.
ijolite A coarse-grained igneous rock, consisting of nepheline, aegirine-augite, with usually melanite garnet as a prominent accessory, occurring in nepheline-syenite complexes in the Kola Peninsula, the Transvaal, and elsewhere.
illite A monoclinic clay mineral, a hydrated silicate of potassium and aluminium. It is the dominant clay mineral in shales and mudstones. The illite group somewhat resembles muscovite.
illumination The quantity of light or luminous flux falling on unit area of a surface. Illumination is inversely proportional to the square of the distance of the surface from the source of light, and proportional to the cosine of the angle made by the normal to the surface with the direction of the light rays. The unit of illumination is the *lux*, which is an illumination of 1 lumen/m^2. Symbol E. Also *illuminance*.
ilmenite An oxide of iron and titanium, crystallizing in the trigonal system; a widespread accessory mineral in igneous and metamorphic rocks, esp. in those of basic composition. A common mineral in detrital sediments, often becoming concentrated in beach sand.
ilvaite Silicate of iron, calcium and manganese. It crystallizes in the orthorhombic system.

image Optical images may be of two kinds: real or virtual. A *real image* is one which is formed by the convergence of rays which have passed through the image-forming device (usually a lens) and can be thrown on to a screen, as in the camera and the optical projector. A *virtual image* is one from which rays appear to diverge. It cannot be projected on to a screen or a sensitive emulsion.

image analysis system A computer image analysis system, initially developed for planetary missions and now applied extensively in geology. A range of spatially related data (geological, geochemical, geophysical etc.) including remotely sensed data, are analysed interactively. Abbrev. *IAS*.

imbibition The absorption or adsorption of a liquid by a solid or a gel, accompanied by swelling of the latter.

imbricate structure (1) A structure produced by thrust faulting leading to the development of numerous small faults and rock slices arranged in parallel like a pack of fallen cards. (2) A sedimentary structure in which pebbles with a flat surface are stacked in the same direction, dipping upcurrent.

immiscibility The property of two or more liquids of not mixing and of forming more than one phase when brought together.

impact crusher Machine in which soft rock is crushed by swift blows struck by rotating bars or plates. The material may break against other pieces of rock or against casing plates surrounding the rotating hammers.

impermeable Not permitting the passage of liquids or gases.

impsonite A member of the asphaltite group.

In Symbol for **indium**.

inactivation The destruction of the activity of a catalyst, serum etc.

inbye The direction from a haulage way to a working face.

inch-penny weight In valuation of gold ore, the width of the lode or reef measured normal to the enclosing rock, multiplied by the assay value in penny weights per ton.

incised meander An entrenched bend of a river, which results from renewed downcutting at a period of rejuvenation.

inclusion A foreign body (gas, liquid, glass or mineral) enclosed by a mineral. Fluid inclusions (e.g. liquid carbon dioxide) may be used to study the genesis of the minerals in which they occur. See **xenolith**.

incompetent rock, bed A rock or bed that yields by plastic flow, folding or shearing during deformation.

incomplete reaction A reversible reaction which is allowed to reach equilibrium, a mixture of reactants and reaction products being obtained.

index fossil A fossil species which characterizes a particular geological **horizon**. It tends to be abundant, with a narrow time range and a wide geographical spread.

index mineral Mineral whose appearance marks a particular grade of metamorphism in progressive regional metamorphism.

Indian topaz See **citrine**. Also a misnomer for yellow corundum.

indicated ore Proved limits of deposit, in the light of known geology of mine and economic factors.

indicator vein In prospecting, one associated with the lode or vein being traced, thus guiding the search.

indices of crystal faces See **Miller indices**.

indicolite, indigolite A blue (either pale or bluish-black) variety of tourmaline.

indigo copper See **covellite**.

indigolite See **indicolite**.

indium A silvery metallic element in the third group of the periodic system. Symbol In, at.no. 49, r.a.m. 114.82, mp 155°C, bp 2100°C, rel.d. 7.28 at 13°C, electrical resistivity 9×10^{-8} ohm metres. Found in traces in zinc ores. The metal is soft and marks paper like lead; it forms compounds with carbon compounds. It has a large cross-section for slow neutrons and so is readily activated. Also used in manufacture of transistors and as bonding material for acoustic transducers.

induced radioactivity Radioactivity induced in non-radioactive elements by neutrons in a reactor, or protons or deuterons in a cyclotron or linear accelerator. X-rays or gamma-rays do not induce radioactivity unless the gamma-ray energy is exceptionally high.

induced reaction A chemical reaction which is accelerated by the simultaneous occurrence in the same system of a second, rapid reaction.

inductance (1) That property of an element or circuit which, when carrying a current, is characterized by the formation of a magnetic field and the storage of magnetic energy. (2) The magnitude of such capability.

induction period The interval of time between the initiation of a chemical reaction and its observable occurrence.

inductor A substance which accelerates a slow reaction between two or more substances by reacting rapidly with one of the reactants.

induration The process of hardening a soft sediment by heat, pressure and cementation. Cf. **diagenesis**.

industrial diamond Small diamond, not of gemstone quality, e.g. **black diamond** and **bort**; used to cut rock in borehole drilling, and in abrasive grinding. Now synthesized on a considerable scale by subjecting carbon to ultra-high pressures and temperature.

inert Not readily changed by chemical means.

inert gases See **noble gases**.

inertinite A carbon-rich **maceral** found in coal.

infrared detection Rays detected and registered photographically with special dyes: photosensitively with a special Cs—O—Ag surface; by photoconduction of lead sulphide and telluride; and, in absolute terms, by bolometer, thermistor, thermocouple or Golay detector.

infrared radiation Electromagnetic radiation in the wavelength range from 0.75 to 1000 μm approximately, i.e. between the visible and microwave regions of the spectrum. The *near* infrared is from 0.75 to 1.5 μm, the *intermediate* from 1.5 to 20 μm and the *far* from 20 to 1000 μm.

infrared spectrometer An instrument similar to an optical spectrometer but employing non-visual detection and designed for use with infrared radiation. The infrared spectrum of a molecule gives information as to the functional groups present in the molecule and is very useful in the identification of unknown compounds.

infusorial earth See **tripolite**.

inherent ash Noncombustible material intimately bound in the original coal-forming vegetation, as distinct from 'dirt' from extraneous sources.

inherent floatability Natural tendency of some mineral species to repel water and to become part of the 'float' without preliminary conditioning in the froth-flotation process.

injection The emplacement of fluid rock matter in crevices, joints or fissures found in rocks. See **intrusion**.

injection complex An assemblage of rocks, partly igneous, partly sedimentary or metamorphic, the former in intricate intrusive relationship to the latter, occurring in zones of intense regional metamorphism.

injection string Pipe run in addition to the *production string* in the borehole to allow the passage of additives.

inlier An outcrop of older rocks surrounded by those of younger age.

inosilicates Those silicate minerals which have an atomic structure in which the SiO_4 groups are linked together in chains, e.g. **pyroxene group**. See **silicates**.

inselberg A steep-sided knob or hill arising from a plain; often found in the semi-arid regions of tropical countries.

insoluble Incapable of being dissolved. Most 'insoluble' salts have a definite, though very limited, solubility.

instantaneous fuse Rapid-burning, as distinct from slow fuse. Ignition proceeds at a few kilometres per second, but is slower than that of **detonating fuse**.

instantaneous specific heat capacity The specific heat capacity at any one temperature level; *true s.h.c.* to distinguish from *mean s.h.c.*.

interface Sharp contact boundary between two phases, either or both of which may be solid, liquid or gaseous. Differs from interphase in lacking a diffuse transition zone.

interference Interaction between two or more waves of the same frequency emitted from coherent sources. The wavefronts are combined, according to the *principle of superposition*, and the resulting variation in the disturbances produced by the waves is the interference pattern. See **interference fringes**.

interference colours See **colours of thin films**.

interference figure More or less symmetrical pattern of concentric rings or lemniscates, cut by a black cross or hyperbola, exhibited by a section of anisotropic mineral when viewed in convergent light between crossed Nicol prisms or polarizers in a polarizing microscope. See **uniaxial**, **biaxial**.

interference fringes Alternate light and dark bands formed when two beams of monochromatic light having a constant phase relation, overlap and illuminate the same portion of a screen.

interference pattern See **interference**.

interfluve A ridge separating two parallel valleys.

interglacial stage A period of milder climate between two glacial stages.

intergranular texture A texture characteristic of holocrystalline basalts and doleritic rocks, due to the aggregation of augite grains between feldspar laths arranged in a network.

interlock Arrangement of switch-gear by which the controlling source prevents premature loading, starting or continuance during partial malfunction where a series of operations is so interlinked as to require smooth on-line operation.

intermediate (1) A general term for any chemical compound which is manufactured from a substance (see **primary**) obtained from raw materials, and which serves as a starting material for the synthesis of another product. (2) A short-lived species in a complex reaction.

intermediate igneous rocks Igneous rocks containing from 52% to 63% silica, and essentially intermediate in composition between the acid (granitic) and basic (gabbroic or basaltic) rocks. See **syenite, syenodiorite, diorite**.

intermolecular forces Term referring to the forces binding one molecule to another. They are very much weaker than the bonding forces holding together the atoms of a molecule.

intermontane basin A basin between mountain ranges often associated with a graben, e.g. Midland Valley of Scotland.

interpenetration twins Two or more crystals united in a regular fashion, according to a fixed plan (the *twin law*), the individual crystals appearing to have grown through one another. Cf. **juxtaposition twins, twinned crystals**.

interphase Transition zone between two phases in a system (solid-liquid, liquid-liquid, liquid-gas). In a solid-liquid system, it is the zone of shear through which the physical or chemical qualities of the contacting surfaces migrate toward one another.

intersertal texture The texture characterized by the occurrence of interstitial glass between divergent laths of feldspar in basaltic rocks.

intrinsic crystal Crystal, the photoelectric properties of which do not depend on impurities.

intrusion The process of emplacement of a magma into pre-existing rock. Also used in the sense of injection of a plastic sediment, e.g. a salt dome.

inverse square law The intensity of a field of radiation is inversely proportional to the square of the distance from the source. Applies to any system with spherical wavefront and negligible energy absorption.

inversion of relief A condition whereby synclinal ridges are separated by anticlinal depressions.

iodine Black lustrous non-metallic halogen solid that sublimes easily. Symbol I, at.no. 53, r.a.m. 126.905, mp 113.5°C, bp 184°C, rel.d. 4.98. Abundance in the Earth's crust 0.46 ppm in seawater 0.05 ppm. Concentrated by biological processes in marine plants and animals. It has a large ionic radius (2.20 Å) and may replace Cl^{1-} and OH^{1-} in some minerals to a small extent. There are a few independent silver iodide minerals. An essential trace element in animals and man.

iolite See **cordierite**.

ion Strictly, any atom or molecule which has resultant electric charge due to loss or gain of valency electrons. Free electrons are sometimes loosely classified as *negative ions*. Ionic crystals are formed of ionized atoms and in solution exhibit ionic conduction. In gases, ions are normally molecular and cases of double or treble ionization may be encountered. When almost completely ionized, gases form a fourth state of matter, known as a **plasma**. Since matter is electrically neutral, ions are normally produced in pairs. See **ionic radius**.

ion concentration That expressed in moles per unit volume for a particular ion. Also *ionic concentration*. See **pH value**.

ion concentration The number of ions of either sign, or of ion pairs, per unit volume. Also *ionization density*.

ion exchange Use of zeolites, artificial resins or specially treated coal to capture anions or cations from solution. Used in separating isotopes, water-softening, de-salting seawater and a wide range of industrial processes such as chemical extraction of uranium from its ores. Instead of solid particles through which liquid runs, immiscible liquid-liquid systems may be employed to transfer ions from one phase to another. Abbrev. *IX*.

ion-exchange capacity Electrical charge surplus to that uniting the framework of the ion-exchange resin or other exchange vehicle.

ion-exchange liquids Those immiscible with water (e.g. kerosine) which are rendered active by addition of suitable chemicals, such as tri-lauryl amine. These liquids can be used in solvent extraction processes in place of solid ion-exchange resins.

ion-exchange resins Term applied to a variety of materials, usually organic, which have the capacity of exchanging the ions in solutions passed through them. Different varieties of resin are used dependent on the nature (cationic or anionic) of the ions to be exchanged. Many of the resins in present-day use are based on polystyrene networks cross-linked with divinyl benzene.

ion flotation Removal of ions or gels from water by adding a surface active agent which forms complexes. These are floated as a scum by the use of air bubbles. See **froth flotation, Gibbs' adsorption theorem**.

ionic crystal Lattice held together by the electric forces between ions, as in a crystalline chemical compound.

ionic radius See panel on p. 122.

ionization The formation of ions by separating atoms, molecules or radicals, or by adding or subtracting electrons from atoms by strong electric fields in a gas, or by

ionic radius

Most minerals are solids with properties largely dependent upon the structural geometry and electrical nature of the ions. Each atom or ion consists of a nucleus surrounded by electrons, so although it is often represented in models by a small sphere, it is not strictly limited like a billiard ball.

The radii of common ions in minerals are shown in the figure and table, and illustrate the large differences between most cations and most anions. The size of any particular ion is related to a number of factors, especially the *atomic number*; the *co-ordination number*, the number of closest surrounding ions; and the *valency*, the combining power of one element with another. This is seen in the different sizes for divalent iron, Fe^{2+}, and trivalent iron, Fe^{3+}, below.

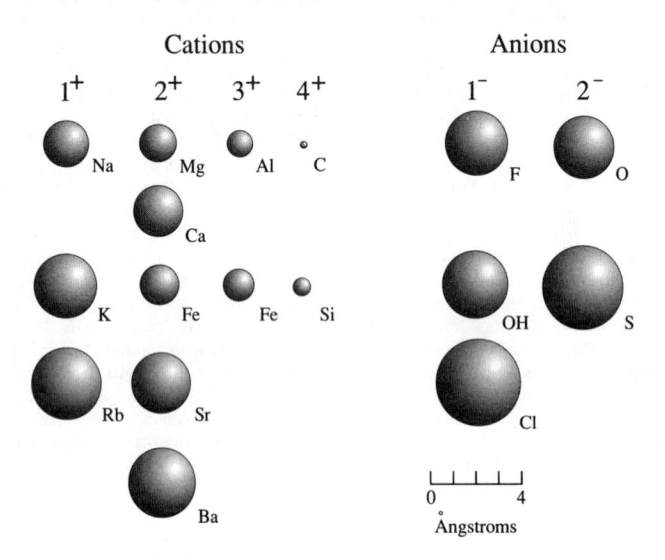

Cations: 1^+ Na, Rb, K; 2^+ Mg, Ca, Fe, Sr, Ba; 3^+ Al, Fe; 4^+ C, Si

Anions: 1^- F, OH, Cl; 2^- O, S

0 — 4 Ångstroms

Ionic radii of some elements present in minerals

Ion	Radius (Å)	Ion	Radius (Å)	Ion	Radius (Å)
C^{4+}	0.15	Na^{1+}	0.98	O^{2-}	1.32
Si^{4+}	0.39	Ca^{2+}	1.06	F^{1-}	1.33
Al^{3+}	0.57	Sr^{2+}	1.27	OH^{1-}	1.40
Fe^{3+}	0.67	K^{1+}	1.33	S^{2-}	1.74
Mg^{2+}	0.78	Ba^{2+}	1.43	Cl^{1-}	1.81
Fe^{2+}	0.82	Rb^{1+}	1.49		

weakening the electric attractions in a liquid, particularly water.

ionization cross-section Effective geometrical cross-section offered by an atom or molecule to an ionizing collision.

ionized (1) Electrolytically dissociated. (2) Converted into an ion by the loss or gain of an electron.

ionized atom Atom with a resultant charge arising from capture or loss of electrons; an ion in gas or liquid.

ionizing radiation See panels on pp.123–4.

Ipswichian A temperate stage of the late Pleistocene. See **Quaternary**.

IPT thermometers Thermometers conforming to the standards laid down by the *Institute of Petroleum Technologists*.

IR Abbrev. for *InfraRed* (spectroscopy).

iridium A brittle, steel-grey metallic element of the platinum family. Symbol Ir,

ionizing radiation: sources and doses

Everyone lives continuously with radiation, which is the dissemination of energy from a source, and the energy thus radiated. The term is applied to electromagnetic waves (radio waves, infrared, light, X-rays, γ-rays etc.) and to acoustic waves. It is also applied to emitted particles (α, β, protons, neutrons etc.).

Ionizing radiation, frequently confused in the public's mind with all forms of radiation, is radiation that produces ionization in matter. Examples are alpha particles, beta particles, gamma rays, X-rays and cosmic rays; examples of non-ionizing radiation are light, infra-red and radiofrequency radiation. *Ionization* is the process by which a neutral atom or molecule acquires or loses an electric charge to produce an electrically charged ion.

The *alpha particle* (α-particle) consists of two protons plus two neutrons emitted by a *radionuclide*, an unstable isotope of an element which undergoes natural radioactive decay; the *beta particle* (β-particle) is an electron emitted by the nucleus of a radionuclide; *gamma rays* are electromagnetic energy without mass or charge emitted by a radionuclide and *cosmic rays* are highly energized radiation from outer space. The time taken for a radionuclide to lose half its activity is its *half-life*, symbol $t_{1/2}$.

Of the ionizing radiation received by everyone continuously, on average, about 87% is *natural radiation*. The remaining 13% is *artificial radiation*. Twelve of these percentage points are from medical X-rays, nuclear medicine investigations and radiotherapy. The remaining 1% is composed of nuclear fallout (0.4%), exposure to radiation from work (0.2%), nuclear discharges (0.1%) and miscellaneous sources (0.4%). Fifty-nine per cent of the *natural ionizing radiation* received on average by the UK population is *radon gas* from the ground (Fig. 1). This is a heavy radioactive element formed

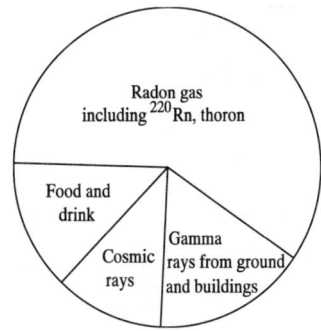

Fig 1. Sources of natural ionizing radiation

by the disintegration of radium and occurs esp. in areas of granitic rocks. Out of doors it disperses but may accumulate in buildings. Uranium, thorium and potassium-40 are the principal sources of natural radioactivity in rocks. Gamma rays, amounting to 16% of natural radiation, are emitted by radioactive materials in the earth and by some natural building materials taken from the earth. Cosmic rays amount to about 11%, and 14% comes from food and drink, where the source of radiation is largely an isotope of potassium (potassium-40 or ^{40}K).

continued on next page

at.no. 77, r.a.m. 192.2, rel.d. at 20°C 22.4, mp 2410°C, electrical resistivity 6×10^{-8} ohm metres. Alloyed with platinum or osmium to form hard, corrosion-resisting alloys, used for pen points, watch and compass bearings, crucibles, standards of length. The radioactive isotope ^{192}Ir is a medium-energy gamma-emitter used for industrial radiography. High concentrations of iridium in clay bands near the Cretaceous–Tertiary boundary worldwide have been ascribed to an impact by an iridium-rich extraterrestrial object like an asteroid.

iridosmine An ore of iridium and osmium, a natural alloy, with Os greater than 35%. Crystallizes in the hexagonal system.

iris A form of quartz showing chromatic reflections of light from fractures, often

ionizing radiation: units of measurement

The unit of measurement of ionizing radiation is the *becquerel*, one atomic disintegration per second (symbol Bq). This is exceedingly small so multiples of becquerels, e.g. *megabecquerels* (MBq), millions of becquerels, are often used. One gram of plutonium-239 has an activity of about 2000 MBq (two thousand million alpha particles emitted per second). A discontinued unit is the *curie*, symbol Ci, equal to 3.7×10^{10} Bq (ten thousand million becquerels).

Alpha particles have very limited powers of penetration and so radionuclides that emit them are not dangerous unless taken into the body. Beta particles may penetrate a cm or so of tissue and the radionuclides that emit them are hazardous. Gamma rays can pass through the body and these and the radionuclides which emit them are hazardous. Cosmic rays also pass through the body in large numbers.

The quantity of energy imparted by ionizing radiation to a unit mass of tissue (the *absorbed dose*) is the *gray*, symbol Gy. One gray is equal to one joule per kilogram. To put all types of source (alpha and beta particles, gamma rays etc) on the same basis of approximately equal biological effect, the *sievert* is used, symbol Sv, the *dose equivalent*. The average total annual dose to the UK population from all radioactive sources, is 2.5 millisieverts (mSv, thousandths of a sievert) per person but there are large variations. From gamma rays the average annual dose per person is 230 microsieverts (μSv, millionths of a sievert). The average dose per person from radon is 1200 microsieverts per year, with a wide range. The amount of gamma rays amd radon received in Britain is shown in Fig. 2 in which the variation is shown by the height of the surface of the map.

Old units for absorbed dose and dose equivalent are the *rad* and *rem*.

Fig. 2. Radon and gamma ray doses in Britain

iron

A metallic element in the eighth group of the periodic system. Symbol Fe, at.no. 26, r.a.m. 55.847, rel.d. at 20°C 7.86, mp 1525°C, bp 2800°C, electrical resistivity 9.8×10^{-8} ohm metres. As the basis metal in steel and cast iron, it is the most widely used of all metals.

It occurs in minerals principally in the ferrous (II), and ferric (III) forms. It is the fourth commonest element of the Earth's crust, with an abundance of 6.2%, and it is thought to make up 80% of the core of the Earth. Very small quantities of metallic iron occur under special conditions in the crust, but most iron is present in a large number of oxide and sulphide minerals, primary and secondary silicates and secondary ores and hydroxides.

The principal oxides and sulphides, all economically important as iron ores, are magnetite (Fe_3O_4), hematite (Fe_2O_3) and pyrite (FeS_2). *Magnetite* or *magnetic iron ore* occurs as an accessory mineral in many igneous rocks but also in large masses probably as a result of magmatic segregation, e.g. in Northern Sweden. *Hematite* occurs as a massive reddish earthy mineral including *kidney iron ore*, but also as black lustrous crystals called *specular iron*, or *micaceous hematite* when it is in fine flakes. Hematite occurs in pockets replacing limestone but also in large deposits as banded Precambrian sediments, e.g. in the Lake Superior region of the US.

Pyrite, or *iron pyrites* is of very common occurrence in sulphide ore bodies, e.g. the Rio Tinto deposits of Spain. *Pyrrhotine, pyrrhotite* or *magnetic pyrites* is another important iron sulphide that commonly contains nickel, as at Sudbury, Canada.

In the primary silicate minerals of igneous rocks, iron occurs esp. in the numerous olivines, pyroxenes, amphiboles and ferromagnesian micas, but very few are silicates of iron alone. The main iron silicates of sedimentary rocks are the hydrous silicates, chamosite, glauconite, and greenalite. The chlorites are important hydrated silicates occurring in many rock types.

Siderite is the ferrous carbonate which occurs esp. in sedimentary rocks that were formerly important, but low-grade, iron ores. With the weathering of almost any iron-bearing minerals, secondary hydroxides of iron may be formed including limonite (a mixture of minerals), goethite and lepidocrocite.

In its biological role, iron is essential for animals and plants, including haemoglobin, the red oxygen carrier in blood.

---

produced artificially by suddenly cooling a heated crystal. Also *rainbow quartz*.

iron See panel on this page.

iron alum See **halotrichite**.

iron bacteria Filamentous bacteria, which can convert iron oxide to iron hydroxide, deposited on their sheaths. Important in formation of **bog iron ore**.

iron-glance From the German *Eisenglanz*, a name sometimes applied to specular iron ore (haematite).

iron meteorites One of the two main categories of meteorites, the other being the *stony meteorites*. They are composed of iron and of iron-nickel alloy, with only a small proportion of silicate or sulphide minerals.

iron-olivine See **fayalite**.

iron ores Rocks or deposits containing iron-rich compounds in workable amounts; they may be primary or secondary; they may occur as irregular masses, as lodes or veins, or interbedded with sedimentary strata. See **iron**.

iron pan A hard layer often found in sands and gravels; caused by the precipitation of iron salts from percolating waters. It is formed a short distance below the soil surface. See **hard pan**.

iron pyrites See **pyrite**.

iron spinel See **spinel**.

ironstone An iron-rich sedimentary rock, found in nodules, layers, or beds.

irreversible reaction A reaction which takes place in one direction only, and therefore proceeds to completion.

island arc A chain of volcanic islands

formed at a convergent plate boundary. Deep ocean trenches occur on the convex side and deep basins on the concave side, e.g. Japan.

isobases Lines drawn through places where equal depression of the land mass took place in Glacial times, as a result of the weight of the ice load.

isochore A curve relating quantities measured under conditions in which the volume remains constant.

isoclinal fold A fold in which both limbs dip in the same direction. See **folding**.

isocline Line on a map, joining points where the angle of dip (or inclination) of the Earth's magnetic field is the same.

isodesmic structure Crystal structure with equal lattice bonding in all directions, and no distinct internal groups.

isodynamic lines Lines on a magnetic map which pass through points having equal strengths of the Earth's field.

iso-electric point Hydrogen ion concentration in solutions, at which dipolar ions are at a maximum. The point also coincides with minimum viscosity and conductivity. At this pH value, the charge on a colloid is zero and the ionization of an ampholyte is at a minimum. It has a definite value for each amino acid and protein.

isogonic line Line on a map joining points of equal magnetic declination, i.e. corresponding variations from true north.

isograd A line joining points where metamorphic rocks have attained the same facies, by being subjected to the same temperature and pressure.

isomagnetic lines The lines connecting places at which a property of the Earth's magnetic field is a constant.

isomerism (1) The existence of more than one substance having a given molecular composition and rel. mol. mass but differing in constitution or structure. (See **optical isomerism**.) The compounds themselves are called *isomers* or *isomerides* (Gk. 'composed of equal parts'). Isobutane and butane have the same formula, C_4H_{10}, but their atoms are placed differently; one type of alkane molecule, $C_{40}H_{82}$, has over 50^{12} possible isomers. Isomerism is frequently met with among organic compounds and complex inorganic salts. (2) The existence of nuclides which have the same atomic number and the same mass number, but are distinguishable by their energy states; that having the lowest energy is stable, the others having varying life-times. If the life-times are measurable the nuclides are said to be in *isomeric states* and undergo *isomeric transitions* to the ground state.

isomers See **isomerism**.

isometric system The cubic system.

isomorphism The name given to the phenomenon whereby two or more minerals, which are closely similar in their chemical constitution, crystallize in the same class of the same system of symmetry, and develop very similar forms. Adjs. *isomorphic, isomorphous*.

isomorphous replacement Replacing atoms at a given position in a crystal structure by others, usually those of a heavy metal. The determination of crystal structure of complex molecules in particular is made much more difficult by the investigator's inability to determine the phase relations of the diffraction pattern. Heavy metal replacement is an important method of overcoming the problem.

isopach A line drawn through points of equal thickness on a rock unit.

isoseismal line A line drawn on a map through places recording the same intensity of earthquake shocks. See **earthquakes**.

isostasy The process whereby areas of crust tend to float in conditions of near equilibrium on the plastic mantle.

isotope One of a set of chemically identical species of atom which have the same **atomic number** but different **mass numbers**. A few elements have only one natural isotope, but all elements have artificially produced radioisotopes. (Gk. *isos*, same; *topos*, place).

isotope geology The study of the relative abundances of radioactive and stable isotopes in rocks to determine radiometric ages and conditions of formation.

isotope separation Process of altering the relative abundance of isotopes in a mixture. The separation may be virtually complete as in a mass spectograph, or may give slight enrichment only as in each stage of a diffusion plant.

isotope structure Hyperfine structure of spectrum lines resulting from mixture of isotopes in source material. The wavelength difference is termed the *isotope shift*.

isotopic abundance The proportion of one isotope to the total amount of an element, as it occurs in nature.

isotopic age determination, dating The calculation of the age in years for geological (or archaeological) materials using the known radioactive decay rates from parent to daughter isotopes. See **uranium-lead**, **potassium-argon**, **rubidium-strontium** and **radiocarbon dating**.

isotopic dilution analysis A method of determining the amount of an element in a specimen by observing the change in isotopic composition produced by the addition

isotopic symbols of a known amount of radioactive allobar.

isotopic symbols Numerals attached to the symbol for a chemical element, with the following meanings: *upper left*, mass number of atom; *lower left*, nuclear charge of atom; *lower right*, number of atoms in molecule, e.g. $_{12}H_2$, $_{12}^{24}Mg$.

isotropic Said of a medium, the physical properties of which, e.g. magnetic susceptibility, elastic constants or refractive indices do not vary with direction.

itacolumite A micaceous sandstone with loosely-interlocking grains, which enable the rock to bend when cut into thin slabs.

Italian asbestos A name often given to tremolite asbestos to distinguish it from Canadian or chrysotile asbestos.

italite A rare coarsely granular plutonic rock composed of leucite and a little glass, a *leucitolite*.

IUGS Abbrev. for *International Union of Geological Sciences*. Under their aegis a classification of igneous rocks has been agreed internationally.

IUPAC Abbrev. for the *International Union of Pure and Applied Chemistry*, a body responsible, among other things, for the standardization of chemical nomenclature, which it alters frequently.

IX Abbrev. for **Ion eXchange**.

J K

J Symbol for joule, the unit of energy, work, quantity of heat.
J Symbol for (1) electric current density; (2) **magnetic polarization**.
jacinth See **hyacinth**.
jackbit Detachable cutting end fitted to shank of miner's rock drill, used to drill short blast holes. Also *rip-bit*.
jacket Erected around an offshore oil well and resting on the seabed, it supports the platform carrying the drilling derrick, operating equipment and living accommodation.
jackhammer A hand-held compressed-air hammer drill for rock drilling.
jack-up rig A prefabricated well-drilling assembly mounted on a barge towed to the drilling site in moderately shallow water (usually less than 100 m). Three or more legs are flooded and lowered to the seabed, and the barge and superstructure are jacked up out of the water and clear of wave action.
jacobsite An oxide of manganese and iron, often with considerable replacement of manganese by magnesium; crystallizes in the cubic system (usually in the form of distorted octahedra). A spinel.
jacupirangite A nepheline-bearing pyroxenite consisting of titanaugite, biotite, iron ores, and nepheline, the last being subordinate to the mafic minerals.
jade A general term loosely used to include various mineral substances of tough texture and green colour used for ornamental purposes. It properly embraces **nephrite** and **jadeite** but is sometimes misapplied to green varieties of minerals such as **amazonstone**, **bowenite**, **hydrogrossular**, **quartz** and **vesuvianite**.
jadeite A monoclinic member of the pyroxene group; sodium aluminium silicate. Usually white, grey or mauve, it occurs only in metamorphic rocks, and is the rare form of jade (*Chinese jade*).
jad, jud Deep groove cut into bed to detach block of natural stone, i.e. an undercut. A *jadder* is a stonecutter and his working tool is a *jadding pick*.
jamesonite See **feather ore**.
jargons, jargoons Names given in the gem trade to the zircons (chiefly colourless, smoky or of golden-yellow colour) from Ceylon. They resemble diamonds in lustre but are less valuable. See **hyacinth**.
jarosite A hydrous sulphate of iron and potassium crystallizing in the trigonal system; a secondary mineral in ferruginous ores.
jasper An impure opaque chalcedonic silica, commonly red owing to the presence of iron oxides.
jaw breaker Heavy-duty rock-breaking machine with fixed vertical, and inclined swing jaw, between which large lumps of ore are crushed. Also *Jaw crusher, Blake crusher, alligator*.
jet A hard coal-black variety of lignite, sometimes exhibiting the structure of coniferous wood; worked for jewellery in the last century.
jet drilling See **fusion drilling**.
jet mill A mill in which particles are pulverized to micrometre size by the collisions occurring among them when they are swept into a small jet of gas at sonic velocity.
jet pump Hydraulic elevator, in which a jet of high-pressure water rises in a pipe immersed in a sump containing the water, sands, or gravels which are to be entrained and pumped. Inefficient but cheap and effective where surplus hydraulic power exists.
jet shales Shales containing 'jet rock', found in the Upper Lias of the Whitby district of England.
jig Device for concentrating ore according to relative density of its constituent minerals.
johannsenite A silicate of calcium, manganese and iron. It is a member of the pyroxene group, crystallizing in the monoclinic system.
Johnson concentrator Machine used to arrest heavy auriferous material flowing in ore pulp. An inclined cylindrical shell rotates slowly, metallic particles being caught in rubber-grooved linings at periphery, lifted and separately discharged.
joint (1) An actual or potential fracture in a rock, in which there is no displacement. See **columnar structure, rift and grain**. (2) A length of drilling pipe or casing, usually 30 to 40 feet long.
jointing See **joint**.
Joosten process Use of chemical reaction between solutions of calcium chloride and sodium silicate to consolidate running soils or gravels when tunnelling. A water-resistant gel is formed.
joule SI unit of work, energy and heat. 1 joule is the work done when a force of 1 newton moves its point of application 1 metre in the direction of the forces. Abbrev. *J*. 1 erg = 10^{-7} J, 1 kWh = 3.6×10^6 J, 1 eV = 1.60×10^{-19} J, 1 calorie = 4.18 J, 1 Btu = 1055 J. See **SI units**.
jumper The borer, steel or bit for a compressed-air rock drill.
Jurassic The middle period of the Mesozoic era covering an approx. time span from 215–

145 million years ago. Named after the type area, the Jura Mountains. The corresponding system of rocks. See **Mesozoic**.

juvenile water Water derived from magma, as opposed to meteoric water (derived from rain or snow), or connate water (trapped in sediments at the time of deposition).

juxtaposition twins Two (or more) crystals united regularly in accordance with a 'twin law', on a plane (the 'composition plane') which is a possible crystal face of the mineral. Cf. **interpenetration twins**.

k Symbol for the velocity constant of a chemical reaction.

k Symbol for (1) **mass transfer coefficient**; (2) Boltzmann constant; (3) radius of gyration.

κ– Symbol for (1) cata-, i.e. containing a condensed double aromatic nucleus substituted in the 1,7 positions; (2) substitution on the tenth carbon atom; (3) electrolytic conductivity; (4) **compressibility**; (5) magnetic susceptibility.

K Symbol for (1) **potassium**; (2) **kelvin**.

K A symbol for (1) **equilibrium constant**; (2) K_s, solubility product; (3) bulk modulus.

[K] A very strong Fraunhofer line in the extreme violet of the solar spectrum. See **[H]**.

kaersutite A hydrated silicate of calcium, sodium, magnesium, iron, titanium and aluminium; a member of the amphibole group crystallizing in the monoclinic system. It occurs in somewhat alkaline igneous rocks.

kainite Hydrated sulphate of magnesium, with potassium chloride, which crystallizes in the monoclinic system. It usually occurs in salt deposits.

Kainozoic See **Cenozoic**.

kaiwekite A volcanic rock containing phenocrysts of olivine, titanaugite, barkevikite and anorthoclase, probably a hybrid between basalt and trachyte.

kaliophilite Silicate of potassium and aluminium, which crystallizes in the hexagonal system. It has a similar composition to kalsilite and is probably metastable at all temperatures at atmospheric pressure.

kalsilite A potassium-aluminium silicate, $KAlSiO_4$; it is related to nepheline and crystallizes in the hexagonal system.

kamacite A variety of nickeliferous iron, found in meteorites; it usually contains about 5.5% nickel. Metallurgically, alpha-iron.

kame A mound of gravel and sand which was formed by the deposition of the sediment from a stream as it ran from beneath a glacier. Kames are thus often found on the outwash plain of glaciers.

kandite A collective term for the kaolin minerals or members of the kaolinite group. These include kaolinite, dickite, nacrite, anauxite, halloysite, meta-halloysite and allophane.

kaolin Clay consisting of the mineral kaolinite. Also *china clay*.

kaolinite A finely crystalline form of hydrated aluminium silicate, $(OH)_4Al_2Si_2O_5$, occurring as minute monoclinic flaky crystals with a perfect basal cleavage; resulting chiefly from the alteration of feldspars under conditions of hydrothermal or pneumatolytic metamorphism, or by weathering. The kaolinite group of clay minerals includes the polymorphs **dickite** and **nacrite**. See **kandite**.

kaolinization The process by which the feldspars in a rock such as granite are converted to kaolinite.

Karbate TN for a form of extremely dense carbon having high enough strength to permit its being made into special shapes, e.g. tubes, pumps, and possessing corrosion resistance and good heat conductivity.

karst Any uneven limestone topography, characterized by joints enlarged into crisscross fissures (*grikes*) and pitted with depressions resulting from the collapse of roofs of underground caverns. It is formed by the action of percolating waters and underground streams.

Kasimovian An epoch of the Pennsylvanian Period.

kataklasis See **cataclasis**.

kataphorite See **katophorite**.

katathermometer Instrument used in mine ventilation survey to assess cooling effect of air current. Thermometer bulb is first exposed dry, then when covered with wetted gauze, and time taken for temperature to fall from 100° to 95°F (38° to 35°C) is observed.

katophorite, kataphorite, cataphorite Hydrated silicate of sodium, calcium, magnesium, iron and aluminium; a member of the amphibole group. It crystallizes in the monoclinic system and occurs in basic alkaline igneous rocks.

Kazanian A stratigraphical stage in the Permian rocks of Russia and eastern Europe.

keatite A high-pressure synthetic form of silica.

Keewatin Group A series of basic pillow lavas associated with sedimentary iron ores (worked in the 'Iron Ranges'); forms part of the Precambrian succession in the Canadian Shield.

keilhauite A variety of sphene containing more than 10% of the rare earths.

kelly The topmost **joint** of a drill string, attached below to the next drill joint and above to the swivel and mudhose connection. It is a

kelly bushing heavy tubular part of square or hexagonal cross-section externally which fits into the corresponding hole of the **kelly bushing**, itself fixed to the **rotary table**. To add a fresh pipe to the drill string, it has to be raised by the **draw works** until the kelly and bushing are clear of the rotary table. The string is then locked by wedge-shaped *slips* to the table and the kelly joint unscrewed and parked in the **rat hole**. After a new pipe has been attached the kelly is rescrewed and the whole string lowered so that drilling can be restarted. See **drilling rig**.

kelly bushing The replaceable bearing with a square or hexagonal hole in which the **kelly** slides and which is attached to the **rotary table** during drilling.

kelvin The SI unit of thermodynamic temperature. It is 1/273.16 of the temperature of the *triple point of water* above **absolute zero**. The temperature interval of 1 kelvin (K) equals that of 1°C (degree Celsius). See **Kelvin thermodynamic scale of temperature**.

Kelvin thermodynamic scale of temperature A scale of temperature based on the thermodynamic principle of the performance of a reversible heat engine. The scale cannot have negative values so *absolute zero* is a well defined thermodynamic temperature. The temperature of the *triple point of water* is assigned the value 273.16 K. The temperature interval corresponds to that of the Celsius scale so that the freezing point of water (0°C) is 273.15 K. Unit is the **kelvin**. Symbol K (not K°).

kelyphitic rim A shell of one mineral enclosing another in an igneous rock, produced by the reaction of the enclosed mineral with the other constituents of the rock. Kelyphitic rims are most common in basic and ultrabasic rocks.

kentallenite A coarse-grained, basic igneous rock, named for the type locality, Kentallen, Argyllshire; it consists essentially of olivine, augite and biotite, with subordinate quantities of plagioclase and orthoclase in approximately equal amounts.

kenyte A fine-grained igneous rock, occurring as lava flows on Mt. Kenya, E. Africa, and in the Antarctic; essentially an olivine-bearing phonolite with phenocrysts of anorthoclase, with or without augite and olivine, in a glassy groundmass.

keratophyre A fine-grained igneous rock of intermediate composition. It is essentially a soda-trachyte, containing albite-oligoclase or anorthoclase in a cryptocrystalline groundmass. The pyroxenes, when present, are often altered to chlorite or epidote.

kerf In coal winning, undercut made by coal-cutting machine to depth of 1 m or more. Also *kirve*.

kermesite Oxysulphide of antimony, which crystallizes in the monoclinic system. It is a secondary mineral occurring as the alteration product of stibnite. Also *pyrostibnite*.

kernite Hydrated sodium borate, $Na_2B_4O_7 \cdot 4H_2O$.

kerogen Fossilized, insoluble, organic material present in a sediment, that yields petroleum products on distillation.

Kerr effect Plane-polarized light at normal incidence on the polished surface of a magnetized material when reflected becomes elliptically polarized. Also *magneto-optical effect*.

kersantite A mica-lamprophyre, named from the type locality Kersanton, near Brest; it consists essentially of biotite and plagioclase feldspar. See **minette**.

ketones Compounds containing a carbonyl group, -CO-, in the molecule attached to two hydrocarbon radicals.

kettle A steep-sided basin in glacial drift, often a lake or swamp, and derived from the melting of a block of stagnant ice.

Keuper The upper series of rocks assigned to the Triassic System in N.W. Europe, lying above the Muschelkalk.

Keweenawan Conglomerates, arkoses, red sandstones and shales of desert origin, associated with great thicknesses of basic lavas and intrusives. This is the youngest of the Precambrian divisions in the Canadian Shield.

key fossil See **index fossil**.

kick Sudden increase in pressure during well drilling, which if not controlled quickly can cause a **blowout**. See **kill line**.

Kick's law Law assuming that the energy required for subdivision of a definite amount of material is the same for the same fractional reduction in average size of the individual particles, i.e. $E = k_k \log_e d_1/drt_2$, where E is the energy used in crushing, k_k is a constant, depending on the characteristics of the material and method of operation of the crusher, and d_1 and d_2 are the average linear dimensions before and after crushing.

kidney ore A form of the mineral haematite, oxide of iron, Fe_2O_3, which occurs in reniform masses, hence the name (L. *ren*, kidney).

kidney stone (1) A pebble or nodule of limestone, resembling a kidney and found in Jurassic rocks. (2) A misnomer for **nephrite**, which was once supposed to be efficacious in diseases of the kidney (Gk. *nephros*, kidney).

kieselguhr See **diatomite**.

kieserite Hydrated magnesium sulphate which crystallizes in the monoclinic system; found in large amounts in some salt deposits.

kieve See **dolly tub**.

kieving See **tossing**.

killas A name used in South West England for Palaeozoic slates or phyllites metamorphosed in contact with granite or mineral veins.

kill line Pipe lines connected through the **blowout preventer** stack to allow denser mud to be pumped into a borehole which has been shut because of the danger of a **blowout**.

kiln Furnace used for: drying ore; driving off carbon dioxide from limestone; roasting sulphide ores or concentrates to remove sulphur as dioxide; reducing iron (II) ores to magnetic state in reducing atmosphere.

kilohertz One thousand hertz or cycles per second. A multiple of the SI unit of frequency. Abbrev. *kHz*.

kimberlite A type of mica-peridotite, occurring in volcanic pipes in South Africa and elsewhere, composed of serpentinized olivine, orthopyroxene, phlogopite, carbonate and chromite, and containing xenoliths of many types of ultramafic rocks, with or without diamonds.

Kimmeridgian A stage in the Upper Jurassic. See **Mesozoic**.

kinematic viscosity The coefficient of viscosity of a fluid divided by its density. Symbol v. Thus $v = \eta/\rho$. Unit in the CGS system is the stokes (cm^2 s^{-1}); in SI m^2 s^{-1}.

klippe An erosional remnant of a thrust sheet. It is essentially an outlier lying on a thrust plane. Pl. *klippen*.

knebelite A silicate of manganese and ferrous iron, in the olivine series. It crystallizes in the orthorhombic system and occurs as the result of metamorphism and metasomatism.

knickpoint A change of slope in the longitudinal profile of a stream as a result of **rejuvenation**.

knockings See **riddle**.

knoll See **reef knolls**.

kobellite A lead, bismuth, antimony sulphide, crystallizing in the orthorhombic system.

komatiite An ultramafic lava with a very high magnesium content and well-developed *spinifex* texture with elongate lacy olivine crystals.

kornerupine A magnesium aluminium borosilicate, crystallizing in the orthorhombic system. Rare, but collected, and cut as a gemstone.

Kr Symbol for **krypton**.

kraton See **craton**.

kriging A statistical technique used in calculating grade and tonnage of ore reserves.

krypton A zero valent element, one of the noble gases. Symbol Kr, at.no. 36, r.a.m. 83.80, mp $-169°C$, bp $-151.7°C$, density 3.743 g/dm^3 at s.t.p. It is a colourless and odourless monatomic gas, and constitutes about 1-millionth by volume of the atmosphere, from which it is obtained by liquefaction. It is used in certain gas-filled electric lamps. It forms few compounds, e.g. KrF$_4$.

kulaite An amphibole-bearing nepheline basalt.

kunzite See **spodumene**.

kupfernickel See **niccolite**.

kX unit See **X-ray unit**.

kyanite, cyanite A silicate of aluminium which crystallizes in the triclinic system. It usually occurs as long-bladed crystals, blue in colour, in metamorphic rocks. See **disthene**.

kylite An olivine-rich theralite.

L

l Abbrev. for *litre*. (1 dm^3 is now used officially).

l (1) Symbol for (a) specific latent heat per gramme; (b) mean free path of molecules; (c) (with subscript) equivalent ionic conductance, 'mobility'. (2) Abbrev. for **laevorotatory**.

λ Symbol for (1) **wavelength**; (2) linear coefficient of thermal expansion; (3) mean free path; (4) radioactive decay constant; (5) thermal conductivity.

L Symbol for (1) molar latent heat; (2) angular momentum; (3) **luminance**; (4) self inductance.

Λ Symbol for **molar conductance**; $Λ_0$, at infinite dilution.

La Symbol for **lanthanum**.

labelled atom Atom which has been made radioactive, in a compound introduced into a flowline in order to trace progress.

labradorescence A brilliant play of colours shown by some labradorite feldspars. Probably due to a fine intergrowth of two phases.

labradorite A plagioclase feldspar containing 50% to 70% of the anorthite molecule; occurs in basic igneous rocks; characterized by a beautiful play of colours in some specimens, due to schiller structure.

laccolith A concordant intrusion with domed top and flat base, the magma having been instrumental in causing the up-arching of the 'roof'. Cf. **phacolith**. See **dykes and sills**.

lacustrine Related to a **lake**.

laevorotatory Said of an optically active substance which rotates the plane of polarization in an anti-clockwise direction when looking against the oncoming light.

lag To protect a shaft or level from falling rock by lining it with timber (*lagging*).

lagoon A shallow stretch of seawater close to the sea but partly or completely separated from it by a low strip of land. Often associated with coral reefs.

lahar A mudflow of volcanoclastic material on the slopes of a volcano.

lake A body of water lying on the surface of a continent, and unconnected (except indirectly by rivers) with the ocean. Lakes may be *freshwater lakes*, provided with an outlet to the sea; or *salt lakes*, occurring in the lowest parts of basins of inland drainage, with no connection with the sea. Lakes act as natural settling tanks, in which sediment carried down by rivers is deposited, containing the shells of molluscs etc. The lakes of former geological periods may thus be recognized by the nature of the sediments deposited in them and the fossils they contain. Lakes occur plentifully in glaciated areas, occupying hollows scooped out by the ice, and depressions lying behind barriers of morainic material.

Lambert's law The illumination of a surface on which the light falls normally from a point source is inversely proportional to the square of the distance of the surface from the source.

lamellibranch See **bivalves**.

lamina See **lamination**.

lamination Stratification on a fine scale, each thin stratum, or *lamina*, frequently being a millimetre or less in thickness. Typically exhibited by shales and fine-grained sandstones.

lamping Use of ultraviolet light to detect fluorescent minerals either when prospecting or in checking concentrates during ore treatment.

lamp man Colliery surface worker in charge of miners' lamps. Works in lamp room or cabin, and controls repairs, recharge, issue, lighting etc. of portable lamps.

lamproite A variety of lamprophyric extrusive rock rich in potassium and magnesium.

lamprophyres Igneous rocks usually occurring as dykes intimately related to larger intrusive bodies; characterized by abnormally high contents of coloured silicates, such as biotite, hornblende and augite, and a correspondingly small amount of feldspar, some being feldspar-free. See **alnöite, camptonite, kersantite, minette, monchiquite, spessartite, vogesite**.

lanarkite A rare monoclinic sulphate of lead, occurring with anglesite and leadhillite (into which it easily alters), as at Leadhills, Lanarkshire, Scotland.

landing Stage in hoisting shaft at which cages are loaded or discharged.

landscape marble A type of limestone containing markings resembling miniature trees etc.; when polished, the surface has the appearance of a sepia drawing.

landslip The sudden sliding of masses of rock, soil, or other superficial deposits from higher to lower levels, on steep slopes. Landslips on a very large scale occur in mountainous districts as a consequence of earthquake shocks, stripping the valley sides bare of all loose material. In other regions landslips occur particularly where permeable rocks, lying on impermeable shales or clays, dip seawards or towards deep valleys. The clays hold up water, becoming lubricated thereby, and the overlying strata, fractured by joints, tend to slip downhill, a movement that is facilitated on the coast by marine erosion.

Langhian A stage in the Miocene. See **Tertiary.**

langite A very rare ore of copper occurring in Cornwall, blue to greenish-blue in colour; essentially hydrated copper sulphate, crystallizing in the orthorhombic system.

Langmuir trough Apparatus in which a rectangular tank is used to measure surface tension of a liquid.

lansfordite Hydrated magnesium carbonate, crystallizing in the monoclinic system. It occurs in some coalmines but is not stable on exposure to the atmosphere and becomes dehydrated.

lanthanide contraction The peculiar characteristic of the lanthanides that the ionic radius decreases as the atomic number increases, because of the increasing pull of the nuclear charge on the unchanging number of electrons in the two outer shells. Thus the elements after lanthanum, e.g. platinum, are very dense and have chemical properties very similar to their higher homologues, e.g. palladium.

lanthanide series Name for the rare earth elements at.nos. 57–71, after lanthanum, the first of the series. Cf. **actinides.** In both series an incomplete f-shell is filling.

lanthanum A metallic element in the third group of the periodic system, belonging to the rare earths group. Symbol La, at.no. 57, r.a.m. 138.91, mp 921°C.

lapilli Small rounded pieces of lava whirled from a volcanic vent during explosive eruptions; lapilli are thus similar to volcanic bombs but smaller in size, with a mean diameter 2–64 mm. Sing. *lapillus.*

lapilli tuff A compact pyroclastic rock composed of lapilli.

lapis lazuli Original sapphire of ancients, a beautiful blue stone used extensively for ornamental purposes. It consists of the deep blue feldspathoid **lazurite**, usually together with calcite and spangles of pyrite.

La Pointe picker Small belt conveyor on which ore is so displayed that radioactive pieces are removed as they pass a Geiger-Müller counter.

Laramide orogeny In the narrow sense, that mountain building movement associated with the production of the Rocky Mountains between late Cretaceous and Palaeocene times. More broadly it is sometimes used to embrace all orogenies that took place during that span of time.

lardalite See **laurdalite.**

larnite A rare calcium silicate, Ca_2SiO_4, formed at very high temperatures in contact metamorphosed limestone.

larvikite See **laurvikite.**

laser *Light Amplification by Stimulated Emission of Radiation.* A source of intense monochromatic light in the ultraviolet, visible or infrared region of the spectrum. It operates by producing a large population of atoms with their electrons in a certain high energy level. By *stimulated emission*, transitions to a lower level are induced, the emitted photons travelling in the same direction as the stimulating photons. If the beam of inducing light is produced by reflection from mirrors or *Brewster's windows* at the ends of a resonant cavity, the emitted radiation from all stimulated atoms is in phase, and the output is a very narrow beam of coherent monochromatic light. Solids, liquids and gases have been used as lasing materials.

laser enrichment The enrichment of uranium isotopes using a powerful laser to ionize atoms of the selected isotope of molecules bearing the uranium isotope. The separation can then be done by chemical or physical means.

lashing A South African term for removing broken rock after blasting. Canadian *mucking, mucking out.*

latent heat More correctly, *specific latent heat.* The heat which is required to change the state of unit mass of a substance from solid to liquid, or from liquid to gas, without change of temperature. Most substances have a latent heat of fusion and a latent heat of vaporization. The specific latent heat is the difference in *enthalpies* of the substance in its two states.

latent magnetization The property possessed by certain feebly magnetic metals (e.g. manganese and chromium) of forming strongly magnetic alloys or compounds.

latent neutrons In reactor theory, the delayed neutrons due from (but not yet emitted by) fission products.

lateral moraine A low ridge formed at the side of a valley glacier. It is composed of material derived from glacial plucking and abrasion, or material that has fallen on to the ice.

lateral shift The displacement of outcrops in a horizontal sense, as a consequence of faulting. Cf. **throw.**

laterite A residual clay formed under tropical climatic conditions by the weathering of igneous rocks, usually of basic composition. Consists chiefly of hydroxides of iron and aluminium. See **bauxite.**

laterization The process whereby rocks are converted into laterite. Essentially, the process involves the abstraction of silica from the silicates. See **laterite**.

lath A term commonly applied to a lath-like crystal.

latite A volcanic rock containing approximately equal amounts of alkali feldspar and sodic plagioclase, i.e. the approx. equivalent of monzonite. See **volcanic rocks**.

lattice A regular space arrangement of points as for the sites of atoms in a crystal.

lattice dynamics The study of the excitations a crystal lattice can experience and their consequences for the thermal, optical and electrical properties of solids.

lattice energy Energy required to separate the ions of a crystal from each other to an infinite distance.

lattice structure One of three types of crystal structure: (1) ionic, with symmetrically arranged ions and good conducting power; (2) molecular, covalent, usually volatile and non-conducting; (3) layer, with large ions each associated with two small ones, forming laminae weakly held by non-polar forces.

lattice vibration Vibration of atoms or molecules in a crystal due to thermal energy.

Laue pattern The pattern of spots produced on photographic film when a heterogeneous X-ray beam is passed through a thin crystal, which acts like an optical grating. Used in the analysis of crystal structure.

laumontite A zeolite consisting essentially of hydrated silicate of calcium and aluminium, crystallizing in the monoclinic system; occurs in cavities in igneous rocks and in veins in schists and slates.

launder Inclined trough for conveying water or crushed ore and water (pulp).

Laurasia The palaeogeographic supercontinent of the northern hemisphere corresponding to Gondwanaland in the southern hemisphere.

laurdalite, lardalite A coarse-grained soda-syenite from S. Norway; it resembles laurvikite but contains rhomb-shaped alkali feldspar crystals and large crystals of nepheline.

Laurentian Granites Precambrian granitic intrusives in the Canadian Shield.

laurvikite, larvikite A soda syenite from S. Norway, very popular as an ornamental stone when cut and polished; widely used for facing buildings, the distinctive feature being a fine blue colour, produced by schiller structure in the anorthoclase feldspars. Synonym *blue granite*.

lautarite Monoclinic iodate of calcium, occurring rarely in caliche in Chile.

lava The molten rock material that issues from a volcanic vent or fissure and consolidates on the surface of the ground (*subaerial lava*), or on the floor of the sea (*submarine lava*). Chemically, lava varies widely in composition; it may be in the condition of glass, or a holocrystalline rock. See **volcano**. Also *basalt, obsidian, pillow structure, pumice*.

lava flows See **extrusive rocks**.

law of Dulong and Petit The atomic heat capacities of solid elements are constant and approximately equal to 26 (when the specific heat capacity is in joules). Certain elements of low atomic mass and high melting point have, however, much lower atomic heat capacities at ordinary temperatures.

law of mass action The velocity of a homogeneous chemical reaction is proportional to the concentrations of the reacting substances.

law of multiple proportions When two elements combine to form more than one compound, the amounts of one of them which combine with a fixed amount of the other exhibit a simple multiple relation. Also *Dalton's law*.

law of rational indices A fundamental law of crystallography which states, in the simplest terms, that in any natural crystal the indices may be expressed as small whole numbers.

laws of reflection (1) When a ray of light is reflected at a surface, the reflected ray is found to lie in the plane containing the incident ray and the normal to the surface at the point of incidence. (2) The angle of reflection equals the angle of incidence.

law of superposition See **superposition, law of**.

lawsonite An orthorhombic hydrated silicate of aluminium and calcium. It occurs in low grade regionally metamorphosed schists, particularly in glaucophane schists.

lay barge Vessel for the storage, welding and laying of pipelines underwater.

layered igneous rocks Igneous rocks which display layers of differing mineral and chemical composition, e.g. Bushveld Complex in South Africa; the Skaergaard Complex of East Greenland.

layering The high-temperature sedimentation feature of igneous rocks.

layer lattice Concentration of bonded atoms in parallel planes, with weaker and non-polar bonding between successive planes. This gives marked cleavage, e.g. in graphite, mica.

layer line Line joining a series of spots on X-ray rotating crystal diffraction photograph. Its position enables the crystal lattice spacing parallel to the axis of rotation to be determined.

lazulite A deep sky-blue, strongly pleochroic mineral, crystallizing in the monoclinic system. In composition essentially a hydrated phosphate of aluminium, magnesium and iron, with a little calcium. Found in aluminous high grade metamorphic rocks and in granite pegmatites.

lazurite An ultramarine blue mineral occurring in cubic crystals or shapeless masses; it consists of silicate of sodium and aluminium, with some calcium and sulphur, and is considered to be a sulphide bearing variety of haüyne. A constituent of **lapis lazuli**.

leached zone See **gossan, secondary enrichment**.

leaching The extraction of a soluble metallic compound from an ore by dissolving in a solvent e.g. cyanide, sulphuric acid. The metal is subsequently precipitated or adsorbed from the solution.

lead A metallic element in the fourth group of the periodic system. Symbol Pb, at.no. 82, r.a.m. 207.19, valency 2 or 4, mp 327.4°C, bp 1750°C, rel.d. at 20°C 11.35. Specific electrical resistivity 20.6×10⁻⁸ ohm metres. Occurs chiefly as **galena**. The naturally occurring stable element consists of four isotopes: Pb 204 (1.5%), Pb 206 (23.6%), Pb 207 (22.6%) and Pb 208 (52.3%). It is used as shielding in X-ray and nuclear work because of its relative cheapness, high density and nuclear properties. Other principal uses: in storage batteries, ammunition, foil and as a constituent of bearing metals, solder and type metal. Abundance in the Earth's crust 13 ppm. Toxic.

lead age The age of a mineral or rock calculated from the ratios of its radiogenic and non-radiogenic lead isotopes.

leader A thin mineralized vein parallel to or otherwise related to the main ore-carrying vein, so aiding its discovery.

lead glance See **galena**.

leadhillite Hydrated carbonate and sulphate of lead, so called from its occurrence with other ores of lead at Leadhills (Scotland).

lean ore Ore of marginal value; low grade ore.

leat Ditch following contour line used to conduct water to working place.

Le Chatelier-Braun principle If any change of conditions is imposed on a system at equilibrium, then the system will alter in such a way as to counteract the imposed change. This principle is of extremely wide application.

lechatelierite A name sometimes applied to naturally fused vitreous silica, such as that which occurs as fulgurites (*lightning tubes*).

Lemberg's stain test A black iron sulphide stain, used to distinguish between calcite and dolomite in limestones.

length-fast, length-slow In optically birefringent minerals, terms used to denote whether the fast or the slow vibration direction is aligned parallel or nearly parallel to the length of prismatic crystals.

lens A portion of a homogeneous transparent medium bounded by spherical surfaces. Each of these surfaces may be convex, concave, or plane. If in passing through the lens a beam of light becomes more convergent or less divergent, the lens is said to be *convergent* or *convex*. If the opposite happens, the lens is said to be *divergent* or *concave*.

lens formula The equation giving the relation between the image and object distances, l' and l and the focal length f of a lens is:

$$\frac{1}{f} = \frac{1}{l'} - \frac{1}{l}$$

in the cartesian *convention of signs* for a thin lens in air.

lenticle A mass of lens-like (lenticular) form. The term may refer to masses of clay in sand, or vice versa, and, in metamorphic rocks, to enclosures of one rock type in another.

lenticular Said of a mineral or rock of this particular shape, embedded in a matrix of a different kind.

lepidocrocite An orthorhombic hydrated oxide of iron (FeO.OH) occurring as scaly brownish-red crystals in iron ores. It is often one of the constituents of limonite, together with its dimorph goethite.

lepidolite A lithium-bearing mica crystallizing in the monoclinic system as pink to purple scaly aggregates. It is essentially a hydrated silicate of potassium, lithium and aluminium, often also containing fluorine, and is the most common lithium-bearing mineral. It occurs almost entirely in pegmatites. Also *lithia mica*.

lepidomelane A variety of biotite, rich in iron, which occurs commonly in igneous rocks.

leptite An even-grained metamorphic rock composed mainly of quartz and feldspars. Approximately synonymous with *granulite* and *leptynite*.

leptynite See **leptite**.

leucite A silicate of potassium and aluminium, related in chemical composition to orthoclase, but containing less silica. At ordinary temperatures, it is tetragonal (pseudocubic), but on heating to about 625°C it becomes cubic. Occurs in igneous rocks, particularly potassium-rich, silica-poor lavas of Tertiary and Recent age, as e.g. at Vesuvius.

leucitite A volcanic rock composed of leucite and clinopyroxene.

leucitophyre A fine-grained igneous rock, commonly occurring as a lava, carrying phenocrysts of leucite and other minerals in a matrix essentially trachytic; a well-known example comes from Rieden in the Eifel.

leucocratic A term used to denote a light colour in igneous rocks, due to a high content of felsic minerals, and a correspondingly small amount of dark, heavy silicates.

leucosapphire See **white sapphire**.

leucoxene An opaque whitish mineral formed as a decomposition product of ilmenite. Normally consists of finely crystalline rutile, but may be composed of finely divided brookite.

levee A long low ridge built up on either side of a stream on its flood plain. It consists of relatively coarse sand and silt deposited by the stream when it overflows its banks.

level An approximately horizontal tunnel in a mine, generally marking a working horizon or level of exploitation, and either in or parallel to the ore body.

levigation Use of sedimentation or elutriation to separate finely-ground particles into fractions according to their movement through a separating fluid. Also used in connection with wet grinding.

levyne, levynite A zeolite; hydrated silicate of calcium and aluminium crystallizing in the trigonal system.

Lewisian complex The oldest rocks in Britain, a division of the Scottish Precambrian. Much of the complex is of Archaean origin, subsequently affected by later Precambrian events.

lherzolite An ultramafic plutonic rock, a peridotite, consisting essentially of olivine, with both ortho- and clino-pyroxene; named from Lake Lherz in the Pyrenees.

Li Symbol for **lithium**.

Lias The oldest epoch of the Jurassic period.

Lias Clay A thick bed of clay found in the Lower Jurassic rocks of Britain. See **Mesozoic**.

liberation First stage in ore treatment, in which comminution is used to detach valuable minerals from gangue.

libethenite An orthorhombic hydrated phosphate of copper, occurring rarely as olive-green crystals in the oxide zone of metalliferous lodes.

libollite A pitch-like member of the asphaltite group.

lifter (1) Projecting rib or wave in lining plates of ball or rod mill. (2) Perforated plate in drum washer or heavy media machine which aids tumbling action or removes floating fraction of ore being treated.

light *Electromagnetic radiation* capable of inducing visual sensation, with wavelengths between about 400 and 800 nm. See **illumination, speed of light**.

lightning tubes See **fulgurites**.

light oils A term for oils with a boiling range of about 100°–210°C, obtained from the distillation of coal tar.

light quanta When light interacts with matter, the energy appears to be concentrated in discrete packets called *photons*. The energy of each photon $E=h\nu$, where ν is the frequency and h is Planck's constant. See **quantum, X-rays, gamma radiation**.

light red silver ore See **proustite**.

lignite A brownish, black coal intermediate between peat and subbituminous coal. It is commoner than coal in Mesozoic and Tertiary deposits. Commonly *brown coal*.

limb The side of a fold.

limburgite A basic, fine-grained igneous rock occurring in lava flows, similar to the dyke rock monchiquite, but having interstitial glass between the dominant olivine and augite crystals. Typically, limburgite is feldspar-free.

lime A substance produced by heating limestone to 825°C or more, as a result of which the carbonic acid and moisture are driven off. Lime, which is much used in the building, chemical, metallurgical, agricultural and other industries, may be classified as *high calcium, magnesian*, or *dolomitic* depending on the composition. Unslaked lime is commonly known as *caustic lime*; also *anhydrous lime, burnt lime, quicklime*.

lime-silicate rocks Rocks resulting from the contact (high-temperature) metamorphism of limestones containing silica in detrital grains, nodules of flint or chert, or siliceous skeletons, the silica combining with the lime to form such silicates as lime-garnet, anorthite, wollastonite.

lime slurry Thick aqueous suspension of finely-ground slaked lime, used to control alkalinity of ore pulps in flotation and cyanide process.

limestone A sedimentary rock consisting of more than 50% by weight of calcium carbonate. Mineralogically, limestone consists of either aragonite or calcite, although the former is abundant only in Tertiary and Recent limestones. Dolomite may also be present, in which case the rock is called *dolostone*.

limestones may be *organic*, formed from the calcareous skeletal remains of living organisms, *chemically precipitated*, or *detrital*, formed of fragments from pre-existing limestones. The majority are organic, and can be classified according to their texture and the nature of the organisms whose skeletons are incorporated, e.g. oölitic limestone, shelly limestone, algal limestone, crinoidal limestone.

limnology The study of lakes.

limonite Although originally thought to be a definite hydrated oxide of iron, now known to consist mainly of cryptocrystalline goethite or lepidocrocite along with adsorbed water; some haematite may also be present. It is a common alteration product of most iron-bearing minerals and also the chief constituent of bog iron ore.

linarite Hydrated sulphate of lead and copper, found in the oxide zone of metalliferous lodes; a deep-blue mineral resembling azurite, also crystallizing in the monoclinic system.

lineament A long feature on the Earth's surface, often more clearly visible by satellite photography. It may be structural or volcanic and can be related to **plate tectonics**.

lineation Any one-dimensional structure in a rock.

line defect See **defect**.

line profile A graph showing the fine structure of a spectral line, the intensity of the line, measured with a microphotometer, being plotted as a function of wavelength.

line spectrum A spectrum consisting of relatively sharp lines, as distinct from a **continuous spectrum**. Line spectra originate in the atoms of incandescent gases or vapours. See **spectrum**.

liparite Synonym for **rhyolite**.

liquefaction The change in packing of the grains of a water-filled sediment, turning it into a fluid mass which can then flow.

liquid crystals Certain pure liquids which are turbid and, like crystals, anisotropic over a definite range of temperature above their freezing points.

liquid-liquid extraction Process, both batch and continuous, whereby two non-mixing liquids are brought together to transfer soluble substance from one to the other for useful recovery of these soluble substances.

lithia *Lithium oxide*, Li_2O.

lithia emerald A misnomer for **hiddenite**.

lithia mica See **lepidolite**.

lithic arenite A sediment of sand-sized particles many of which are rock fragments rather than mineral grains.

lithic tuff A volcanic tuff in which rock fragments are more abundant than crystals or vitric fragments.

lithification Processes which convert an unconsolidated sediment into a **sedimentary rock**.

lithiophilite Orthorhombic phosphate of lithium and manganese, forming with triphylite a continuously variable series.

lithium An element. Symbol Li, r.a.m. 6.939, at.no. 3, mp 186°C, bp 1360°C, rel.d. 0.585. It is the least dense solid, chemically resembling sodium but less active. Abundance in the Earth's crust 20 ppm. There are numerous minerals mainly occurring in pegmatites associated with granitic rocks. *Spodumene* ($LiAlSi_2O_6$), lithium-bearing micas, *lepidolite* and *zinnwaldite*, *amblygonite* ($LiAlPO_4(OH,F)$) and *petalite* ($LiAlSi_4O_{10}$) are the commonest lithium minerals. It is used in alloys and in the production of tritium; also as a basis for lubricant grease with high resistance to moisture and extremes of temperature; and as an ingredient of high-energy fuels. Its salts are used medically as an anti-depressant.

lithographic stone A compact, porous fine-grained limestone, often dolomitic, formerly employed in lithography. Pale creamy-yellow in colour, but occasionally grey. Fair samples may be obtained from the Jurassic rocks in Britain, but the finest material comes from Solenhofen and Pappenheim in Bavaria. See **Solenhofen stone**.

lithography The systematic description of rocks, more esp. sedimentary rocks. See **petrology**.

lithology The character of a rock expressed in terms of its mineral composition, its structure, the grain size and arrangement of its component parts, i.e. all those visible characters that in the aggregate impart individuality to the rock. The term is most commonly applied to sedimentary rocks in hand specimen and outcrop.

lithophile An element that is concentrated in the silicate minerals rather than in sulphide or metal phases, and occurs in stony meteorites or the Earth's crust. Also *oxyphile*.

lithosphere The outer, rigid shell of the Earth lying over the **asthenosphere**. It includes the *crust, continents* and *plates*. See **continental crust, oceanic crust**.

lithostratigraphy Stratigraphy based on the observable petrographical characters of the rock successions.

lit-par-lit injection The injection of fluid or molten material, mostly granitic, along bedding, cleavage or schistosity planes in a rock. The product is an alternation of apparently igneous and non-igneous material, known as a *migmatite*. Some of the rocks previously considered to have been formed by the process are now thought to be the products of partial melting, the igneous layers representing the fraction of the rock which became liquid.

Little Ice Age Variously applied to cool periods in otherwise warm stages, including periods in the 16th and 18th centuries.

littoral The area between high and low tide marks. Cf. **neritic zone**.

live crude oil Oil arriving at the well head and containing gas. Cf. **dead crude oil**.

liver opal A form of opaline silica, resembling liver in colour.

lizardite A mineral of the **serpentine** group.

Llandeilo An epoch of the Ordovician period. See **Palaeozoic**.

Llandovery The oldest series of the Silurian Period. See **Palaeozoic**.

Llanvirn An epoch of the Ordovician period.

load cast A *sole mark* appearing as an irregular bulge on the underside of a sedimentary rock. It lacks evidence of current direction. See **flute cast**.

loader A mechanical shovel or other device for loading trucks underground.

loading Adsorption by resins of dissolved ionized substances in ion-exchange process. In uranium technology, the loading factor is the mass of U_3O_8 adsorbed by unit volume of the resin.

loading capacity In ion exchange, saturation limit of resin.

loadstone See **lodestone**.

loam A rich friable soil containing more or less equal amounts of sand, silt, clay and usually organic matter.

locked test In preliminary tests on unknown ores, retention of a potentially troublesome fraction of the test product for addition to a new batch of ore, to ascertain whether in continuous work such a build-up would be upsetting. Also *cyclic test*.

lode (1) A mineral deposit composed of a zone of veins. (2) Steeply inclined fissure of non-alluvial mineral enclosed by walls of country rock of different origin.

lodestone, loadstone Iron oxide. A form of magnetite exhibiting polarity, behaving, when freely suspended, as a magnet.

loess Homogeneous unstratified, usually calcareous, blanket deposit consisting mainly of quartz silt with a particle size 0.01–0.05 mm. It originates as wind blown dust and vast accumulations cover large areas of China, eastern Europe and North and South America.

log washer Trough or tank set at a slope, in which ore is tumbled with water by means of one or more box girders with projecting arms set helically, so that as they rotate coarse lumps are cleaned and delivered upslope while mud and sand overflow downslope.

löllingite, loellingite Arsenide of iron, $FeAs_2$, occurring as steel-grey crystals, prismatic in habit, belonging to the orthorhombic system.

longitudinal dune A long, narrow sand dune parallel to the direction of the prevailing wind.

longshore current A current which flows parallel to the shoreline.

longshore drift The movement of material along the shore by a **longshore current**.

long tom Portable sluice in which rough concentrates made during treatment of alluvial sands or gravels are sometimes worked up to a better grade.

longwall coal-cutting machine A machine which severs coal in mechanized mining.

longwall working Method of mining bedded deposits, notably coal, in which the whole seam is removed, leaving no pillars. If advancing, work goes outwards from the shaft and roads must be maintained through excavated areas. If retreating, ground behind workers can be allowed to cave if surface rights are not thus affected.

loparite A titanate of sodium, calcium and the rare earth elements; a variety of **perovskite**.

loss of vend Difference in weight between raw (run-of-mine) coal and saleable product leaving washery.

loughlinite A variety of sepiolite (see **meerschaum**) containing sodium.

love arrows See **flèches d'amour**.

löweit A hydrated sulphate of sodium and magnesium, crystallizing in the trigonal system, occurring in salt deposits.

LPG *Liquefied Petroleum Gases* such as propane and butane, used for fuels.

Lu Symbol for **lutetium**.

Ludhamian An early stage of the Pleistocene. See **Quaternary**.

Ludlow The youngest series of the Silurian Period. See **Palaeozoic**.

lumen Unit of *luminous flux*, being the amount of light emitted in a unit solid angle by a small source of one **candela**. In other words, the lumen is the amount of light which falls on unit area when the surface area is at unit distance from a source of one candela. Abbrev. *lm*.

luminance Measure of brightness of a surface, e.g. candela per square metre of the surface radiating normally. Symbol *L*.

luminescence Emission of light (other than from thermal energy causes) such as *bioluminescence* etc. Thermal luminescence in excess of that arising from temperature occurs in certain minerals. See **fluorescence, phosphorescence**.

lustre This depends upon the quality and amount of light that is reflected from the surface of a mineral. The highest degree of lustre in opaque minerals is *splendent*, the comparable term for transparent minerals being *adamantine* (i.e. the lustre of diamond). *Metallic* and *vitreous* indicate less brilliant lustre, while *silky, pearly, resinous*, and *dull* are self-explanatory terms covering other degrees of lustre.

Lutetian A stage in the Eocene. See **Tertiary**.

lutetium A metallic element, the densest member of the rare earth group. Symbol Lu, at.no. 71, r.a.m. 174.97. It occurs in monazite and xenotime.

lutite A consolidated rock composed of silt and/or clay, e.g. **shale, mudstone**.

lux Unit of illuminance or illumination in SI system, 1 lm/m^2. Abbrev. *lx*.

Lw Symbol for lawrencium.

Lydian stone, lydite *Touchstone*. Black flinty jasper, also other silicified fine-grained rocks. The name touchstone has reference to the use of lydite as a streak plate for gold: the colour left after rubbing the metal across it indicates to the experienced eye the amount of alloy.

lyosorption The adsorption of a liquid on a solid surface, esp. of solvent on suspended particles.

lyotropic Concentration determines the phase of lyotropic materials.

M

m Symbol for (1) electromagnetic moment; (2) **mass**; (3) mass of electron; (4) **molality**.

m- Abbrev. for (1) *meta-*, i.e. containing a benzene nucleus substituted in the 1,3 positions; (2) *meso-* (1).

μ Symbol for (1) chemical potential; (2) dipole moment; (3) index of refraction; (4) magnetic permeability; (5) micron (obsolete; replaced by μm); (6) the prefix micro-.

$\mu-$ Symbol signifying: (1) meso- (2) ; (2) meso- (3); (3) a bridging ligand.

M General symbol for a metal or an electropositive radical. Symbol for (1) magnetization per unit volume; (2) mafic and related minerals.

M Abbrev. for *mega-*, i.e. 10^6. Symbol for (1) **relative molecular mass**; (2) luminous emittance; (3) moment of force; (4) mutual inductance.

Ma Symbol for a million years. Sometimes *my*.

maar The explosion vent of a volcano.

Maastrichtian The highest stage in the Cretaceous. See **Mesozoic**.

maceral The organic material which comprises coal. It includes **exinite**, **inertinite** and **vitrinite**.

macle The French term for **twinned crystals**; in the diamond industry, more commonly used than *twin*, esp. for twinned octahedra.

MacLeod's equation The surface tension of a liquid is given as the difference between the densities of the liquid and its vapour raised to the fourth power, multiplied by a constant.

macro-axis The long axis in orthorhombic and triclinic crystals.

macrophyric A textural term for medium- to fine-grained igneous rocks containing phenocrysts 2 mm long. Cf. **microphyric**.

Madagascar aquamarine A strongly dichroic variety of the blue beryl obtained, as gemstone material, from the Malagasy Republic.

Madagascar topaz See **citrine**.

Madeira topaz A form of *Spanish topaz*. See **citrine**.

mafic A mnemonic term for the ferromagnesian and other non-felsic minerals actually present in an igneous rock.

maghemite An iron oxide with the crystal structure of magnetite but the composition of hematite. A cation-deficient spinel, produced by the oxidation of magnetite.

magma Molten rock, including dissolved water and other gases. It is formed by melting at depth and rises either to the surface, as lava, or to whatever level it can reach before crystallizing again, in which case it forms an igneous intrusion.

magmatic cycle See **igneous cycle**.

magnesia alba Commercial basic magnesium carbonate.

magnesia alum See **pickeringite**.

magnesian spinel See **spinel**.

magnesite Carbonate of magnesium, crystallizing in the trigonal system. Magnesite is a basic refractory used in open-hearth and other high-temperature furnaces; it is resistant to attack by basic slag. It is obtained from natural deposits and is calcined at high temperature to drive off moisture and carbon dioxide, before being used as a refractory.

magnesium See panel on p. 141.

magnesium carbonate $MgCO_3$. See **magnesite**.

magnesium orthodisilicate Occurs in nature as **serpentine**.

magneson 4-(4-nitrophenolazo)resorcinol. Used as a reagent for detection and determination of magnesium, with which it forms a characteristic blue colour in alkaline solution.

magnet A mass of iron or other material which possesses the property of attracting or repelling other masses of iron, and which also exerts a force on a current-carrying conductor placed in its vicinity.

magnetic anomaly The value of the local magnetic field remaining after the subtraction of the dipole portion of the Earth's field. The local deviation measured in magnetic prospecting.

magnetic anomaly patterns See panel on p. 142.

magnetic axis A line through the effective centres of the poles of a magnet.

magnetic dip See **dip**.

magnetic epoch A long period of geological time in which the Earth's magnetic field was essentially of one polarity.

magnetic events Geologically short periods during **magnetic epochs** when the magnetic field had a reversed polarity.

magnetic field strength A measure of the strength of a magnetic field. Unit ampere per metre, symbol Am^{-1}; the CGS unit is the *oersted*. 1 $Am^{-1} = 4 \times 10^{-3}$ oersted. Also *magnetic field intensity*, *magnetizing force*.

magnetic iron ore See **magnetite**.

magnetic map A map showing the distribution of the Earth's magnetic field.

magnetic moment Vector such that its product with the magnetic induction gives the torque on a magnet in a homogeneous magnetic field. Also *moment of a magnet*.

magnesium

A light metallic element in the second group of the periodic system. Symbol Mg, at.no. 12, r.a.m. 24.312, mp 651°C, rel.d. 1.74, electrical resistivity 42×10^{-8} ohm metres, bp 1120°C at 1 atm, specific latent heat of fusion 377 kJ/kg. Found in nature only as compounds. The metal is a brilliant white in colour, and magnesium ribbon burns in air, giving an intense white light. The sixth most abundant element of the Earth's crust, which contains 2.76% Mg but the figure is much higher in the Earth as a whole. There is 0.13% in sea water. Magnesium forms a large number of minerals including the silicates *olivine*, of which *forsterite* is the pure magnesian end-member, *pyroxene* (enstatite, hypersthene, diopside), *amphiboles* (tremolite), *mica* (phlogopite), *garnet* (pyrope), as well as the hydrated minerals, *serpentine* and *talc*. The olivines occur particularly in ultrabasic and basic igneous rocks, the pyroxenes in basaltic and gabbroic types and amphiboles and mica in many igneous and metamorphic rocks. Serpentine minerals occur mainly in altered ultrabasic rocks; talc is typically a mineral of crystalline schists and garnet, occurring mainly in metamorphic rocks. In evaporite deposits there are several magnesian salts such as sulphates, chlorides and borates.

Dolomite, calcium-magnesium carbonate, occurs as a gangue in mineral veins and extensively in bedded sedimentary deposits. It is used as a refractory for furnaces, manufacture of magnesium compounds and magnesium metal and other industrial processes. *Magnesite*, magnesium carbonate, is an earthy mineral that occurs in veins and as an alteration of dolomite.

Magnesium plays an essential but rather specialized role biologically. It takes part in many biochemical reactions and is a constituent of chlorophyll, essential for photosynthesis in all green plants. *Guano*, the excreta of sea birds, contains magnesium phosphate minerals, as does bone.

Magnetic North The direction in which the north pole of a pivoted magnet will point. It differs from the Geographical North by an angle called the *magnetic declination*.

magnetic oxide of iron See **magnetite**.

magnetic polarization The production of optical activity by placing an inactive substance in a magnetic field.

magnetic prospecting Form of **geophysical prospecting** which measures either the distortion of the Earth's magnetic field by local accumulation of ferromagnetic materials, e.g. magnetite, pyrrhotite, chromite or ilmenite, or the magnetic susceptibility of rocks. Methods include the use of a simple dip needle, instrumental magnetometer in field studies or of airborne magnetometers (*aeromagnetic survey*).

magnetic pyrites See **pyrrhotite**.

magnetic reversal A change of the Earth's magnetic field, to the opposite polarity. See **magnetic anomaly patterns**.

magnetic separator A device for separating, by means of an electromagnet, any magnetic particles in a mixture from the remainder of the mixture, e.g. for separating iron filings from brass filings.

Magnetic South The direction in which the south pole of a pivoted magnet will point. It differs from the Geographical South by an angle called the *magnetic declination*.

magnetic stability A term used to denote the power of permanent magnets to retain their magnetism in spite of the influence of external magnetic fields, vibration etc.

magnetic units Units for electric and magnetic measurements in which μ_0, the permeability of free space, is taken as unity. Now replaced by SI units in which μ_0 is given the value $4\pi \times 10^{-7}$ Hm^{-1}.

magnetism Science covering magnetic fields and their effect on materials, due to unbalanced spin of electrons in atoms. See **Coulomb's law, paramagnetism, diamagnetism, ferromagnetism.**

magnetite, magnetic iron ore An oxide of iron, crystallizing in the cubic system. It has the power of being attracted by a magnet, but it has no power to attract particles of iron to itself, except in the form of lodestone.

magnetization Orientation from randomness of saturated *domains* in a ferromagnetic material. The magnetization per unit volume **M** is related to the field strength **H** and the magnetic induction **B** by

magnetic anomaly patterns

When new basaltic rock wells up along mid-oceanic ridges it spreads out symmetrically on either side away from the ridge. See **plate tectonics**. As this new rock cools it assumes the direction of magnetism of the prevailing Earth's magnetic field, displaying *remanent magnetism*. However, the polarity of the Earth's field reverses from time to time and this is recorded in successive stripes of rock, which then show *normal* and *reversed magnetization* as in the simplified diagram below. The ages of these stripes and their distances away from the ridge allow the rate of sea-floor spreading to be calculated.

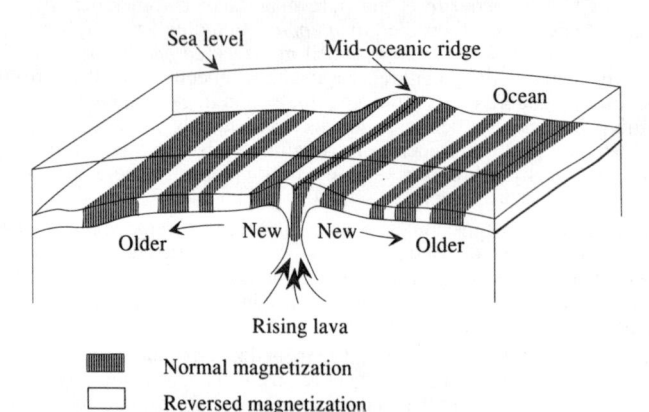

Normal and reversed magnetization produced at a mid-oceanic ridge

$$B = \mu_0 (H + M) \text{ tesla},$$

where μ_0 is the permeability of free space. The SI unit is ampere per metre.

magnetochemistry The study of the relation of magnetic properties to chemical structure. Particularly, the extent of paramagnetism in transition metal compounds may be related to the type of ligand bonded to the metal.

magneto-optic rotation See **magnetic polarization**.

magnification The ratio of the size of the image to that of the object in an optical system where the object is plane and lies perpendicular to the axis of the system. Also *transverse magnification*. See **magnifying power**.

magnifying power The ratio of the apparent size of the image of an object formed by an optical instrument to that of the object seen by the naked eye. For a microscope, it is necessary to assume that the object would be examined by the naked eye at the least distance of distinct vision, 25 cm. Unless otherwise stated, the *linear* magnification is assumed to be indicated.

main airway In colliery, the airway directly connected with point of entry to mine.

main and tail Single-track underground rope haulage by means of the main rope to draw out the full wagons and the tail rope to draw back the empties.

main line Pipeline connecting producing areas to a refinery. Also *trunk line*.

main rope See **main and tail**.

malachite Basic copper carbonate $CuCO_3Cu(OH)_2$, crystallizing in the monoclinic system. It is a common ore of copper, and occurs typically in green botryoidal masses in the oxidation zone of copper deposits.

malignite An alkaline igneous rock, a *mesocratic* variety of nepheline syenite.

Malm The youngest epoch of the Jurassic period. See **Mesozoic**.

Malus' law For a plane-polarized beam of light incident on a polarizer, $I = I_0 \cos^2 \theta$, where I = intensity of transmitted beam; I_0 = intensity of incident beam; θ = angle between the plane of vibrations of the beam and the plane of vibrations which are transmitted by the polarizer.

mammals See **vertebrates**.

mammoth Extinct Pleistocene ancestor of the elephant, with long tusks curling upwards and hairy skin which helped their survival along the borders of continental glaciers. They had a much wider distribution than present-day elephants which are restricted to Africa and Asia.

manganepidote See **piedmontite**.

manganese See panel on p. 144.

manganese epidote See **piedmontite**.

manganese garnet See **spessartine**.

manganese nodules Irregular small concretions with high concentrations of Mn and Fe, Cu, Ni minerals. They are most common on the deep ocean floors, notably the North Pacific where there is little sedimentation. See **todorokite**, **birnessite**.

manganese spar See **rhodochrosite**.

manganite A grey or black hydrated oxide of manganese, crystallizing in the monoclinic system (*pseudo-orthorhombic*). It is a minor ore of manganese, and also occurs in deep-sea **manganese nodules**.

manganophyllite A phlogopite or biotite mica containing appreciable manganese.

manganosite An oxide of manganese (MnO) which crystallizes in the cubic system.

mangerite An intermediate member of the charnockite rock series, the equivalent of hypersthene monzonite.

man-riding Said of equipment on oil rigs which is used by personnel and is built to a higher standard of safety than material handling systems.

mantle That part of the Earth between **crust** and **core**, ranging from depths of approx. 40 km to 2900 km.

mantled gneiss dome A dome-like structure consisting of granite at the centre, surrounded by gneiss, and intruded into low-grade regionally metamorphosed rocks. Such structures are found in Precambrian shield areas.

marble A term strictly applying to a granular crystalline metamorphosed limestone, but in a loose sense it includes any calcareous or other rock of similar hardness that can be polished for decorative purposes.

marcasite (1) A disulphide of iron which crystallizes in the orthorhombic system. It resembles iron pyrites, but has a rather lower density, is less stable, and is paler in colour when in a fresh condition. (2) In the gemstone trade, *marcasite* is either pyrites, polished steel (widely used in ornamental jewellery in the form of small 'brilliants'), or even white metal.

mare A lunar plain filled with mafic volcanic rocks.

marekanite A rhyolitic perlite broken down into more or less rounded pebbles, named from the type locality, Marekana river, eastern Siberia.

margarite An aggregate of minute sphere-like crystallites, arranged like beads, found as a texture in glassy igneous rocks.

margarite Hydrated silicate of calcium and aluminium, crystallizing in the monoclinic system, often as lamellae with a pearly lustre; one of the so-called **brittle micas**.

marginal ore Ore which, at current market value of products from its excavation and processing, just repays the cost of its treatment.

marialite Silicate of aluminium and sodium with sodium chloride, crystallizing in the tetragonal system. It is one of the end-members in the isomorphous series of the scapolite group.

marine erosion The processes, both physical and chemical, which are responsible for the wearing-away and destruction of coastlines.

marker horizon A layer of rock in a sedimentary sequence which, because of its distinctive appearance or fossil content, is easily recognized over large areas, facilitating correlation of strata.

markfieldite A variety of diorite with a porphyritic structure and a granophyric groundmass.

marl A general term for a very fine-grained rock, either clay or loam, with a variable admixture of calcium carbonate.

marmorization The recrystallization of a limestone by heat to give a marble.

martite A variety of **hematite** (Fe_2O_3) occurring in dodecahedral or octahedral crystals believed to be pseudomorphous after magnetite, and in part perhaps after pyrite.

maskelynite A glass which occurs in colourless isotropic grains in meteorites and has the composition of a plagioclase. It probably represents re-fused feldspar.

mass The quantity of matter in a body. Two principle properties are concerned with mass: (1) the *inertial mass*, i.e. the mass as a measure of resistance of a body to changes in its motion; (2) the *gravitational mass*, i.e. the mass as a measure of the attraction of one body to another. The general theory of relativity shows inertial and gravitational mass are equivalent. By a suitable choice of units they can be made numerically equal for a given body, which has been confirmed to a high degree of precision by experiment. Mass is always conserved. Cf. **mass-energy equation**.

manganese

A hard, brittle metallic element, in the seventh group of the periodic system. It is brilliant white in colour, with reddish tinge. Symbol Mn, at.no. 25, r.a.m. 54.94, valency 2,3,4,6,7, rel.d. at 20°C 7.39, mp 1220°C, bp 1900°C, electrical resistivity 5.0×10^{-8} ohm metres, hardness in Mohs' scale 6. Manganese is mainly used in steel manufacture, as a deoxidizing and desulphurizing agent.

It is the twelfth commonest element of the Earth's crust with an abundance of 0.11%. It exists in several oxidation states, of which Mn (II), Mn (III) and Mn (IV) are importamt geochemically. The ionic radius of Mn^{2+} is close to that of Fe^{2+} which it replaces in silicates, and it also substitutes for Mg in ferromagnesian silicates as a minor element. Mn occurs in the silicates *garnet* (spessartine, almandine, pyrope), *epidote* and *tourmaline*. Manganese has an affinity for sulphur, resulting in sulphides that are found in mineral veins, but there is very little Mn in the iron sulphides, *pyrite* and *pyrrhotite*. There are numerous independent sedimentary oxide, hydroxide and carbonate minerals, most in the higher oxidation state, and a number of other manganese minerals, some very rare: many are found in hydrothermal deposits.

In the magmatic rocks the manganous ion, Mn^{2+} predominates. The manganous minerals typically have a pink colour, e.g. *rhodochrosite* ($MnCO_3$) and *rhodonite* ($MnSiO_3$), that give ornamental stones. The oxidate ores of manganese are dark brown or black and concentrate or scavenge many uncommon elements. (As, Ba, Cu, Mo, V, W etc).

The principal oxides-hydroxides (which are also the main ores) are *pyrolusite*, MnO_2; *hausmannite*, Mn_3O_4; *braunite*, Mn_2O_3 with silica present; and *manganite*, $Mn_2O_3.H_2O$. *Wad*, or *bog manganese*, is an earthy dark brown or black impure mixture of manganese oxides and hydroxides with other elements present (e.g. Cu, Co, Ni). *Psilomelane* is dark grey hydrated oxide of manganese often with Ba or K oxides. Manganese is an essential trace element for humans. See **manganese nodules**.

mass-energy equation $E = mc^2$. Confirmed deduction from Einstein's special theory of relativity that all energy has mass. If a body gains energy E, its inertia is increased by the amount of mass $m = E/c^2$, where c is the speed of light. Derived from the assumption that all conservation laws must hold equally in all frames of reference and using the principle of conservation of *momentum*, of *energy* and of *mass*.

massicot Yellow lead oxide. A rare mineral of secondary origin, associated with galena.

mass number The total number of photons and neutrons in an atomic nucleus, each being taken as a unit of mass.

mass spectrograph A vacuum system in which positive rays of various charged atoms are deflected through electric and magnetic fields so as to indicate, in order, the charge-to-mass ratios on a photographic plate, thus measuring the atomic masses of isotopes with precision. System used for the first separation for analysis of the isotopes of uranium.

mass spectrometer A *mass spectograph* in which the charged particles are detected electrically instead of photographically.

mass spectrum See **spectrum**.

mass transfer The transport of molecules by convection or diffusion, as in the operations of extraction, distillation and absorption.

mass transfer coefficient The molecular flux per unit driving force. Symbol k, K.

mass wasting Downhill gravity movement of rock material, e.g. *landslides*, *rockfall*.

matte Fusion product consisting of mixed sulphides produced in the smelting of sulphide ores. In the smelting of copper, for example, a slag containing the gangue oxides and a matte consisting of copper and iron sulphides are produced. The copper is subsequently obtained by blowing air through the matte, to oxidize the iron and sulphur.

matter The substances of which the physical universe is composed. Matter is characterized by gravitational properties (on Earth by weight) and by its indestructibility under normal conditions.

Matura diamonds Colourless (fired) zircons from Ceylon, which on account of their brilliancy are useful as gemstones. A misleading name.

maturity (1) A stage of the geomorphological cycle characterized by maximum relief and well-developed drainage. (2) The ultimate stage in the development of a sediment, characterized by stable minerals and rounded grains.

McCabe-Thiele diagram A graphical method, based on vapour-liquid equilibrium properties, for establishing the theoretical number of separation stages in a continuous distillation process.

M-discontinuity See **Mohorovičić discontinuity**.

meander Sharp sinuous curves in a stream particularly in the mature part of its course. The meanders are accentuated by continuing erosion on the convex side and deposition on the concave side of the stream course.

measured ore Proved quantity and assay grade of ore deposit as ascertained by competent measurement of exposures and an adequate sampling campaign.

mechanical deposits Those of sediments which owe their accumulation to mechanical or physical processes.

mechanics The study of forces on bodies and of the motions they produce.

meerschaum A hydrated silicate of magnesium. It is clay-like, and is shown microscopically to be a mixture of a fibrous mineral called parasepiolite and an amorphous mineral β-sepiolite. It is used for making pipes, and was formerly used in Morocco as a soap. Also *sepiolite*.

megaripple A sandwave with a wavelength of < 1 m.

meimechite An ultramafic volcanic rock containing phenocrysts of olivine.

meionite Silicate of aluminium and calcium, together with calcium carbonate, which crystallizes in the tetragonal system. It is an end-member of the isomorphous series forming the scapolite group.

melaconite Cupric oxide crystallizing in the monoclinic system. It is a black earthy material found as an oxidation product in copper veins, and represents a massive variety of **tenorite**.

melange A heterogeneous mixture of rock materials mappable as a rock unit. Mostly of tectonic origin but some are caused by large-scale slumping of sediments.

melanite A dark-brown or black variety of andradite garnet containing appreciable titanium.

melanocratic A term applied to rocks which are abnormally rich in dark and heavy ferro-magnesian minerals (to the extent of 60% or more). See **leucocratic**.

melanterite Hydrated ferrous sulphate which crystallizes in the monoclinic system. It usually results from the decomposition of iron pyrites or marcasite. Also *copperas, green vitriol.*

melilite Calcium magnesium aluminium silicate, crystallizing in the tetragonal system, and occurring in alkaline igneous and contact metamorphosed rocks and slags.

melilitite An ultramafic volcanic rock composed essentially of melilite and pyroxene.

melilitolite An ultramafic plutonic rock composed essentially of melilite, pyroxene and olivine.

melteigite A melanocratic intrusive rock with 10%–30 % nepheline, part of the *ijolite* series.

melting point The temperature at which a solid begins to liquefy. Pure metals, eutectics, and some intermediate constituents melt at a constant temperature. Alloys generally melt over a range. Abbrev. *mp.* Also *fusing point.*

membrane filter Thin layer filter made by fusing cellulose ester fibres or by β-bombardment of thin plastic sheets, so that they are perforated by tiny uniform channels. Also *molecular filter.*

Mendeleev's table See **periodic system**.

menilite An alternative and more attractive name for **liver opal**; it is a grey or brown variety of that mineral.

Mercalli scale A scale of intensity used to measure earthquake shocks. The various observable movements are graded from 1 (very weak) to 12 (total destruction). See **earthquakes**.

Mercator's projection A derivative of the simple *cylindrical projection.* The meridian scale is adjusted to coincide with the latitudinal scale at any given point, allowing true representation of shape but not of area, regions towards the poles being exaggerated in size. Long used by navigators, for straight lines drawn on it represent a constant bearing. A transverse form of Mercator is sometimes used in large-scale maps of limited E.–W. extent such as many of the UK Continental Shelf.

mercury A white metallic element which is liquid at atmospheric temperature. Chemical symbol Hg, at.no. 80, r.a.m. 200.59, rel.d. at 20°C 13.596, mp −38.9°C, bp 356.7°C, electrical resistivity 95.8×10^{-8} ohm metres. A solvent for most metals, the products being called *amalgams.* Its chief uses are in the manufacture of batteries, drugs, chemicals, fulminate and vermilion. Used as metal in mercury-vapour lamps, arc rectifiers, power-control switches, and in many scientific and electrical instruments. Metallic mercury occurs rarely but its principal ore is *cinnabar*, Hg S, which occurs in brilliant red acicular crystals. Organic forms of mercury are toxic. Also *quicksilver*.

Merioneth The youngest epoch of the Cambrian period.

mesa A flat-topped, steep-sided tableland. Small mesas are called *buttes*.

meso- (1) Optically inactive by intramolecular compensation. Abbrev. *m-*. (2) Substituted on a carbon atom situated between two hetero-atoms in a ring. (3) Substituted on a carbon atom forming part of an intramolecular bridge.

mesocolloid A particle whose dimensions are 25 to 250 nm containing 100–1000 molecules.

mesolite A zeolite intermediate in composition between **natrolite** and **scolecite**. Crystallizes in the monoclinic system, and occurs in amygdaloidal basalts and similar rocks.

Mesolithic The middle division of the Stone Age. Cf. **Palaeolithic, Neolithic**.

Mesozoic See panel on p. 147.

Messinian A stage in the Miocene. See **Tertiary**.

metallic bond In metals, the valence electrons are not even approximately localized in discrete covalent bonds, but are delocalized and interact with an indefinite number of atomic nuclei. This gives rise both to the opacity and lustre of metals and to their electrical conductivity.

metallic lustre A degree of lustre exhibited by certain opaque minerals, comparable with that of polished steel.

metalloid An element having both metallic and non-metallic properties, e.g. arsenic.

metallurgical balance sheet Report in equation form of the products from treatment of a known tonnage of ore of specified assay value, yielding a known weight of concentrate and tailing, to which the head value can be attributed for economic and technical control.

metallurgical coke Coke of high strength to resist pressure and breakage, and of high purity. Used for smelting mineral ores.

metamict Mineral which has been exposed to natural radioactivity so that its crystalline structure breaks down to a glassy amorphous state, e.g. zircon.

metamorphic facies All those rocks that have reached chemical equilibrium under the same pressure-temperature range of metamorphism.

metamorphism Change in the mineralogical and structural characteristics of a rock as a consequence of heat and/or pressure.

metasomatism Change in the bulk chemical composition of a rock by the introduction of liquid or gaseous material from elsewhere.

meteor A 'shooting star'. A small body which enters the Earth's atmosphere from interplanetary space and becomes incandescent by friction, flashing across the sky and generally ceasing to be visible before it falls to Earth. See **bolide**.

meteor crater A circular unnatural crater of which Meteor Crater in Arizona is best known; believed to be caused by the impact of meteorites. Also *meteorite crater*.

meteoric water Ground water of recent atmospheric origin.

meteorite crater See **meteor crater**.

meteorites Mineral aggregates of cosmic origin which reach the Earth from interplanetary space. Cf. **meteor, bolide**. See **achondrite, aerolites, chondrite, iron meteorites, pallasite, siderite, stony meteorites**.

methane CH_4. The simplest *alkane*, a gas, mp $-186°C$, bp $-164°C$; occurs naturally in oil wells and as marsh gas. *Fire damp* is a mixture of methane and air; coal gas contains a large proportion of methane.

methanol CH_3OH. A colourless liquid, bp 66°C, rel.d. 0.8. It may be produced by the destructive distillation of wood; it is nowadays synthesized from CO and H_2 in the presence of catalysts. It is an important intermediate for numerous chemicals, and is used as a solvent and for denaturing ethanol. Also *methyl alcohol, wood alcohol*.

methyl alcohol See **methanol**.

metre The Système International (SI) fundamental unit of length. The metre is defined (1983) in terms of the velocity of light. The metre is the length of path travelled by light in vacuum during a time interval of 1/299 792 458 of a **second**. Originally intended to represent 10^{-7} of the distance on the Earth's surface between the North Pole and the Equator, formerly it has been defined in terms of a line on a platinum bar and later (1960) in terms of the wavelength from ^{86}Kr. See **SI units**.

metre-kilogram(me)-second-ampere See **MKSA**.

-metry A suffix denoting a method of analysis or measurement, e.g. acidimetry, iodimetry, nephelometry.

Mexican onyx A translucent, veined and parti-coloured aragonite found in Mexico and in the south-western US.

Mg Symbol for **magnesium**.

Mesozoic

The Mesozoic ('middle life') was characterized by ammonites and reptiles, together with brachiopods, lamellibranchs, gastropods and corals. The first period, the Triassic, had however an impoverished fauna and flora that followed the extinctions at the end of the Palaeozoic era. During the second period, the Jurassic, there was a rich flora in the warm climate, when reptiles, notably dinosaurs, were dominant on land. In the Cretaceous, flowering plants spread and many large reptiles, ammonites, most belemnites and many brachiopod species became extinct and **Chalk** was the most important formation.

Era	Period	Series	Stage		Age, Ma
Mesozoic	Cretaceous	Upper		Maastrichtian	74
			Senonian	Campanian	83
				Santonian	87
				Coniacian	89
			Turonian		90
			Cenomanian		97
		Lower	Albian		112
			Aptian		125
			Barremian		132
			Hauterivian		135
			Valanginian		141
			Ryazanian		146
	Jurassic	Upper (Malm)	Portlandian		152
			Kimmeridgian		155
			Oxfordian		157
		Middle (Dogger)	Callovian		161
			Bathonian		166
			Bajocian		174
			Aalenian		178
		Lower (Lias)	Toarcian		187
			Pliensbachian		194
			Sinemurian		204
			Hettangian		208
	Triassic	Upper	Rhaetian		210
					235
		Middle			241
		Lower			245

mg Abbrev. for *milligram*.

miarolitic structure A structure found in an igneous rock, consisting of irregularly shaped cavities into which the constituent minerals may project as perfectly terminated crystals.

miaskite A leucocratic biotite nepheline monzosyenite. See **plutonic rocks**.

mica A group of silicates which crystallize in the monoclinic system; they have similar chemical compositions and highly perfect basal cleavage. Mica is one of the best electrical insulators.

micaceous iron ore A variety of specular hematite (Fe_2O_3) which is foliated or which simulates mica in the flakiness of its habit.

micaceous sandstone A sandstone containing conspicuous flakes of mica.

mica-schist Schist composed essentially of micas and quartz, the foliation being mainly due to the parallel disposition of the mica flakes. See **schist**.

micelle A colloidal sized aggregate of molecules, such as those formed by surface active agents.

micrite The microcrystalline matrix of fine-grained limestones seen to have a semi-opaque appearance when viewed microscopically, due to its small grain size. Also *micritic limestone*.

micro Applied to names of rocks, it indicates the medium-grained form, e.g. *microdiorite*, *microsyenite*, *microtonalite*.

micro-analysis A special technique of both qualitative and quantitative analysis, by means of which very small amounts of substances may be analysed.

microbalance Sensitive to one microgram, for use in micro-analysis; may be either beam or quartz fibre.

microbiological mining (1) The use of natural or *genetically engineered* strains of bacteria to enhance or induce acid leaching of metals from ores (*bacterial leaching*), either in situ within an ore deposit, or to promote leaching of metals from mine waste. (2) The use of bacteria to recover useful or toxic metals from natural drainage waters, mine drainage or waste water from tips (*bacterial recovery*). Also *biomining*.

microchemistry Preparation and analysis of very small samples, usually less than ten milligrams but, in radioactive work, down to a few atoms.

microcline A silicate of potassium and aluminium which crystallizes in the triclinic system. A feldspar, it resembles orthoclase, but is distinguished by its optical and other physical characters. See **potassium feldspar**.

microcrystalline texture A term applied to a rock or groundmass in which the individual crystals can be seen only under the microscope.

microfelsitic texture A term applied to the cryptocrystalline texture seen, under the microscope, in the groundmass of quartz-felsites and similar rocks; due to the devitrification of an originally glassy matrix.

microfossils Fossils or fossil fragments too small to be studied without using a microscope.

microgranite A medium-grained, microcrystalline, acid igneous rock having the same mineral composition and texture as a granite.

micrographic texture A distinctive rock texture in which the simultaneous crystallization of quartz and feldspar has led to the former occurring as apparently isolated fragments, resembling runic hieroglyphs, set in a continuous matrix of feldspar.

microlite A general term for minute crystals of tabular or prismatic habit found in microcrystalline rocks. These give a reaction with polarized light.

microlux A unit for very weak illuminations, equal to one-millionth of a lux.

micrometre One-millionth of a metre. Symbol μm. Formerly *micron*.

micropalaeontology The study of **microfossils**.

microperthite A feldspar which consists of intergrowths of potassium feldspar and albite in a microscopic scale.

microphyric A textural term descriptive of medium- to fine-grained igneous rocks containing phenocrysts < 2 mm long. Cf. **macrophyric**.

microscope An instrument used for obtaining magnified images of small objects. The *simple microscope* is a convex lens of short focal length, used to form a virtual image of an object placed just inside its principal focus. The *compound microscope* consists of two short-focus convex lenses, the objective and the eyepiece mounted at opposite ends of a tube. For most microscopes, the magnifying power is roughly equal to $450/f_o f_e$, where f_o and f_e are the focal lengths of objective and eyepiece in centimetres. The *petrological* or *polarizing microscope* is widely used for the examination of *thin sections* of rocks and minerals. A polarizer (formerly a *Nicol prism*) polarizes the incident light below the stage, which may be rotated. A second polarizer, the *analyser*, can then be used to examine the light from the section to determine *extinction angles*, *interference fringes* etc.

microspherulitic texture A texture in

Micum test

which spherulites on a microscopic scale are distributed through the groundmass of an igneous rock.

Micum test A standard laboratory test for determining the mechanical strength of coke.

middlings In ore dressing, an intermediate product left after the removal of clean concentrates and rejected tailings. It consists typically of interlocked particles of desired mineral species and gangue, or of by-product minerals not responding to the treatment used. See **riddle**.

mid-ocean ridge A major, largely submarine, mountain range where two *plates* are being pulled apart and new volcanic **lithosphere** is being created. See **magnetic anomaly patterns, plate tectonics**.

migmatite A rock with both igneous and metamorphic characteristics; generally consisting of a host metamorphic rock injected by granitic material.

Milankovitch theory of climatic change An astronomical theory proposed by the Yugoslav mathematician M. Milankovitch, that the changes in solar radiation received by the Earth as a result of irregularities in its movements accounted for the glacial and interglacial stages of the Pleistocene. The variations were (1) the variations in eccentricity of the Earth's orbit (periods of 10^5 and 4×10^5 years); (2) variations in the obliquity of tilt of the Earth's axis (period 4×10^4 years); (3) the precession of the equinoxes (period 2×10^4 years). There is evidence for the two shorter cycles in data obtained from analysis of deep-sea bottom cores.

milarite A hydrated silicate of aluminium, beryllium, calcium and potassium, crystallizing in the hexagonal system.

mill In UK, a crushing and grinding plant. In US, the whole equipment for comminuting and concentrating an ore.

Miller indices Integers which determine the orientation of a crystal plane in relation to the three crystallographic axes. The reciprocals of the intercepts of the plane on the axes (in terms of lattice constants) are reduced to the smallest integers in ratio. Also *crystal indices*.

millerite, capillary pyrite Nickel sulphide, crystallizing in the trigonal system. It usually occurs in very slender crystals and often in delicately radiating groups.

Miller process Purification of bullion by removal of base metals as chlorides. Chlorine gas is bubbled through the molten metal.

millet-seed sandstone A sandstone consisting essentially of small spheroidal grains of quartz; typical of deposits accumulated under desert conditions.

mill-head Ore accepted for processing after removal of waste rock and detritus. Also, *assay grade* of such ore.

millidarcy See **darcy**.

millilambert A unit of brightness equal to 0.001 lambert; more convenient magnitude than the lambert.

millilux Unit of illumination equal to one-thousandth of a lux.

millimass unit Equal to 0.001 of atomic mass unit. Abbrev. *mu*.

milling Removing valueless material and harmful constituents from an ore, in order to render marketing more profitable. Also *comminution, dressing*.

milling grade The grade at which an ore is sufficiently rich to repay cost of processing.

milling width The extent of the lode which will determine the tonnage sent daily from mine to mill.

million tons oil equivalent Abbrev. *mtoe*. A standard basis for comparing between countries the annual production and consumption of different fuels (with different **calorific value**) by relating each fuel to crude oil. Hydroelectric and nuclear power are also expressed in *oil equivalence* terms by calculating the amount of oil needed to produce an equivalent quantity of electrical energy. One mtoe is equivalent to about 1.5 million tons of coal.

Millstone Grit A long established name for the middle coarse sandstone division of the Carboniferous, roughly corresponding to the Namurian.

mimetite A chloride-arsenate-phosphate of lead with AsP; cf. **pyromorphite**. It crystallizes in the hexagonal system, often in barrel-shaped forms, and is usually found in lead deposits which have undergone a secondary alteration.

Mindel The second major glacial stage of the Pleistocene epoch of the Alps. See **Günz, Riss, Würm**.

mine See **mining**.

mineral A naturally-occurring substance of more or less definite chemical composition and physical properties. It has a characteristic atomic structure frequently expressed in the crystalline form or other properties.

mineral caoutchouc See **elaterite**.

mineral deposit A naturally occurring body containing minerals of economic value.

mineral dressing See **mineral processing**.

mineralization The process by which minerals are introduced into a rock.

mineral oils Petroleum and other hydrocarbon oils obtained from mineral sources.

mineral processing, dressing

mineral processing, dressing Crushing, grinding, sizing, classification or separation of ore into waste, and value by chemical, electrical, magnetic, gravity and physicochemical methods. First-stage extraction metallurgy. Also *ore dressing*, *beneficiation*, *préparation mécanique* (French).

mineral vein A fissure or crack in a rock which has been subsequently lined or filled with minerals. See **lode**.

miner's dip needle A portable form of *dip needle* used for indicating the presence of magnetic ores.

miner's lamp A portable lamp specially designed to be of robust construction and adequate safety for use in mines.

minette (1) A lamprophyre composed essentially of biotite and orthoclase, occurring in dykes associated with major granitic intrusions. (2) Jurassic ironstones of Briey and Lorraine.

mining See panel on p. 151.

mining dial See **dial**.

minnesotaite The iron-bearing equivalent of talc. A major constituent of the siliceous iron ores of the Lake Superior region.

minor intrusions Igneous intrusions of relatively small size, compared with **plutonic (major) intrusions**. They comprise dykes, sills, veins and small laccoliths. The injection of the minor intrusions constitutes the dyke phase of a volcanic cycle.

minverite A basic intrusive rock, in essentials a dolerite, containing a brown, soda-rich hornblende; named from the type locality, St Minver, Cornwall.

Miocene An epoch of the Neogene sub-period of the Cenozoic era. See **Tertiary**.

miospore A spore or pollen grain arbitrarily defined in palaeopalynology as less than 200 micrometres (200 μm) in diameter.

mirabilite See **Glauber salt**.

mirror A highly-polished reflecting surface capable of reflecting light rays without appreciable diffusion. The commonest forms are plane, spherical (convex and concave) and paraboloidal (usually concave). The materials used are glass silvered on the back or front, speculum metal, or stainless steel.

miscibility The property enabling two or more liquids to dissolve when brought together and thus form one phase.

miscibility gap The region of composition and temperature in which two liquids form two layers or phases when brought together.

mispickel See **arsenopyrite**.

Mississippian A period of the Upper Palaeozoic era, lying between the Devonian and Pennsylvanian. It covers an approx. time span from 360–320 million years and is the North American equivalent of the *Lower Carboniferous* in Europe. The corresponding system of rocks. See **Palaeozoic**.

missourite A melanocratic plutonic rock composed of clinopyroxene, olivine and leucite.

mist A suspension, often colloidal, of a liquid in a gas.

Mitscherlich's law of isomorphism Salts having similar crystalline forms have similar chemical constitutions.

mixed crystal A crystal in which certain atoms of one element are replaced by those of another.

mizzonite One of the series of minerals forming the scapolite group, consisting of a mixture of the meionite and marialite molecules. Mizzonite includes those minerals with 50%–80% of the meionite end-member molecule. Found in metamorphosed limestones and in some altered basic igneous rocks.

MKSA *Metre-Kilogram(me)-Second-Ampere* system of units, now superseded by the S.I. system. See **SI units**.

mobile belt A long zone of the Earth's crust associated with igneous activity and deformation. Traditionally associated with geosynclinal development. See **plate tectonics**.

Mocha stone See **moss agate**.

mode The actual mineral composition of a rock expressed quantitatively in percentages. Cf. **norm**.

modifier Modifying agent used in froth flotation to increase *either* the wettability *or* the water-repelling quality of one or more of the minerals being treated.

modulus of elasticity For a substance, the ratio of stress to strain within the elastic range, i.e. where Hooke's law is obeyed. Measured in units of stress, e.g. GN/m^2. See **elasticity**.

mofette A volcanic opening through which emanations of carbon dioxide, nitrogen and oxygen pass. It marks the last phase of volcanic activity.

Moho Abbrev. for **Mohorovičić discontinuity**.

Mohorovičić discontinuity The boundary, at which there is a marked change in seismic velocity, separating the Earth's crust above from the mantle below. Abbrev. *Moho*. See **Earth**.

Mohr-Coulomb theory The resistance of rock to crushing is due to internal friction *plus* cohesion of bonding materials.

Mohs' scale of hardness A scale introduced by Mohs to measure the hardness of minerals. See **hardness**.

Moinian This group, with various names

mining

The process of extracting metallic or non-metallic mineral deposits, including gemstones, from below the Earth's surface. The mine is a subterranean excavation made in connection with exploitation of, or search for, minerals of economic interest. Terms *quarry*, *pit* and *opencast* are reserved for workings open to daylight; in the North of England *mine* is sometimes applied to any coal seam irrespective of thickness and grade. The methods of mining are devised for each deposit, so there are many variations. Some of the terms used are as shown in the figure; since mining has been practised for many hundreds of years these terms may be of considerable antiquity.

An *adit* is a horizontal tunnel or passage into a mine, for access or drainage, and a *shaft* is a vertical or highly inclined excavation (sometimes down a lode) made for access, ventilation etc. A tunnel more-or-less parallel to an ore zone for exploration or exploitaion is a *drift* or *drive*; one that cuts across is a *cross-cut*. The ore may be removed by *stoping*, leaving a *stope* where the ore is worked out. An internal shaft that goes down into an ore body is a *winze*, one that goes up into it a *raise* or *rise*. The *footwall* is the lower wall of country rock in contact with a vein or lode. The upper wall is the *hanging wall*. Depths to and between levels may be measured in feet, metres or fathoms (i.e. 6 feet; abbrev. *fm*). The inner end of a drift is a *dead end*. Drift is also used for a heading obliquely through a coal seam, and *drift mining* uses an inclined haulage road to the surface. *Glory hole* mining is where ore is worked in a conical excavation in an open cut and dropped down a shaft for removal in a haulage tunnel or adit below the orebody. *Opencast* mining is surface mining, often of bedded deposits such as coal or sedimentary ironstone, by the removal of overburden in an open pit.

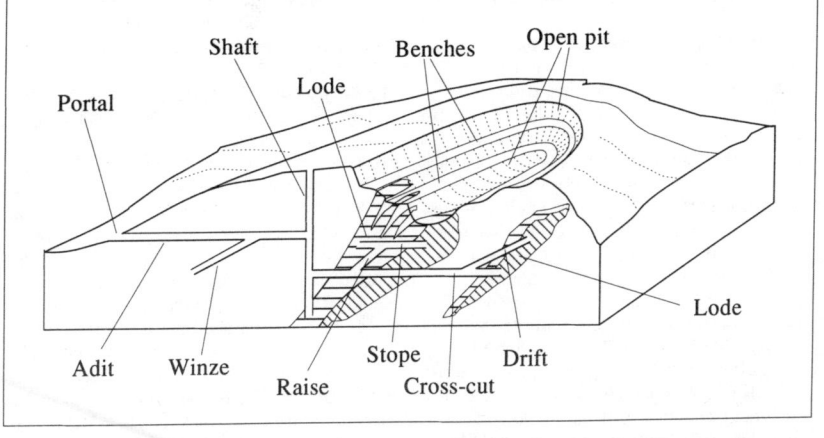

such as 'Moine Schists', consists of late Precambrian metamorphosed sediments, and forms much of the Northern Highlands of Scotland.

mol See **mole**.

molality The concentration of a solution expressed as the number of moles of dissolved substance per kilogram of solvent.

molal specific heat capacity The *specific heat capacity* of 1 mole of an element or compound. Also *volumetric heat* (for gases).

molar absorbance The **absorbance** of a solution with a concentration of 1 mol dm^{-3} measured in a cell of a thickness of 1 cm.

molar conductance The conductance which a solution would have if measured in a cell large enough to contain one mole of solute between electrodes 1 cm apart.

molar heat See **molar heat capacity**.

molar heat capacity The heat required to raise the temperature of a substance by 1 K. The symbol for that measured at constant

molarity

volume is C_V and for that at constant pressure C_P.

molarity The concentration of a solution expressed as the number of moles of dissolved substance per dm^3 of solution.

molar surface energy The surface energy of a sphere containing 1 mole of liquid; equal to $\gamma V^{2/3}$, where γ is the surface tension and V the molar volume. It is zero near the critical point and its temperature coefficient is often a colligative property. See **Eötvös equation**.

molar volume The volume occupied by one mole of a substance under specified conditions. That of an ideal gas at s.t.p. is 2.241×10^{-2} m^3 mol^{-1}.

molasse Sediments produced by erosion of mountain ranges following the final stages of an *orogeny*. Sandstones and other detrital rocks are the dominant products of this type of sedimentation. Cf. **flysch**.

moldavite A type of **tektite**.

mole The amount of substance that contains as many entities (atom, molecules, ions, electrons, photons etc.) as there are atoms in 12 g of ^{12}C. It replaces in SI the older terms *gram-atom*, *gram-molecule* etc., and for any chemical compound will correspond to a mass equal to the relative molecular mass in grams. Abbrev. *mol*. See **Avogadro number**.

molecular (1) Pertaining to a molecule or molecules. (2) Pertaining to 1 mole.

molecular elevation of boiling point The rise in the boiling point of a liquid which would be produced by the dissolution of 1 mole of a substance in 1 kg of the solvent, if the same laws held as in dilute solution.

molecular formula A representation of the atomic composition of a molecule. When no structure is indicated, atoms are usually given in the order C, H, other elements alphabetically. Functional groups may be written separately, thus sulphanilamide may be written $C_6H_8N_2O_2S$ or $H_2N.C_6H_4.SO_2NH_2$.

molecular models Three dimensional models of the structures of many molecules including complex molecules like proteins and DNA have been used as an aid both in determining their structure and understanding their function. *Space filling models* use truncated spheres of diameters corresponding to the atomic radius which can be fitted together to give the proper bond angles. *Stick models* show only the positions of a special repeating feature in the structure such as the *alpha carbon* of a peptide. These positions can be represented by a marker on a rod (*stick*) which can be set to the appropriate three-dimensional co-ordinates. These

molybdenum

mechanical models are becoming redundant as computer modelling has become widely available.

molecular rotation One-hundredth of the product of the specific rotation and the relative molecular mass of an optically-active compound.

molecular sieve Framework compound, usually a synthetic zeolite, used to absorb or separate molecules. The molecules are trapped in 'cages', the sizes of which can be selected to suit solvent.

molecular structure The way in which atoms are linked together in a molecule.

molecular volume The volume occupied by 1 mole of a substance in gaseous form at standard temperature and pressure (approx. 2.241×10^{-2} m^3).

molecular weight (1) The mass of a molecule of a substance referred to that of an atom of ^{12}C taken as 12.000. (2) The sum of the relative atomic masses of the constituent atoms of a molecule. More accurately *relative molecular mass*.

molecule An atom or a finite group of atoms which is capable of independent existence and has properties characteristic of the substance of which it is the unit. Molecular substances are those which have discrete molecules, such as water, benzene or haemoglobin. Diamond, sodium chloride and zeolites are examples of non-molecular substances.

mole fraction Fraction, of the total number of molecules in a phase, represented by a given component.

mollusc A member of the phylum Mollusca. Unsegmented coelomate invertebrates with a head (usually well developed), a ventral muscular foot and a dorsal visceral hump; the skin over the visceral hump (the *mantle*) often secretes a largely calcareous shell and encloses a mantle cavity. Includes the ammonites, belemnites, bivalves, cephalopods and gastropods, all of which have made important contributions to the fossil record.

molybdenite Disulphide of molybdenum, crystallizing in the hexagonal system. It is the most common ore of molybdenum, and occurs in lustrous lead-grey crystals in small amounts in granites and associated rocks.

molybdenum A metallic element in the sixth group of the periodic system. Symbol Mo, at.no. 42, r.a.m. 95.94, rel.d. at 20°C 10.2, hardness 147 (Brinell), mp 2625°C, bp 3200°C, resistivity c. 5×10^{-8} ohm metres. Its physical properties are similar to those of iron, its chemical properties to those of a non-metal. Used in the form of wire for filament supports, hooks etc. in electric lamps and radio valves, for electrodes of mercury-vapour lamps, and for winding electric resistance furnaces. It is added to a number of types of alloy steels. Its main ore is *molybdenite*, MoS_2, a lead-grey mineral. It is also obtained from *wulfenite*, $PbMoO_4$. Molybdenum is an essential trace element in plants and animals.

molybdite Strictly, molybdenum oxide. Much so-called molybdite is ferrimolybdite, a hydrated ferric molybdate which crystallizes in the orthorhombic system. It is commonly impure and occurs in small amounts as an oxidation product of molybdenite. Also *molybdic ochre*.

monazite Monoclinic phosphate of the rare earth metals, $CePO_4$, containing cerium as the principal metallic constituent, and also some thorium. Monazite is exploited from beach sands, where it may be relatively abundant; one of the principal sources of rare earths and thorium.

monchiquite An alkaline lamprophyre with phenocrysts of olivine, pyroxene and usually mica or amphibole, in a groundmass of glass and feldspathoids.

monitor In hydraulic mining, high-pressure jet of water used to break down loosely consolidated ground in open-cast work. Also *giant*. In ore treatment, a device which checks part of the process and sounds a signal, makes a record, or initiates compensating adjustment if the detail monitored requires it.

mon-, mono- Containing one atom, group etc., e.g. monobasic.

monobasic Containing one hydrogen atom replaceable by a metal with the formation of a salt.

monochromatic By extension from *monochromatic light*, any form of oscillation or radiation characterized by a unique or very narrow band of frequency.

monochromatic light Light containing radiation of a single wavelength only. No source emits truly monochromatic light, but a very narrow band of wavelengths can be obtained, e.g. the cadmium red spectral line, wavelength 643.8 nm with a *half-width* of 0.0013 nm. Light from some lasers have extremely narrow line widths.

monochromatic radiation Electromagnetic radiation (originally only the visible) of one single frequency component. By extension, a beam of particulate radiation comprising particles all of the same type and energy. *Homogeneous* or *monoenergic* is preferable in this sense.

monochromator Device for converting heterogeneous radiation (electromagnetic or particulate) into a homogeneous beam by absorption, refraction or diffraction processes.

monocline A fold with one limb which dips steeply; the beds, however, soon approximate to horizontality on either side of this flexure.

monoclinic system One of six crystal systems in which the 3 crystal axes are of unequal lengths, having 1 of their intersections oblique and the other 2 at right angles. Also *oblique system*. See **Bravais lattices**.

monodisperse system A colloidal dispersion having particles all of effectively the same size.

monolayer See **monomolecular layer**.

monomer A substance considered as a single unit, in contrast with *dimers* and *polymers*. In particular, a molecule from which a polymer may be built, either by *addition* or *condensation polymerization*.

monominerallic rocks Rocks consisting essentially of one mineral, e.g. dunite and anorthosite.

monomolecular layer A film of a substance one molecule in thickness.

monomolecular reaction A reaction in which only one species is involved in forming the activated complex of the reaction.

monomorphous Existing in only one crystalline form.

monopod platform Drilling or production rig with one central leg, used in arctic conditions because of risk of ice damage to conventional designs.

monosymmetric system See **monoclinic system**.

monovalent Capable of combining with 1 atom of hydrogen or its equivalent, thus having an oxidation number or co-ordination number of one.

montasite An asbestiform variety of the amphibole grunerite. It differs from **amosite** in having less harsh and more silky fibres.

montebrasite A variety of **amblygonite** in which the amount of hydroxyl exceeds that of fluorine.

monticellite A silicate of calcium and magnesium which crystallizes in the orthorhombic system. It occurs in metamorphosed dolomitic limestones and, more rarely, in some ultrabasic igneous rocks.

montmorillonite A hydrated silicate of aluminium, one of the important clay minerals and the chief constituent of bentonite and fuller's earth. The montmorillonite group of clay minerals are also collectively termed *smectites*.

monzodiorite A plutonic rock intermediate between monzonite and diorite. Synonym *syenodiorite*. See **plutonic rocks**.

monzogabbro A plutonic rock intermediate between monzonite and gabbro. Synonym *syenogabbro*. See **plutonic rocks**.

monzogranite A variety of granite with roughly equal amounts of alkali feldspar and plagioclase. See **adamellite, plutonic rocks**.

monzonite A coarse-grained igneous rock of intermediate composition, characterized by approximately equal amounts of orthoclase and plagioclase (near andesine in composition) together with coloured silicates in variety. Named from Monzoni in the Tyrol. Often referred to as *syenodiorite*. See **plutonic rocks**.

monzonorite A norite containing some orthoclase.

moonstone A variety of alkali feldspar or sometimes plagioclase, which possesses a bluish-pearly opalescence attributed to lamellar micro- or crypto-perthitic intergrowth. It is used as a gemstone.

moraine Material laid down by moving ice. Moraines are found in all areas which have been glaciated, and are of several types. *Terminal moraines* are irregular ridges of material marking the farthest extent of the ice and representing debris pushed along in front of the ice. *Lateral moraines* are found along the sides of present-day and former glaciers. *Medial moraines* form by the combination of the two lateral moraines when two glaciers join. *Ground moraine* is an irregular sheet of **till** laid down beneath an ice sheet.

mordenite A hydrated sodium, potassium, calcium, aluminium silicate. A zeolite crystallizing in the orthorhombic system and occurring in amygdales in igneous rocks and as a hydration product of volcanic glass.

morganite A pink or rose-coloured variety of beryl, used as a gemstone.

morion A variety of smoky quartz which is almost black in colour.

mortar structure A cataclastic structure resulting from dynamic metamorphism in which small grains produced by granulation occupy the interstices between larger grains.

Moscovian An epoch of the Pennsylvanian period.

Moseley's law For one of the series of characteristic lines in the X-ray spectrum of atoms, the square root of the frequency of the lines is directly proportional to the *atomic number* of the element. This result stresses the importance of atomic number rather than atomic weight.

moss agate A variegated cryptocrystalline silica containing visible impurities, as manganese oxide, in moss-like or dendritic form. Also *Mocha stone*.

Mössbauer effect When an atomic nucleus emits a γ-ray photon it must recoil to conserve linear momentum. Consequently there is a change of frequency of the radiation due to the movement of the source (*Doppler effect*). If the atom is firmly bound in a crystal lattice so that it may not recoil, the momentum is taken up by the whole lattice; an effect much used in the study of the structure of solids.

mother liquor The solution remaining after a solute has been crystallized out.

mother-of-emerald A variety of *prase*, a leek-green quartz owing its colour to included fibres of actinolite; thought at one time to be the mother rock of emerald.

mother-of-pearl The iridescent nacreous material from the shells of molluscs, used as a gemstone and for small ornamental objects. See **bivalves**.

motion compensation Automatic machinery which maintains a constant downhole pressure on the drill bit bored from a floating platform.

mottramite Descloizite in which the zinc is almost entirely replaced by copper.

mould The impression of an original shape. Usually refers to fossils, but may be minerals or sedimentary structures. Cf. **cast**.

mountain cork A variety of asbestos which consists of thick interlaced fibres. It is light and will float, and is of a white or grey colour.

mountain leather A variety of asbestos which consists of thin flexible sheets made of interlaced fibres.

mountain wood A compact fibrous variety of asbestos looking like dry wood.

mp Abbrev. for **melting point**.

mtoe Abbrev. for **Million Tons Oil Equivalent**.

muck See **dirt**.

mud A fine-grained unconsolidated rock, of the clay grade, often with a high percentage of water present. It may consist of several minerals.

mud See **drilling mud**.

mud acid Inorganic acids and chemicals used to promote flow in oil wells by acid treatment.

mud column The column of mud from the top to the bottom of a borehole.

mud flow Mass movement of fine-grained material in a highly fluid state. When abundant coarse material is also carried it is known as a *debris flow*.

mudline Separating line between clear overflow water and settled slurry in a **de-watering** or thickening plant.

mud motor Hydraulic motor situated downhole and actuated by the pressure of the **drilling mud** forced down the borehole. Can be used for the main drilling, when the drill pipe does not rotate or to actuate subsidiary drills, reamers etc. See **drain holes**.

mud pipe An outer casing which is run down through semi-liquid material found during drilling until it rests on firmer strata. It protects the casing proper which contains the **drilling mud**.

mud pump Pump used in the drilling of deep holes (e.g. oil wells) to force thixotropic mud to bottom of hole and flush out rock chips.

mudstone An argillaceous sedimentary rock characterized by the absence of obvious stratification. Cf. **shale**.

mud volcano A conical hill formed by the accumulation of fine mud which is emitted together with various gases, from an orifice in the ground. It is generally derived from a volcanic hot spring but sedimentary mud volcanoes can also be produced by earthquake activity, which generates the extrusion of liquid mud.

mugearite A dark, finely-crystalline, basic volcanic rock which contains oligoclase, orthoclase, and usually olivine in greater amount than augite. Occurs typically at Mugeary in Skye.

muller Pestle, dragstone, iron wearing plate or shoe used to crush and/or abrade rock entrained between it and a baseplate over which it is moved.

Müller's glass See **hyalite**.

mullion structure A linear structure found in severely folded sedimentary and metamorphic rocks in which the harder beds form elongated fluted columns. Mullions are parallel to fold axes.

mullite A silicate of aluminium, rather similar to sillimanite but with formula close to $Al_6Si_2O_{13}$. It occurs in contact-altered argillaceous rocks.

multiple intrusions Minor intrusions formed by several successive injections of approximately the same magma.

mundic See **pyrite**.

Murex process Magnetic separation of desired mineral from pulp, in which the ore is mixed with oil and magnetite, and then pulped in water. Magnetite clings selectively to the desired mineral.

muscovite The common or white mica; a hydrous silicate of aluminium and potassium, crystallizing in the monoclinic system. It occurs in many geological environments but deposits of economic importance are found in granitic pegmatites. Used as an insulator and a lubricant.

Muscovy glass Formerly a popular name for **muscovite**.

my Same as *Ma*, a million years.

mylonite A banded, chertlike, cataclastic rock produced by the shearing and granulation of rocks associated with intense folding and thrusting.

mylonitization The process by which rocks are granulated and pulverized and formed into mylonite.

myrmekite An intergrowth of plagioclase and vermicular quartz.

N O

n Symbol for (1) amount of substance; (2) **neutron**.

n- Abbrev. for *normal*, i.e. containing an unbranched carbon chain in the molecule.

ν Symbol for (1) frequency; (2) neutrino; (3) Poisson's ratio; (4) **kinematic viscosity**.

N Symbol for (1) **nitrogen**; (2) newton.

N Symbol for (1) **Avogadro number**; (2) number of molecules; (3) neutron number.

N- Symbol indicating substitution on the nitrogen atom.

N_A Symbol for **Avogadro number**.

Na Symbol for **sodium**.

NA Abbrev. for **Numerical Aperture**.

nacreous A term applied to the lustre of certain minerals, usually on crystal faces parallel to a good cleavage, the lustre resembling that of pearls.

nacrite A species of clay mineral, identical in composition with kaolinite, from which it differs in certain optical characters and in atomic structure.

naked-light mine Non-fiery mine, where safety lamps are not required.

Namurian A stratigraphical series name in the Upper Carboniferous of Europe. Approximately corresponding to the Millstone Grit of England and Wales. See **Palaeozoic**.

naphtha A mixture of light hydrocarbons which may be of coal tar, petroleum or shale oil origin. *Coal tar naphtha* (boiling range approx. 80°–170°C) is characterized by the predominant aromatic nature of its hydrocarbons. Generally *petroleum naphtha* is a cut between gasoline and kerosine with a boiling range 120°–180°C, but much wider naphtha cuts may be taken for special purposes, e.g. feedstock for high temperature cracking for chemical manufacture. The hydrocarbons in petroleum naphthas are predominantly aliphatic.

naphthalene $C_{10}H_8$; consists of two condensed benzene rings. Glistening plates, insoluble in water, slightly soluble in cold ethanol and ligroin, readily soluble in hot ethanol and ethoxyethane; mp 80°C, bp 218°C; sublimes easily and is volatile in steam. It occurs in the coal tar fraction boiling between 180°C and 200°C. It forms an additive compound with picric acid (trinitrophenol). Naphthalene is more reactive than benzene, and substitution occurs in the first instance in the *alpha* position. It is an important raw material for numerous derivatives, many of which play a rôle in the manufacture of dyestuffs.

naphthalene derivatives Substitution products of naphthalene.

naphthols $C_{10}H_7OH$. There is a 1-naphthol, mp 95°C, bp 282°C, and a 2-naphthol, mp 122°C, bp 288°C. Both are present in coal tar and can be prepared from the respective naphthalene sulphonic acids or by diazotizing the naphthylamines. 1-Naphthol is prepared on a large scale by heating α-naphthylamine under pressure with sulphuric acid. They have a phenolic character. 2-Naphthol is an antiseptic and can be used as a test for primary amines.

napoleonite A diorite containing spheroidal structures, about 2.5 cm in diameter, which consist of alternating shells essentially of hornblende and feldspars.

nappe A major structure of mountain chains such as the Alps, consisting essentially of a great recumbent fold with both limbs lying approximately horizontally. It is produced by a combination of compressional earth movements and sliding under gravity, resulting in translation of the folded strata over considerable horizontal distances.

native Said of naturally occurring metal, e.g. *native* gold, *native* copper.

native hydrocarbons A series of compounds of hydrogen and carbon formed by the decomposition of plant and animal remains, including the several types of coal, mineral oil, petroleum, paraffin, the fossil resins, and the solid bitumens occurring in rocks. Many which have been allotted specific names are actually mixtures. By the loss of the more volatile constituents as natural gas, the liquid hydrocarbons are gradually converted into the solid bitumens, such as *ozocerite*. See **asphalt**, **bitumen**, **coal**, **mineral oils**.

natrojarosite Hydrated sulphate of sodium and iron crystallizing in the trigonal system.

natrolite Hydrated silicate of sodium and aluminium crystallizing in the orthorhombic system. A sodium-zeolite. It usually occurs in slender or acicular crystals, and is found in cavities in basaltic rocks and as an alteration product of nepheline or plagioclase.

natron Hydrated sodium carbonate, occurring in soda-lake deposits.

natural background In the detection of nuclear radiation, the radiation due to *natural radioactivity* and to cosmic rays, enhanced by contamination and fallout.

natural gas Any gas found in the Earth's crust, including gases generated during volcanic activity (see **pneumatolysis**, **solfatara**). The term, however, is particularly applied to natural hydrocarbon gases which are associated with the production of petroleum. These gases are principally methane and ethane, sometimes with propane, butane, nitrogen, CO_2 and sulphur compounds (notably H_2S). The gas is found both above the petroleum and dissolved in it, but many very large gas fields are known which produce little or no petroleum. Natural gas has largely replaced **town gas** in many countries where it occurs abundantly (US, Canada, Algeria, W. Europe); since 1967 important finds in the North Sea have provided supplies for the British gas-grid system.

natural glass Magma of any composition is liable to occur in the glassy condition if cooled sufficiently rapidly. Acid (i.e. granitic) glass is commoner than basic (i.e. basaltic) glass; the former is represented among igneous rocks by pumice, obsidian and pitchstone; the latter by tachylyte. Natural quartz glass occurs in masses lying on the surface of certain sandy deserts (e.g. Libyan Desert); while both clay rocks and sandstones are locally fused by basic intrusions. See **buchite, fulgurite,tektites**.

natural radioactivity Radioactivity found in nature. Such radioactivity indicates that the isotopes involved have a half-life comparable with the age of the Earth, or result from the decay of such isotopes. Most such nuclides can be grouped in one of three **radioactive series**. It accounts for 86% of all the radiation received by humans and originates from space, from natural radioactive elements, e.g. uranium (^{238}U), present in small amounts in rocks and soil, and from food, water and buildings. Uranium decays in a sequence of products (forming lead ultimately), one of which is the **noble gas** radon: radioactive radon seeps through soil and may accumulate in poorly ventilated buildings. Radioactive elements naturally present in fossil fuels are released when the fuels are burned. See **radiation, ionizing radiation**.

natural uranium Uranium with its natural isotopic abundance, not depleted by the removal of ^{235}U.

Nauta mixer A proprietary mixer comprising a stationary cone, with point vertically downwards, within which a screw arm rotates on its own axis parallel to the conical surface. The lower end of the screw is fixed in a universal joint at its lower end, and to an arm at its upper end, by means of which the whole screw, while rotating, is moved round the interior surface of the cone. Much used for mixing small quantities of one constituent with large quantities of another and mostly used on dry solids.

Nautiloidea A subclass of the Cephalopoda, having a wide central siphuncle and a planospiral chambered shell. Abundant from early Cambrian to late Cretaceous, but now represented by one genus, *Nautilus*, which lives in tropical seas. All chambers except the terminal living chamber contain gas which buoys up the heavy shell.

Nb Symbol for **niobium**.

Nd Symbol for **neodymium**.

Ne Symbol for **neon**.

neck A plug of volcanic rock representing a former feeder channel of an extinct volcano.

needle stone A popular term for clear quartz containing acicular inclusions, usually of rutile, but in some specimens, of actinolite. Also *rutilated quartz*. The name has also been used for various acicular zeolites.

Néel temperature The temperature at which the magnetic susceptibility of an anti-ferromagnetic material has a maximum value.

negative crystal Birefringent material for which the velocity of the extraordinary ray is greater than that of the ordinary ray.

negative mineral A doubly refracting mineral in which the ordinary refractive index is greater than the extraordinary. Calcite is a negative mineral, for which the values of ω and ε are 1.658 and 1.485 respectively. See **optic sign**.

nektonic Free-swimming *pelagic* organisms.

Neocomian The oldest epoch of the Cretaceous period.

neodymium A metallic element, a member of the rare earth group. Symbol Nd, at.no. 60, r.a.m. 144.24, rel.d. 6.956, mp 840°C. The metal is found in cerite, monazite and orthite. Neodymium glass is used for solid state lasers and light amplifiers.

Neogene A period of the Tertiary covering a time span from approx. 23 – 2 million years ago. The corresponding system of rocks. See **Tertiary**.

Neolithic Period The later portion of the Stone Age, characterized by well-finished, polished stone implements, agriculture and domesticated farm animals. Cf. **Palaeolithic Period**.

neon Light, gaseous, inert element, recovered from atmosphere. Symbol Ne, at.no. 10, r.a.m. 20.179, mp −248.67°C, bp −245.9°C. Historically important in that J. J. Thomson, through his parabolas for charge/mass of particles, found two isotopes in neon, the first non-radioactive isotopes to be recognized. Used in many types of lamp, particularly to start up sodium vapour discharge lamps. Pure neon was the first gas to be used for high voltage display lighting, being bright orange in colour. Much used in cold-cathode tubes.

nepheline, nephelite Silicate of sodium and aluminium, NaAlSiO$_4$, but generally with some potassium partially replacing sodium, which crystallizes in the hexagonal system. It is frequently present in igneous rocks with a high sodium content and a low percentage of silica, i.e. the undersaturated rocks. See **elaeolite**.

nepheline-syenite Coarse-grained igneous rock of intermediate composition, undersaturated with regard to silica, and consisting essentially of nepheline, a varying content of alkali-feldspar, with soda amphiboles and/or soda-pyroxenes. Common hornblende, augite, or mica are present in some varieties. See **foyaite, laurdalite**.

nephelinite A fine-grained igneous rock normally occurring as lava flows, and resembling basalt in general appearance; it consists essentially of nepheline and pyroxene, but not of olivine or feldspar. The addition of the former gives *olivine-nephelinite*, and of the latter, *nepheline-tephrite*.

nephelometric analysis A method of quantitative analysis in which the concentration or particle size of suspended matter in a liquid is determined by measurement of light absorption. Also *photoextinction method*, *turbidimetric analysis*.

nephrite One of the minerals grouped under the name of *jade* ('New Zealand Jade'); consists of compact and fine-grained tremolite or actinolite. It has been widely used for ornaments in the Americas and the East.

neptunean dyke An intrusive sheet of sedimentary rock.

neptunism The obsolete theory that all rocks including granite, basalt etc. are deposited from water. Synonym *Wernerism*. Cf. **plutonism**.

neptunium Element. Symbol Np, at.no. 93; named after planet Neptune; produced artificially by nuclear reaction between uranium and neutrons. Has principal isotopes 237 and 239.

neptunium series Series formed by the decay of artificial radioelements, the first member being *plutonium-241* and the last *bismuth-209*, of which *neptunium-237* is the longest lived (half-life 2.2×10^6 years).

neritic zone That portion of the sea floor lying between low water mark and the edge of the continental shelf, at a depth of about 180 m. Sediments deposited here are of *neritic facies*, showing rapid alternations of the clay and sand grades; ripple marks etc. indicate accumulation in shallow water.

nesosilicate A silicate mineral whose atomic structure contains isolated groups of silicon-oxygen tetrahedra. See **silicates**.

Neumann principle Physical properties of a crystal are never of lower symmetry than the symmetry of the external form of the crystal. Consequently tensor properties of a cubic crystal, such as elasticity or conductivity, must have cubic symmetry, and the behaviour of the crystal will be isotropic.

neutralization The interaction of an acid and a base with the formation of a salt. In the case of strong acids and bases, the essential reaction is the combination of hydrogen ions with hydroxyl ions to form water molecules.

neutral point A point in the field of a magnet where the Earth's magnetic field (usually the horizontal component) is exactly neutralized.

neutral solution An aqueous solution which is neither acidic nor alkaline. It therefore contains equal quantities of hydrogen and hydroxyl ions and has a pH value of 7.

neutral state Said of ferromagnetic material when completely demagnetized. Also *virgin state*.

neutrino A fundamental particle, a *lepton*, with zero charge and zero mass. A different type of neutrino is associated with each of the four charged leptons. Its existence was predicted by Pauli in 1931 to avoid β-decay infringing the laws of conservation of energy and angular momentum. As they have very weak interactions with matter, neutrinos were not observed experimentally until 1956.

neutron Uncharged subatomic particle, mass approximately equal to that of the proton, which enters into the structure of atomic nuclei. Interacts with matter primarily by collisions. Spin quantum number of neutron = +D , rest mass = 1.008 665 a.m.u., charge is zero and magnetic moment is −1.9125 nuclear Bohr magnetons. Although stable in nuclei, isolated neutrons decay by β-emission into protons, with a half-life of 11.6 minutes.

Nevadan orogeny An **orogeny** of Jurassic age affecting western US.

névé A more or less compacted snow-ice occurring above the snowline; it consists of small rounded crystalline grains formed from snow crystals. Also *firn*.

New Red Sandstone A name frequently applied to the combined Permian and Triassic Systems, and particularly applicable in N. Europe, where the palaeontological evidence is insufficient to allow of their separation. The term reflects the general resemblance between the rocks comprising these two systems and the Old Red Sandstone of Devonian age.

Newton's laws of motion (1) Every body continues in a state of rest or uniform motion in a straight line unless acted upon by an external impressed force. (2) The rate of change of momentum is proportional to the impressed force and takes place in the direction of the force. (3) Action and reaction are equal and opposite, i.e. when two bodies interact the force exerted by the first body on the second body is equal and opposite to the force exerted by the second body on the first. These laws were first stated by Newton in his *Principia*, 1687. Classical mechanics consists of the applicaton of these laws.

Newton's rings Circular concentric interference fringes seen surrounding the point of contact of a convex lens and a plane surface. Interference occurs in the air film between the two surfaces. If r_n is the radius of the nth ring, R is the radius of curvature of the lens surface and λ the wavelength. $r_n = \sqrt{nR\lambda}$. See **contour fringes**.

New Zealand greenstone Nephritic 'jade' of gemstone quality, from New Zealand.

Ni Symbol for **nickel**.

niccolite Nickel arsenide, crystallizing in the hexagonal system. It is one of the chief ores of metallic nickel. Also *copper nickel*, *kupfernickel*.

nickel Silver-white metallic element. Symbol Ni, at.no. 28, r.a.m. 58.71, mp 1450°C, bp 3000°C, electrical resistivity at 20°C, 10.9×10^{-8} ohm metres, rel.d. 8.9. Used for structural parts of valves. Its ores are *nickeliferous pyrrhotite*, Fe_nS_{n+1} with Ni; *garnierite*, a nickeliferous serpentine, and *pentlandite*, $(Fe,Ni)S$. Primary nickel minerals are oxidized to many *nickel blooms*, green hydrated nickel salts. The biological role of nickel is uncertain.

nickel antimony glance See **ullmanite**.

nickel arsenic glance See **gersdorffite**.

nickel bloom See **annabergite**.

Nicol prism A device for obtaining plane polarized light, it consists of a crystal of Iceland spar which has been cut and cemented together in such a way that the ordinary ray is totally reflected out at the side of the crystal, while the extraordinary plane polarized ray is freely transmitted. Largely superseded by Polaroid.

nigrite A pitch-like member of the asphaltite group.

niobite See **columbite**.

niobium Rare metallic element. Symbol Nb, at.no. 41, r.a.m. 92.9064, mp 2500°C, used in high temperature engineering products (e.g. gas turbines and nuclear reactors) owing to the strength of its alloys at temperatures above 1200°C. Combined with tin (Nb_3Sn) it has a high degree of superconductivity. Occurs with tantalum in *columbite*, niobate and tantalite of iron and manganese.

nitrates Salts or esters of nitric (V) acid. Metal nitrates are soluble in water; decompose when heated. The nitrates of polyhydric alcohols and the alkyl radicals explode with violence. Uses: explosives, fertilizers, chemical intermediates.

nitre *Potassium nitrate* (V). Also *saltpetre*. See **Chile nitre, soda nitre**.

nitric (V) acid NHO_3, a fuming unstable liquid, bp 83°C, mp −41.59°C, rel.d. 1.5, miscible with water. Old name *aqua fortis*. Prepared in small quantities by the action of conc. sulphuric acid on sodium nitrates and on large scale by the oxidation of nitrogen or ammonia. An important intermediate of fertilizers, explosives, organic synthesis, metal extraction and sulphuric acid manufacture. See **nitrates, chamber process**.

nitrification The treatment of a material with nitric acid.

nitrocelluloses *Cellulose nitrates*. They are the nitric acid esters of cellulose formed by action of a mixture of nitric and sulphuric acids on cellulose. The cellulose can be nitrated to a varying extent ranging from 2 to 6 nitrate groups in the molecule. Nitrocelluloses with a low nitrogen content, up to the tetranitrate, are not explosive, and are used in lacquer, artificial silk and imitation leather book cloth manufacture. They dissolve in ether-alcohol mixtures, and in so-called lacquer solvents. A nitrocellulose with a high nitrogen content is gun-cotton, an explosive.

nitrogen Gaseous element, colourless and odourless. Symbol N, at.no. 7, r.a.m. 14.0067, mp −209.86°C, bp −195°C. It was the determination by Rayleigh and Ramsay of its rel. at. mass that led to the discovery of the inert gases in the atmosphere, argon etc. Approx. 80% of the normal atmosphere is nitrogen, which is also widely spread in minerals, the sea, and in all living matter. Its abundance in the Earth's crust is only 19 ppm. It occurs in rare independent minerals, mainly nitrates in evaporite deposits; there is some in coal and petroleum and it is present in volcanic gases. Used as a neutral filler in filament lamps, in sealed relays, and in Van de Graaff generators; and in high voltage cables as an insulant. Liquid is a coolant. Used extensively in inorganic and organic chemistry.

nitrogen oxides Consisting mainly of nitric oxide, nitrogen dioxide and nitrous oxide, they are produced by natural processes (lightning, bacterial **nitrification** and decomposition) and by burning fossil fuels. In coal-fired power stations, NOx are derived mainly from the coal, but in oil-fired stations NOx come mainly from the nitrogen in the air supporting combustion. As atmospheric pollutants they contribute to **acid rain**. Control is achieved by changes in boiler and burner design, and by catalytic reduction to nitrogen using ammonia as a reductant. Abbrev. *NOx*

nitroglycerine $C_3H_5(ONO_2)_3$, a colourless oil; mp 11°–12°C; insoluble in water; prepared by treating glycerine with a cold mixture of concentrated nitric and sulphuric acids. It solidifies on cooling, and exists in two physical crystalline modifications. In thin layers it burns without explosion, but explodes with tremendous force when heated quickly or struck. See **dynamite, explosive**. Used medically in solution and tablets for angina.

NMR Abbrev. for **Nuclear Magnetic Resonance**.

No Symbol for nobelium.

noble gases Elements helium, neon, argon, krypton, xenon and radon-222, much used (except the last) in gas-discharge tubes. (Radon-222 has short-lived radioactivity, half-life less than 4 days.) Their outer (valence) electron shells are complete, thus rendering them inert to all the usual chemical reactions; a property for which argon, the most abundant, finds increasing industrial use. The heavier ones, Rn, Xe, Kr, are known to form a few unstable compounds, e.g. XeF_4. Also *inert gases*, *rare gases*.

noble metals Metals, such as gold, silver, platinum etc., which have a relatively positive electrode potential, and which do not enter readily into chemical combination with non-metals. They have high resistance to corrosive attack by acids and corrosive agents, and resist atmospheric oxidation. Cf. **base metal**.

nodular structures Spheroidal, ovoid, or irregular bodies often encountered in both igneous and sedimentary rocks, and formed by segregation about centres. See **clay ironstone, doggers, flint, septaria**.

nodulizing Aggregation of finely divided material such as mineral concentrates, by aid of binder and perhaps kilning, into nodules sufficiently strong and heavy to facilitate subsequent use, such as charging into blast furnaces.

non-aqueous solvents May be classed broadly into (1) water-like or levelling solvents, which are highly polar and form strong electrolytic solutions with most ionizable solutes, (2) differentiating solvents, which bring out differences in the strength of electrolytes. Examples of (1) are $NH_2(NH_4$—$NH_{2-})$, $SO_2(SO^{++}$—$SO_{2-})$, $N_2O_4(NO^+$—$NO_{3-})$. Examples of (2) are weak amines or acids, ethers, halogenated hydrocarbons.

non-ionizing radiation Radiation which does not produce ionization in matter. See **ionizing radiation**.

non-metal An element which readily forms negative ions, often in combination with other non-metals. Non-metals are generally poor conductors of electricity.

non-sequence A break in the stratigraphical record, less important and less obvious than an unconformity, and deduced generally on palaeontological evidence.

non-spectral colour Colour outside the range which contributes to white light, but which affects photocells etc.

non-specular reflection Wave reflection of light or sound from rough surfaces, resulting in scattering of wave components, depending on relation between wavelength and dimensions of irregularities. Also *diffuse reflection*.

non-stoichiometric compounds Some solid compounds do not possess the exact compositions which are predicted from Daltonic or electronic considerations alone (e.g. iron (II) sulphide is $FeS_{1.1}$), a phenomnon which is associated with the so-called **defect structure** of *crystal lattices*. Often show semi-conductivity, fluorescence and centres of colour.

nontronite A clay mineral in the montmorillonite group (smectites), containing appreciable ferric iron replacing aluminium.

norbergite Magnesium silicate with magnesium fluoride or hydroxide. A member of the humite group, it crystallizes in the orthorhombic system and occurs in metamorphosed dolomitic limestones.

nordmarkite An alkali quartz-bearing syenite, described originally from Nordmarken in Norway; consists essentially of microperthite, aegirine, soda amphibole and accessory quartz.

norite A coarse-grained igneous rock of basic composition consisting essentially of plagioclase (near labradorite in composition) and orthopyroxene. Other coloured minerals are usually present in varying amount, notably clinopyroxene.

norm The theoretical composition of an igneous rock expressed in terms of standard mineral molecules, calculated from the

normal fault A fracture in rocks along which relative displacement has taken place under tensional conditions, the *fault* hading to the downthrow side. Cf. **reversed fault**.

normality An obsolescent concentration unit, abbreviated *N*. It is used mainly for acids or bases and for oxidizing or reducing agents. In the first case, it refers to the concentration of titratable H^+ or OH^- in a solution. Thus a solution of sulphuric acid with a concentration of 2 mol l^{-1} (2 M) will be a 4 N solution.

normally magnetized crust See **magnetic anomaly patterns**.

normal pressure Standard pressure, 101.325 kN/m^2 = 760 torr, to which experimental data on gases are referred.

normal salts Salts formed by the replacement by metals of all the replaceable hydrogen of the acid.

normal solution See **normality**.

normal temperature and pressure Earlier term for **s.t.p.**, **STP**.

normative composition The theoretical mineral chemical composition of a rock expressed in terms of the standard minerals of the **norm**.

North Sea gas See **natural gas**.

nosean, noselite Silicate of sodium and aluminium with sodium sulphate, crystallizing in the cubic system. Occurs in extrusive igneous rocks which are rich in alkalis and deficient in silica, e.g. phonolite.

novaculite A fine-grained or cryptocrystalline rock composed of quartz or other forms of silica; a form of chert. Used as a whetstone.

NOx See **nitrogen oxides**.

Np Symbol for **neptunium**.

NTP *Normal Temperature and Pressure*. Previous term for **s.t.p.**, **STP**, i.e. 0°C and 101.325 kN/m^2.

nuclear chemistry The study of reactions involving the transmutation of elements, either by spontaneous decay or by particle bombardment.

nuclear magnetic resonance The nuclei of certain isotopes, e.g. ^1H, ^{19}F, ^{31}P, behave like small bar magnets and will line up when placed in a strong d.c. magnetic field. The direction of alignment of the nuclei in a sample may be altered by radiofrequency irradiation in a surrounding coil whose axis is at right angles to the magnetic field. The frequency at which energy is absorbed and the direction of alignment changes is known as the resonant frequency and varies with the type of nucleus, e.g. in a magnetic field of 14 000 gauss ^1H resonates at 60 MHz, ^{19}F at 56 MHz and ^{31}P at 24.3 MHz. Nuclei with even mass and atomic number, e.g. ^{12}C, ^{16}O, have no magnetic properties and do not exhibit this behaviour. Abbrev. *NMR*. NMR gives invaluable information on the structure of molecules. It has also been developed to provide a non-invasive clinical imaging of the human body, magnetic resonance imaging (MRI); multiple projections are combined to form images of sections through the body, providing a powerful diagnostic aid.

nuclear magnetic resonance spectroscopy The resonant frequencies for identical nuclei in a constant external field vary slightly with the chemical environment of the nucleus (about 1–10 ppm for ^1H). By keeping the external magnetic field constant and varying the radiofrequency radiation over a small range, 1000 Hz, for a solution of a given molecule, an absorption spectrum for the compound under examination is obtained for which information may be deduced on the structure of the molecule.

nucleus Composed of protons (positively charged) and neutrons (no charge), and constitutes practically all the mass of the atom. Its charge equals the atomic number; its diameter is from 10^{-15} to 10^{-14} m. With protons, equal to the atomic number, and neutrons to make up the atomic mass number, the positive charge of the protons is balanced by the same number of extra-nuclear electrons.

nuclide A species of atom as characterized by its atomic number, its mass number and nuclear energy state.

nuée ardente A glowing cloud of gas and volcanic ash that moves rapidly downhill, as a density flow.

nugget A lump of native precious metal, esp. *gold*.

numerical aperture Product of the refractive index of the object space and the sine of the semi-aperture of the cone of rays entering the entrance pupil of the objective lens from the object point. The resolving power is proportional to the numerical aperture. Abbrev. *NA*.

numerical index of efficiency of screening See **efficiency of screening, numerical index of**.

nummulites A group of extinct Foraminifera which were important rock-forming organisms in the early Tertiary period. A nummulitic limestone is one which is composed mainly of their skeletal remains. See **protozoa**.

nunatak An isolated mountain peak which projects through an ice sheet.

o- Abbrev. for **ortho-**, i.e. containing a benzene nucleus substituted in the 1.2 positions.

ω Symbol for (1) angular frequency; (2) angular velocity; (3) **dispersive power**; (4) pulsatance; (4) specific magnetic rotation.

ω– Symbol indicating (1) substitution in the side chain of a benzene derivative; (2) substitution on the last carbon atom of a chain, farthest from a functional group.

O Symbol for **oxygen**.

O- Symbol indicating that the radical is attached to the oxygen atom.

Ω Symbol for (1) **ohm**; (2) angular velocity.

--Ω Symbol for the ultimate disintegration product of a radioactive series.

obduction The process during collisions between **tectonic plates** whereby a piece of the subducted plate is broken off and pushed on to the overriding plate.

objective *Objective lens*. Usually the lens of an optical system nearest the object. Abbrev. *OG*, objective glass for microscope work.

oblique system See **monoclinic system**.

obsequent drainage See **drainage patterns**.

observation well As oil extraction proceeds from a reservoir, special observation wells may be drilled, or allocated, for monitoring the changing fluid levels or conditions of pressure in the reservoir.

obsidian A volcanic glass of granitic composition, generally black with vitreous lustre and conchoidal fracture; occurs at Mt. Hecla in Iceland, in the Lipari Isles, and in Yellowstone National Park, US. A green silica glass found in ploughed fields in Moravia is cut as a gemstone and sold under the name *obsidian*. True obsidian is used as a gemstone and is often termed *Iceland agate*.

occlusion The retention of a gas or a liquid in a solid mass or on the surface of solid particles, esp. the retention of gases by solid metals.

ocean-floor spreading See **plate tectonics**, **magnetic anomaly patterns**.

oceanic crust That part of the Earth's crust which is normally characteristic of oceans. In descending vertical section, it consists of approximately 5 km of water, 1 km of sediments and 5 km of basaltic rocks.

oceanic ridge See **mid-ocean ridge**.

oceanite A type of basaltic igneous rock occurring typically in the oceanic islands as lava flows; characterized by a higher percentage of coloured silicates (olivine and pyroxene), and a lower percentage of alkalis, than in normal basalt.

ochre (1) Naturally occurring red, yellow and brown iron oxides, or clays strongly coloured by iron oxides (limonite), formed by residual weathering and used as pigments. (2) Highly coloured alteration products from other metals, e.g. chrome ochre. See **umber**.

octahedral system See **cubic system**.

octahedrite A form of *anatase*, crystallizing in tetragonal bipyramids.

octahedrites A class of iron meteorites showing an octahedral internal structure.

octahedron A form of the cubic system which is bounded by 8 similar faces, each being an equilateral triangle with plane angles of 60°. Pl. *octahedra*.

octane number The percentage, by volume, of *iso*-octane (2,2,4-trimethylpentane) in a mixture of *iso*-octane and *normal* heptane which has the same knocking characteristics as the motor fuel under test; it serves as an indication of the knock-rating of a motor fuel.

OD Abbrev. for **Ordnance Datum**.

odontolite See **bone turquoise**.

oersted CGS electromagnetic unit of magnetic field strength, such that 2π oersteds is a field at the centre of a circular coil 1 cm in radius carrying a current of 1 abampere (10 ampere). Now replaced by the SI unit ampere per metre. $1 \text{ A m}^{-1} = 4\pi \times 10^{-3}$ oersted.

off-lap The dispositional arrangement of a series of conformable strata laid down in the waters of a shrinking sea, or on the margins of a rising landmass, so that the successive strata cover smaller areas than their predecessors. Cf. **overlap**.

offset well Well drilled near to an existing well to further explore or exploit a field.

offstream Term for large scale process plant that is not in production due to maintainance, development work or other circumstances, e.g. oil wells, refineries or chemical plant.

ohm SI unit of electrical resistance, such that 1 ampere through it produces a potential difference across it of 1 volt. Symbol Ω.

ohm-cm The CGS unit of *resistivity*.

ohm-metre SI unit of *resistivity*.

Ohm's law In metallic conductors, at constant temperature and zero magnetic field, the current I flowing through a component is proportional to the potential difference V between its ends, the constant of proportionality being the *conductance* of the component. So $I = V/R$ or $V = IR$, where R is the *resistance* of the component. Law is strictly applicable only to electrical components carrying direct current and for practical purposes to those of negligible reactance carrying alternating current. Extended by analogy to any physical situation where a pressure difference causes a flow through an impedance, e.g. heat through walls, liquid through pipes.

oil-based mud Mud used when drilling very hot formations or through water absorbing strata. Also faster but dirtier drilling.

oilgas A gas of high energy value, obtained by the destructive distillation of high-boiling mineral oils. It consists chiefly of methane, ethene, ethyne, benzene and higher homologues.

oils A group of neutral liquids comprising three main classes: (1) *fixed (fatty) oils*, from animal, vegetable and marine sources, consisting chiefly of glycerides and esters of fatty acids; (2) *mineral oils*, derived from petroleum, coal, shale etc., consisting of hydrocarbons; (3) *essential oils*, volatile products, mainly hydrocarbons, with characteristic odours, derived from certain plants.

oil shale An argillaceous sediment containing diffused **kerogen** in a state suitable for distillation into paraffin and other mineral oils by the application of heat. See **shale oils**.

oil string See **production string**.

old age The final stage in the cycle of erosion in which base-level is nearly attained and the landscape has little relief.

oldhamite Sulphide of calcium, usually found as cubic crystals in meteorites.

Old Red Sandstone The continental facies of the **Devonian** System in Britain, comprising perhaps 12 000 m of red, brown or chocolate sandstones, red and green marls, cornstones, breccias, flags, conglomerates and volcanic rocks. They yield on certain horizons the remains of archaic fishes, eurypterids, plants and rare shelly fossils. Abbrev. *ORS*.

Oligocene The youngest epoch of the **Palaeogene** period, the time succession being Palaeocene, Eocene and Oligocene. See **Tertiary**.

oligoclase One of the plagioclase feldspars, consisting of the Albite (Ab) and Anorthite (An) molecules combined in the proportions of $Ab_9 An_1$ to $Ab_7 An_3$. It is found esp. in the more acid igneous and metamorphic rocks.

olistostrome A sediment which consists of a jumbled mass of heterogeneous blocks of material, generated by gravity sliding.

olivenite A hydrated arsenate of copper which crystallizes in the orthorhombic system. It is a rare green mineral of secondary origin found in copper deposits.

oliver filter Drum filter used in large-scale dewatering or filtration of mineral pulps, usually after thickening. Pulp is drawn by vacuum to filtering membrane as drum rotates slowly through a trough, and filtrate is drawn off while moist filter cake is scraped from down-running side of drum.

olivine (1) Orthosilicate of iron and magnesium, crystallizing in the orthorhombic system, which occurs widely in the basic and ultramafic igneous rocks, including olivine-gabbro, olivine-dolerite, olivine-basalt, peridotites etc. See **chrysolite**. The clear-green variety is used as a gemstone under the name *peridot*. For *iron olivine*, see **fayalite**; for *magnesium olivine*, see **forsterite**. (2) As a prefix for many rocks which contain olivine, e.g. *olivine-basalt*, *olivine-gabbro*.

omphacite An aluminous sodium-bearing pyroxene, occurring in eclogites as pale-green mineral grains.

on-lap An **unconformity** above which beds are successively pinched out by younger beds. Largely synonymous with **overlap**, and the reverse of **off-lap**.

onstream Term describing the functional status of process plant in full production after the commissioning phase. Used usually of large scale plant such as oil wells, refineries or continuous chemical plant.

onyx A cryptocrystalline variety of silica with layers of different colour, typically whitish layers alternating with brown or black bands. Used in cameos, the figure being carved in relief in the white band with the dark band as background.

onyx marble A banded form of calcite. *Oriental alabaster* is a beautifully banded form.

oolite A sedimentary rock composed of ooliths. In most cases ooliths are composed of calcium carbonate, in which case the rock is an oolitic limestone, but they can also be made of chamosite or limonite, in which case the rock is an oolitic ironstone. Written by itself, the word can be assumed to refer to an oolitic limestone. Also *oölite*.

oolith A more or less spherical concretion of calcium carbonate, chamosite, or dolomite, not exceeding 2 mm in diameter, usually showing a concentric-layered and/or a radiating fibrous structure. Also *oölith*.

oolitic Pertaining to an **oolite**.

ooze (1) A fine-grained, soft, deep-sea deposit, composed of shells and fragments of foraminifera, diatoms and other organisms. (2) A soft mud.

opal A cryptocrystalline or colloidal variety of silica with a varying amount of water. The transparent coloured varieties, exhibiting opalescence, are highly prized as gemstones.

opal agate A variety of opal, of different shade of colour and agate-like in structure.

opalescence The milky, iridescent appearance of a solution or mineral, due to the reflection of light from very fine, suspended particles.

opalescence The play of colour exhibited by precious opal, due to interference at the surfaces of minutely thin films, the thicknesses of the latter being of the same order of magnitude as the wavelength of light.

opaque A substance totally absorbent of rays of a specified wavelength, e.g. wood is opaque to visible light but slightly transparent to infrared rays, and completely transparent to X-rays and waves for radio communication.

OPEC Abbrev. for **Organization of Petroleum Exporting Countries**.

opencast Quarry. Open cut. Mineral deposit worked from surface and open to daylight. See **mining**.

open-circuit grinding Size reduction of solids in which the material to be crushed is passed only once through the equipment, so that all the grinding has to be done in a single step. Generally less efficient than *closed-circuit grinding*.

open flow Running an oil well without any valves or constrictions at the casing head.

ophicalcite See **forsterite-marble**.

ophiolite A group of mafic and ultramafic igneous rocks ranging from extrusive spilites to intrusive gabbros, associated with deep-sea sediments. Ophiolites are commonly found in **convergence zones**.

ophitic texture A texture characteristic of dolerites in which relatively large pyroxene crystals completely enclose smaller, lath-shaped plagioclases. See **poikilitic texture**.

optical activity A property possessed by many substances whereby plane-polarized light, in passing through them, suffers a rotation of its plane of polarization, the angle of rotation being proportional to the thickness of substance traversed by the light. In the case of molten or dissolved substances it is due to the possession of an asymmetric molecular structure, e.g. no mirror plane of symmetry in the molecule. See **chirality**.

optical isomerism The existence of isomeric compounds which differ in their **chirality**.

optical pyrometer An instrument which measures the temperatures of furnaces by estimating the colour of the radiation, or by matching it with that of a glowing filament.

optical rotary dispersion The change of optical rotation with wavelength. Rotatory dispersion curves may be used to study the configuration of molecules. Abbrev. *ORD*.

optical rotation The rotation of the plane of polarization of a beam of light when passing through certain materials.

optical spectrum The visible radiation emitted from a source separated into its component frequencies.

optic axial angle The angle between the two optic axes in biaxial minerals, usually denoted as 2V (when measured in the mineral) or 2E (in air).

optic axis The direction(s) in a doubly refracting crystal for which both the ordinary and the extraordinary rays are propagated with the same velocity. Only one exists in uniaxial crystals, two in biaxial.

optics The study of light. *Physical optics* deals with the nature of light and its wave properties; *geometrical optics* ignores the wave nature of light and treats problems of reflection and refraction from the ray aspect.

optic sign Anisotropic minerals are either optically positive or negative, indicated by + or − in technical descriptions. See **negative mineral, positive mineral**.

orbicular structure A structure exhibited by those plutonic igneous rocks which contain spherical orbs up to several centimetres in diameter, each showing a development of alternating concentric shells of different minerals, so deposited by rhythmic crystallization.

ORD Abbrev. for **Optical Rotatory Dispersion**.

Ordnance Datum The level based on the mean sea level determined by tidal measurements at Newlyn, Cornwall, from which heights on British maps are measured. Abbrev. *OD*.

Ordovician The second oldest period of the Palaeozoic era, covering an approx. time span from 510–440 million years. Named after the *Ordovices*, an ancient British tribe of the Welsh borders. Also, the corresponding system of rocks. See **Palaeozoic**.

ore A term applied to any metalliferous mineral from which the metal may be profitably extracted. It is extended to non-metals and also to minerals which are potentially valuable.

ore bin Storage system (usually of steel or concrete) which receives ore intermittently from the mine. A fine ore bin holds material crushed to centimetric size and keeps from one to three days' milling supply.

ore body Deposit, seam, bed, lode, reef, placer, lenticle, mass, stockwork, according to geological genesis.

ore dressing See **mineral processing**.

ore reserves Ore whose grade and tonnage has been established by drilling etc. with reasonable assurance. See **resources**.

Organization of Petroleum Exporting Countries An organization formed in 1960 with Iran, Iraq, Kuwait, Saudi Arabia and Venezuela as member states. It was joined by Qatar in 1961, Indonesia and Libya in 1962, United Arab Emirates in 1967, Algeria in 1969, Nigeria in 1971, and Ecuador and Gabon in 1973. It aims to co-ordinate the policies of these countries in relation to the production and distribution of oil, and to maximize their income from it. Abbrev. *OPEC*.

Oriental alabaster See **onyx marble**.

Oriental almandine A name sometimes used for *corundum*, of gemstone quality, which is deep-red in colour, resembling true almandine (a garnet) in this, but no other, respect.

Oriental amethyst A misnomer for purple corundum or sapphire. Also *false amethyst*.

Oriental cat's eye See **cymophane**.

Oriental emerald A name sometimes used for *corundum*, of gemstone quality, resembling true emerald in colour.

Oriental ruby See **ruby**.

Oriental topaz A variety of *corundum*, resembling topaz in colour.

orogenesis See **orogeny**.

orogenic belt A region of the Earth's crust, usually elongated, which has been subjected to an *orogeny*. Recently formed orogenic belts correspond to mountain ranges, but older belts have often been eroded flat.

orogeny The tectonic process whereby large areas are folded, faulted, metamorphosed and subject to igneous activity. Different periods of orogeny are given specific names, e.g. *Alpine, Caledonian, Laramide*.

orpiment Arsenic trisulphide, which crystallizes in the monoclinic system; commonly associated with realgar; golden-yellow in colour and used as a pigment.

ORS Abbrev. for **Old Red Sandstone**.

orthite See **allanite**.

ortho- (1) Derived from an acid anhydride by combination with the largest possible number of water molecules, e.g. orthophosphoric acid. (2) Consisting of diatomic molecules with parallel nuclear spins and an odd rotational quantum number, e.g. orthohydrogen.

orthoclase Silicate of potassium and aluminium, $KAlSi_3O_8$, crystallizing in the monoclinic system; a feldspar, occurring as an essential constituent in granitic and syenitic rocks, and as an accessory in many other rock types. See also **microcline, sanidine**.

orthoferrosilite The ferrous iron end-member of the orthopyroxene group of silicates.

orthogneiss Term applied to gneissose rocks which have been derived from rocks of igneous origin. Cf. **paragneiss**.

orthophyre A textural term applied to medium- and fine-grained syenitic rocks consisting of closely packed orthoclase crystals of stouter build than in the typical trachytic texture. The term actually implies the presence of porphyritic orthoclase crystals.

orthopyroxene A group of pyroxene minerals crystallizing in the orthorhombic system, e.g. enstatite, hypersthene.

orthoquartzite A pure quartz sandstone.

orthorhombic system The style of crystal architecture which is characterized by three crystal axes, at right angles to each other and all of different lengths. It includes such minerals as olivine, topaz and barytes. See **Bravais lattices**.

Os Symbol for **osmium**.

os See **esker**.

oscillatory zoning The compositional variation within a crystal which consists of alternating layers rich in the two end-members of an isomorphous solid-solution series.

oscilloscope (1) Equipment incorporating a cathode-ray tube, time-base generators, triggers etc., for the display of a wide range of waveforms by electron beam. (2) Formerly, mechanical or optical equipment with a corresponding function, e.g. Duddell oscilloscope.

osmium A metallic element, a member of the platinum group. Symbol Os, at.no. 76, r.a.m. 190.2, mp 2700°C. Osmium is the densest element, rel.d. at 20°C 22.48. Like platinum it is a powerful catalyst for gas reactions, and is soluble in aqua regia, but unlike platinum when heated in air it gives an oxide, volatile OsO_4. Alloyed with iridium it forms an extremely hard material.

osmosis Diffusion of a solvent through a semi-permeable membrane into a more concentrated solution, tending to equalize the concentrations on both sides of the membrane.

osmotic coefficient The quotient of the van't Hoff factor and the number of ions produced by the dissociation of one molecule of the electrolyte.

osmotic pressure The pressure which must be applied to a solution, separated by a semi-permeable membrane from pure solvent, to prevent the passage of solvent through the membrane. Symbol π. For substances which do not dissociate, it is related to the concentration (c) of the solution and the absolute temperature by the relationship $\pi = cRT$, where R is the gas constant.

ostracod, ostracode Small arthropods (0.4–1.5 mm long) belonging to the subclass Ostracoda having a bivalve shell. They range from the Lower Cambrian to the present day and are used for zoning in the Jurassic. The name should not be confused with *ostracoderms* (fossil fishes).

ostracoderms Fossil *agnathan* fishes. See **vertebrates**.

Ostwald's dilution law The application of the law of mass action to the ionization of a weak electrolyte, yielding the expression

$$\frac{\alpha^2}{(1-\alpha)V} = K,$$

where a is the degree of ionization, V the **dilution** (2), and K the ionization constant, for the case in which two ions are formed.

ottrelite A manganese-bearing chloritoid mineral occurring in schists, a product of the metamorphism of certain argillaceous sedimentary rocks.

outcrop An occurrence of a rock at the surface of the ground.

outgassing (1) Removal of occluded, absorbed or dissolved gas from a solid or liquid. For metals and alloys, done by heating in vacuo. (2) The release of juvenile gases from molten rocks, leading to the development of the Earth's atmosphere and oceans.

outlier A remnant of a younger rock which is surrounded by older strata.

outwash fan A sheet of gravel and sand, lying beyond the margins of a sheet of till, deposited by meltwaters from an ice sheet or glacier.

overburden Earth or rock overlying the valuable deposit. In smelting, a furnace is *overburdened* when the ratio of ore to flux or fuel is too high.

overfold A fold with both limbs dipping in the same direction, but more steeply inclined than the other. Cf. **isoclinal fold**.

overgrowth See **crystalline overgrowth**.

overhand stopes Stopes in which severed ore from an inclined seam or lode gravitates downward to tramming level. Also *overhead stopes*. Cf. **back stopes, underhand stopes**.

overlap The relationship between conformable strata laid down during an extension of the basin of sedimentation (e.g. on the margins of a slowly sinking landmass), so that each successive stratum extends beyond the boundaries of the one lying immediately beneath. Cf. **off-lap, overstep**.

overman (1) An underground manager of one or more ventilating districts in a coalmine. (2) An umpire appointed to an arbitration board in a mine dispute.

oversaturated Refers to an igneous rock in which excess silica crystallizes as a separate silica mineral or as a glass. See **undersaturated**.

overshot tool See **fishing tool**.

overstep The structural relationship between an unconformable stratum and the outcrops of the underlying rocks, across which the former transgresses. Cf. **overlap**.

overthrust A fault of low hade along which one slice or block of rock has been pushed bodily over another, during intense compressional earth movements. The horizontal displacement along the **thrust plane** may amount to several kilometres.

oxbow lake A meander loop which has been cut off.

Oxfordian A stage in the Upper Jurassic. See **Mesozoic**.

oxidates Those sedimentary rocks and weathering products whose composition and geochemical behaviour are mainly determined by the oxidation process. This category includes mainly those minerals and rocks containing iron and manganese.

oxidation The addition of oxygen to a compound. More generally, any reaction involving the loss of electrons from an atom. It is always accompanied by reduction.

oxidation number Abbrev. *o.n.* For simple atoms or ions, it is equal to the charge. For more complex groups, a formal oxidation number is often applied to specific atoms, particularly the central atom of a co-ordination compound. Thus, assuming that the ligands are chloride ions (o. n. -1), the o. n. of copper in the complex ion $[CuCl_4]^-$ may be deduced to be $+2$.

oxidation potential A measure of the electron concentration of a system in internal equilibrium. Symbol Eh. Also *redox potential*.

oxides Compounds of oxygen with another element. Oxides are formed by the combination of oxygen with most other elements, particularly at elevated temperatures, with the exception of the noble gases and some of the noble metals.

oxidizing agent A substance which is capable of bringing about the chemical change known as oxidation in another substance. It is itself reduced.

oxygen A non-metallic element. Symbol O, at.no. 8, r.a.m. 15.9994, valency 2. It is a colourless, odourless gas which supports combustion and is essential for the respiration of most forms of life. Mp $-218.4°C$, bp $-183°C$, density 1.429 04 g/dm^3 at s.t.p., formula O_2. An unstable form is ozone, O_3. Oxygen is the most abundant element, forming 21% by volume of the atmosphere, 89% by weight of water, and nearly 50% by weight of the rocks of the Earth's crust. It is

oxygen-isotope determinations

manufactured from liquid air, for use in hot welding flames, in steel manufacture, in medical practice and in anaesthesia; liquid oxygen is much used in rocket fuels.

oxygen-isotope determinations A method of using the ^{16}O and ^{18}O isotope ratio measurements from cores taken from Greenland and Antarctic ice sheets or from oxygen-bearing geological materials, e.g. carbonates from shells of marine organisms. The results may be used to estimate (1) the temperature at which the original snow fell before turning to ice; (2) the sea temperature at the time of deposition of marine fossils and (3) the global ice volume, thus giving a chronology of the ice ages during the Pleistocene.

oxyhornblende See **basaltic hornblende**.

oxyphile Descriptive of elements which have an affinity for oxygen, and therefore occur in the oxide and silicate minerals of rocks rather than in the sulphide minerals or as native elements. Also *lithophile*.

ozocerite, ozokerite A mineral paraffin wax, of dark yellow, brown, or black colour.

ozokerite See **ozocerite**.

ozone O_3. Produced by the action of ultra-violet radiation or electrical corona discharge on oxygen or air. It is a powerful oxidizing agent which absorbs harmful shortwave ultraviolet radiation in the atmosphere which would otherwise harm life on Earth.

PQ

p Symbol for (1) electric dipole moment; (2) impulse; (3) momentum; (4) pressure.
γ Symbol for (1) heat flow rate; (2) luminous flux; (3) magnetic flux; (4) work function.
ϑ Symbol for (1) the phenyl radical C_6H_5; (2) amphi-, i.e. containing a condensed double aromatic nucleus substituted in the 2.6 positions; (3) phase displacement.
P Symbol for (1) **phosphorus**; (2) **poise**.
P Symbol for (1) electric polarization; (2) power; (3) pressure.
Π Symbol for pressure, esp. osmotic pressure.
ψ Symbol for **pseudo-**.
Ψ Symbol for (1) electric flux; (2) **magnetic field strength**.
[*P*] Symbol for **parachor**.
Pa Symbol for **protactinium**.
Pachuca tank Large vertically-set cylindrical vessel used in chemical treatment of ores, in which pulp is reacted with suitable solvents for long periods, while the contents are agitated by compressed air. Also *Brown tank*.
pack Waste rock, mill tailings etc., used to support excavated stopes. Also *fill*.
packed-hole assembly See **bottom-hole assembly**.
packer Expansible plugs sent down the borehole to seal off a section, often prior to making a more permanent seal with cement. Can be used outside the casing, between the casing and drill tube or within the drill tube.
pagodite Like ordinary massive *pinite* in its amorphous compact texture and other physical characters, but contains more silica. The Chinese carve the soft stone into miniature pagodas and images. Also *agalmatolite*.
pahoehoe Lava with a glassy, smooth, ropy surface.
paisanite A sodic microgranite containing *riebeckite* as the principal coloured mineral.
palaeo- Prefix from Gk. *palaios*, ancient. Also *paleo*.
palaeobotany The study of fossil plants. See **palaeontology**.
Palaeocene The name given to the oldest epoch of the **Tertiary** period.
palaeoclimatology The study of climatic conditions in the geological record, using evidence from fossils, sediments and their structures, geophysics and geochemistry.
palaeocurrent An ancient current whose direction can often be worked out by examination of sedimentary structures (e.g. *ripple marks*, *cross bedding*, *sole structures*).
palaeoecology The study of fossil organisms in terms of their mode of life, their interrelationships, their environment, their manner of death and their eventual burial.
Palaeogene A period lying above the Cretaceous and below the *Neogene*, containing the **Palaeocene**, **Eocene** and **Oligocene** epochs. It covers a time span from approx. 65–25 million years. The corresponding system of rocks. See **Tertiary**.
palaeogeography The study of the relative positions of land and water at particular periods in the geological past.
Palaeolithic Period The oldest stone age, characterized by successive 'cultures' of stone implements, made by extinct types of men. Cf. **Neolithic Period**.
palaeomagnetism The study of the Earth's ancient magnetism. *Remanent magnetization* in both igneous and sedimentary rocks provides a means of determining former magnetic poles. See **magnetic anomaly patterns**.
palaeontology The study of fossil animals and plants, including their morphology, evolution and mode of life.
Palaeozoic See panel on p. 169.
palaeozoology The study of fossil animals. See **palaeontology**.
palagonite A hydrous altered basaltic glass. It occurs as infillings in rocks, and is a soft-brown or greenish-black crypto-crystalline substance. Named from Palagonia, Sicily.
paleo- See **palaeo-**.
palingenesis The production of new magma by the complete or partial melting of previously existing rocks. See **granitization**.
palladium A metallic element. Symbol Pd, at.no. 46, r.a.m. 106.4. The metal is white, mp 1549°C, bp 2500°C, rel.d. 11.4. Used as a catalyst in hydrogenation. Native palladium is mostly in grains, and is frequently alloyed with platinum and iridium. It is extracted from copper-nickel ores.
pallasite A group name for stony meteorites which contain fractured or rounded crystals of olivine in a network of nickel-iron.
palygorskite A group of clay minerals, hydrated magnesium aluminium silicates, in appearance resembling cardboard or paper, having a fibrous structure. Also *attapulgite*.
palynology The study of fossil spores and pollen. They are very resistant to destruction and in many sedimentary rocks are the only fossils that can be used for stratigraphical correlation.
Pangaea A hypothetical supercontinent that

Palaeozoic

The oldest of the Phanerozoic eras. The Palaeozoic ('ancient life') began with the Cambrian period, when there was a great expansion of animal life, now recorded by the fossils, esp. trilobites, brachiopods, graptolites and molluscs, as well as early plant life. Graptolites and trilobites reached their acme in the Ordovician, and in the Silurian lamellibranchs became abundant. Amphibians evolved by the end of the Devonian when the graptolites had become extinct. During the Carboniferous there was a rich flora in Coal Measure forests but a glacial climate existed in Gondwana continents. The trilobites died out in the Permian, a period of desert conditions in Britain.

Era	Period		Series	Stage	Age, Ma
Palaeozoic	Permian		Zechstein		256
			Rotliegendes		290
	Carboniferous	Pennsylvanian, Silesian	Stephanian		
			Westphalian		318
			Namurian		
					333
		Mississippian, Dinantian	Viséan		350
			Tournaisian		
					362
	Devonian		Upper	Famerinian	367
				Frasnian	
					377
			Middle	Givetian	381
				Eifelian	
					386
			Lower	Emsian	390
				Siegenian	
					398
				Gedinnian	
					408
	Silurian		Ludlow		424
			Wenlock		430
			Llandovery		
					439
	Ordovician		Ashgill		443
			Caradoc		
					464
			Llandeilo		469
			Llanvirn		
					476
			Arenig		493
			Tremadoc		
					510
	Cambrian		Upper		517
			Middle		
					536
			Lower		570

existed in the geological past and consisted of all the present continents before they split up.

panidiomorphic A term applied to igneous rocks with well-developed crystals.

panning Use by prospector or plant worker of gold pan, batea, plaque, or dulong to concentrate heavier minerals in a crushed sample by washing away the lighter ones.

pantellerite A peralkaline leucocratic rhyolite.

paper chromatography A type of **chromatography** using a sheet of special grade filter paper as the adsorbent. Advantages: microgram quantities; bands can be formed in two dimensions and cut with scissors.

para-. A prefix from Gk. *para*, beside. In chemistry (1) a polymer of .... ; (2) a compound related to .... .

para- (1) Containing a benzene nucleus substituted in the 1.4 positions. (2) Consisting of diatomic molecules with anti-parallel nuclear spins and an even rotational quantum number.

parachor A quantity which may be regarded as the molecular volume of a substance when its surface tension is unity; in most cases it is practically independent of temperature. Its value is given by the expression

$$\frac{M\gamma^{\frac{1}{4}}}{\rho_L - \rho_V},$$

where M is the molecular mass, γ the surface tension, ρ_L and ρ_V are the densities of the liquid and vapour respectively.

paraffins A term for the whole series of saturated aliphatic hydrocarbons of the general formula C_nH_{2n+2}. Also *alkane hydrocarbons*. They are indifferent to oxidizing agents, and not reactive, hence the name *paraffin* (L. *parum affinis*, little allied).

paraffin wax Higher homologues of alkanes, wax-like substances obtained as a residue from the distillation of petroleum; mp 45°–65°C, rel.d. 0.9, resistivity 10^{13} to 10^{17} ohm metres, permittivity 2–2.3.

paragneiss A term given to gneissose rocks which have been derived from detrital sedimentary rocks. Cf. **orthogneiss**.

paragonite A hydrated sodium aluminium silicate. It is a sodium mica, has a yellowish or greenish colour, and is usually associated with metamorphic rocks. Differs from **muscovite** chiefly in containing sodium rather than potassium.

parallax Generally, the apparent change in the position of an object seen against a more distant background when the viewpoint is changed. Absence of parallax is often used to adjust two objects, or two images, at equal distances from the observer.

Parallel Roads The strandlines of a glacial lake which occupied a glacial valley during the Pleistocene Period, when the lower part of the valley was blocked by ice, e.g. as at Glen Roy, Scotland.

paramagnetism Phenomenon in materials in which the susceptibility is positive and whose permeability is slightly greater than unity. An applied magnetic field tends to align the magnetic moments of the atoms or molecules and the material acquires magnetization in the direction of the field; it disappears when the field is removed. Used to obtain very low temperatures by adiabatic demagnetization.

parameter The *parameters* of a plane consist of a series of numbers which express the relative intercepts of that plane upon the crystallographic axes. Given in terms of the established unit lengths of those axes.

paramorph The name given to a mineral species which has changed its molecular structure without any change of chemical constitution, e.g. aragonite altered to calcite. Cf. **pseudomorph**.

paramudras Flint nodules of exceptionally large size and doubtful significance occurring in the Chalk exposed on the east coast of England.

parent In radioactive particle decay of A into B, A is the parent and B the daughter product.

pargasite A monoclinic amphibole of the hornblende group, particularly rich in magnesium, sodium, calcium and aluminium. It occurs chiefly in metamorphic rocks.

parrot coal See **boghead coal**.

pathfinder elements Chemical elements such as Ag, As and Sb are trace elements that are enriched in almost all types of gold deposits and have higher average crustal abundances than gold. Consequently, identification of anomalous concentrations of these is a valuable tool in gold exploration. Other pathfinder elements associated with some gold deposits are B, Ba, Bi, C, Cd, Cu, Hg, Pb, Te, Ti and Zn.

partial melting The process by which a rock, subjected to high temperature and pressure, is partly melted and the liquid removed, to solidify to a rock of different composition from the parent.

partial roasting Roasting carried out to eliminate some but not all of the sulphur in an ore or sulphidic concentrate.

particle Single piece of solid material, usually defined (when small) by its mesh, or size passing through a specified size of sieve.

particle size The general dimensions of

grains in a rock, esp. a sediment. Many definitions have been used. One of the more common is the **Wentworth scale**:

> 256 mm	boulder
64–256 mm	cobble
4–64 mm	pebble
2–4 mm	gravel
1/16–2 mm	sand
1/256–1/16 mm	silt
< 1/256 mm	clay

Field definitions are a little less precise: 'If the grains can be distinguished then it is at least silt grade; if it doesn't feel gritty on the teeth, then it is clay'. Also *grain size*.

particulates Microscopic air-borne material such as sand and volcanic ash but also man-made industrial dust from power stations and other industrial processes.

partition chromatography See **chromatography**.

partition coefficient The ratio of the equilibrium concentrations of a substance dissolved in two immiscible solvents. If no chemical interaction occurs, it is independent of the actual values of the concentrations. Also *distribution coefficient*.

passage beds The general name given to strata laid down during a period of transition from one set of geographical conditions to another; e.g. the Downtonian Stage consists of strata intermediate in character (and in position) between the marine Silurian rocks below and the continental Old Red Sandstone above.

passive margin A continental margin characterized by thick, relatively undeformed sediments, deposited at the trailing edge of a *lithospheric plate*. See **active margin**.

passivity Lack of response of metal or mineral surface to chemical attack such as would take place with a clean, newly exposed surface. Due to various causes, including insoluble film produced by ageing, oxidation, or contamination; run-down of surface energy at discontinuity lattices; adsorbed layers. Phenomenon prevents use of cyanide process to dissolve large particles of gold, but is sometimes used to aid froth-flotation by rendering specific minerals passive to collector agents.

paste Glass used for imitation gemstones.

Pastonian A temperate stage of the Pleistocene. See **Quaternary**.

patronite Ore of vanadium, perhaps VS_4.

pay Pay dirt, ore, rock, streak. Any mineral deposit which will repay efficient exploitation.

pay string The pipe through which oil or gas passes from the *pay zone* to the well head.

Pb Symbol for **lead**.

Pd Symbol for **palladium**.

peacock ore A name given to **bornite** or sometimes **chalcopyrite**, because they rapidly become iridescent from tarnish.

pearl A concretion of nacre from the inside of a mollusc shell around a foreign body such as a sand particle or parasite, and prized as a gemstone. See **gems and gemstones**.

pearl spar See **dolomite**.

peas See **coal sizes**.

peat The name given to the layers of dead vegetation, in varying degrees of alteration, resulting from the accumulation of the remains of marsh vegetation in swampy hollows in cold and temperate regions. Geologically, peat may be regarded as the youngest member of the series of coals of different rank, including brown coal, lignite and bituminous coal, which link peat with anthracite. Peat is very widely used as a fuel, after being air-dried, in districts where other fuels are scarce and in some areas, e.g. in Russia and Ireland, it is used to fire power stations. It is low in ash, but contains a high percentage of moisture and is bulky; specific energy content about 16 MJ/kg or 7000 Btu/lb.

pebble A small rounded fragment of rock between 4 and 64 mm diameter. Adj. *pebbly*, 'containing scattered pebbles'. See **particle size**, **Wentworth scale**.

pebble mill A **ball mill** in which selected pebbles or large pieces of ore are used as grinding media.

pectolite A silicate of calcium and sodium, with a variable amount of water, which crystallizes in the triclinic system. It occurs in aggregations like the zeolites in the cavities of basic eruptive rocks, and as a primary mineral in some alkaline igneous rocks.

pediment A broad and relatively flat rock surface abutting a mountain range, in an arid environment. It may be covered by a veneer of alluvium.

pedion A crystal form consisting of a single plane; well shown by some crystals of tourmaline which may be terminated by a pedion, with or without pyramid faces.

pedology The scientific study of soils.

pegmatite A term originally applied to granitic rocks characterized by intergrowths of feldspar and quartz, as in graphic granite; now applied to igneous rocks of any composition but of particularly coarse grain, occurring as offshoots from, or veins in, larger intrusive rock bodies, representing a flux-rich residuum of the original magma.

pelagic deposits A term applied to any accumulation of sediments under deep-water conditions.

Peléan eruption A type of eruption characterized by lateral explosions generating nuées ardentes.

Pélé's hair Long threads of volcanic glass which result from jets of lava being blown aside by the wind in the volcano of Kilauea, Hawaii.

pelitic gneiss A gneissose rock derived from the metamorphism of argillaceous sediments.

pelitic schist A schist of sedimentary origin, formed by the dynamothermal metamorphism of argillaceous sediments such as clay and shale.

pelletization Treatment of finely divided ore, concentrate or coal to form aggregates some 1–1.5 cm in diameter, for furnace feed, transport, storage or use (e.g. as coal briquettes). Powder, with suitable additives, is rolled into aggregates (green balls) in pelletizing drum and then hardened in a furnace by specialized baking methods.

pencatite A crystalline limestone which contains brucite and calcite in approximately equal molecular proportions.

pencil stone The name given to the compact variety of pyrophyllite, used for slate-pencils. The term *pencil ore* has been used for the broken splinters of radiating massive hematite, as they give a red streak.

penecontemporaneous Describing a process occurring in a rock very soon after its formation.

peneplain A gently rolling lowland, produced after long-continued denudation.

penetrant (1) Substance which increases the penetration of a liquid into porous material or between contiguous surfaces, e.g. alkylaryl sulphonate. (2) A wetting agent.

penetrating shower Cosmic-ray shower containing mesons and/or other penetrating particles.

penetration theory Theory that mass transfer across an interphase into a stirred liquid takes place by diffusive penetration of the solute into the liquid surface, which is continually being renewed, hence the rate of mass transfer is proportional to the square root of the *diffusion coefficient*.

penetration twins See **interpenetration twins**. Cf. **juxtaposition twins**.

pennine, penninite A silicate of magnesium and aluminium with chemically combined water. It crystallizes in the monoclinic system and is a member of the chlorite group.

Pennsylvanian The period of the Upper Palaeozoic era lying between the Mississippian and Permian. It covers an approx. time span from 320–290 million years and is the North American equivalent of the Upper Carboniferous in Europe. The corresponding system of rocks. See **Palaeozoic**.

Pensky-Martens test Standard test for determining the flash and fire points of oils. Based on closed or open cups depending on the nature of the oil under test.

pentagonal dodecahedron A form of the cubic system comprising twelve identical pentagonal faces.

pentanes C_5H_{12}. Low-boiling paraffin hydrocarbons. Pentane has a bp 36°C, rel.d. 0.63.

penthouse Protective covering for workmen at bottom of shaft. Also *pentice*.

pentlandite A sulphide of iron and nickel which crystallizes in the cubic system. It commonly occurs intergrown with pyrrhotite.

peralkaline An igneous rock in which the molecular proportion of $Na_2O + K_2O$ exceeds that of Al_2O_3. This usually produces alkaline pyroxenes or amphiboles in the rock.

percussion drilling System in which a string of tools falls freely on the rock being penetrated. Also, pneumatic drilling in which hammer blows are struck on drill shank. See **cable tool drilling**.

percussion figure A figure produced on the basal pinacoid or cleavage face of mica when it is sharply tapped with a centre punch. It consists of a 6-rayed star, 2 rays more prominent than the others, lying in the unique plane of symmetry.

perfect crystal A single crystal in which the arrangement of the atoms is uniform throughout.

perforating After a well has reached the producing zone the base is sealed with cement and the sides need to be pierced to allow ingress. Often done by a special gun with radial bores from which charges fire projectiles through the casing.

periclase Native magnesia. Oxide of magnesium, which crystallizes in the cubic system. It is commonly found in metamorphosed magnesian limestones, but readily hydrates to the much commoner brucite.

pericline A variety of **albite** which usually occurs as elongated crystals. The name is also used for a type of twinning in feldspars (the *pericline law*).

peridot See **olivine**.

peridotite A coarse-grained ultramafic igneous rock consisting essentially of olivine, with other mafic minerals such as hypersthene, augite, biotite and hornblende, but free

from plagioclase. See **dunite, kimberlite**.
period A major unit of geological time, e.g. *Silurian*. See **geological column**.
periodic law See **periodic system**.
periodic system Classification of chemical elements into periods (corresponding to the filling of successive electron shells) and groups (corresponding to the number of valence electrons). Original classification by relative atomic mass (Mendeleev, 1869). Formerly *periodic law*.
periodic table The most common arrangement of the **periodic system**.
peristerite A whitish variety of albite, or oligoclase, which is beautifully iridescent.
perknite A family of coarse-grained ultramafic igneous rocks which consist essentially of pyroxenes and amphiboles, but contain no feldspar.
perlite An acid and glassy igneous rock which exhibits perlitic structure.
perlitic structure A structure found in glassy igneous rocks, which consists of systems of spheroidal concentric cracks produced during cooling.
permafrost In arctic and subarctic regions, the permanently frozen soil.
permanent hardness Of water, the hardness which remains after prolonged boiling is *permanent hardness*. Due to the presence of calcium and magnesium chlorides or sulphates.
permeability The ability of a rock to transmit fluids, esp. water, oil and gas.
permeameter Instrument for measuring static magnetic properties of ferromagnetic sample, in terms of magnetizing force and consequent magnetic flux.
Permian The youngest geological period of the Palaeozoic era, lying between the Carboniferous and the Triassic. It covers a time span of approx. 290–245 million years and is named after the type area of Perm in the USSR. The corresponding system of rocks. See **Palaeozoic**.
permitted explosives Explosives which may be used in mines, under specified conditions, where a danger of explosion from flammable gas exists.
Permo-Trias The Permian and Triassic systems considered together, as is commonly done in areas such as the British Isles where the rocks are of similar facies.
perovskite Calcium titanate, with rare earths, which crystallizes in the monoclinic system but is very close to being cubic. An accessory mineral in melilite-basalt, and in contact metamorphosed impure limestones. The crystal form of artificial ceramics which are **superconductors** at around 80 K.

perthite The general name for megascopic intergrowths of potassium and sodium feldspars, both components having been miscible to form a homogeneous compound at high temperatures, but the one having been thrown out of solution at a lower temperature, thus appearing as inclusions in the other. Perthite may also be formed by the replacement of potassium feldspar by sodium feldspar. See **microperthite**.
perthosite A type of soda-syenite consisting to a very large extent of perthitic feldspars, occurring at Ben Loyal and Loch Ailsh in Scotland.
petalite A silicate of lithium and aluminium which crystallizes in the monoclinic system. It typically occurs in granite pegmatites.
petrifaction, petrified Terms applied to any organic remains which have been changed in composition by molecular replacement but whose original structure is in large measure retained. *Petrified wood* is wood which has had its structure replaced by e.g. calcium carbonate or silica. Many of the original minute structures are preserved.
petrochemicals Chemicals derived from crude oil or natural gas. They include light hydrocarbons such as butene, ethene and propene, obtained by fractional distillation or catalytic cracking.
petrographic province Region characterized by a group of genetically related rocks, e.g. Andes. See **comagmatic assemblage**.
petrography Systematic description of rocks, based on observations in the field, on hand specimens, and on thin microscopic sections. Cf. **petrology**.
petroleum Naturally-occurring green to black coloured mixtures of crude hydrocarbon oils, found as earth seepages or obtained by boring. Petroleum is widespread in the Earth's crust, notably in the US, USSR and the Middle East. In addition to hydrocarbons of every chemical type and boiling range, petroleum often contains compounds of sulphur, vanadium etc. Commercial petroleum products are obtained from crude petroleum by distillation, cracking, chemical treatment etc.
petroleum coke Nearly pure carbon formed during the refining of crude oil by high temperature carbonization of the heavy residues.
petroleum reservoirs See panels on pp. 174–5.
petrology The study of rocks which includes consideration of their mode of origin, present conditions, chemical and mineral composition, their alteration and decay.

petroleum reservoirs

Natural subsurface porous and permeable rocks to which oil or gas has migrated and accumulated under adequate trap conditions. The trap is formed by an overlying or up-dip impermeable cap. There are various kinds of structural traps in which petroleum can occur, some of considerable complexity. In *anticlinal traps* (Fig. 1, below) the oil or gas migrates to the top of an anticlinal structure beneath an impervious cap rock.

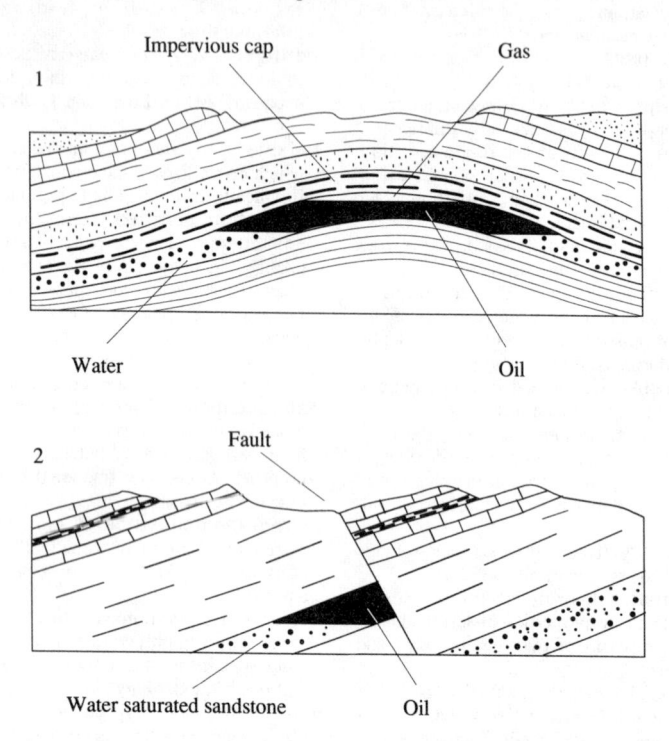

Fault traps (Fig. 2, above) occur when oil in a pervious bed is prevented from escaping by an impervious bed across a fault.

In *diapiric* or *piercement salt dome* (Fig. 3, opposite) the plastic salt has squeezed up and bent or ruptured the overlying beds, to produce potential reservoirs. These occur on the flanks of the salt dome or above it and in small associated fault traps.

The variation in sediments as originally deposited, and subsequent compaction may allow oil pools in *stratigraphic traps* to occupy lenticular bands of porous sandstone that pass laterally as well as vertically into clay or shale (Fig. 4, opposite). Stratigraphic traps are more difficult to locate than structural ones.

The distinction of these various types of reservoir is not always clear cut. In Figs. 2, 3 and 4 gas has been omitted for clarity.

continued on next page

petroleum reservoirs (contd)

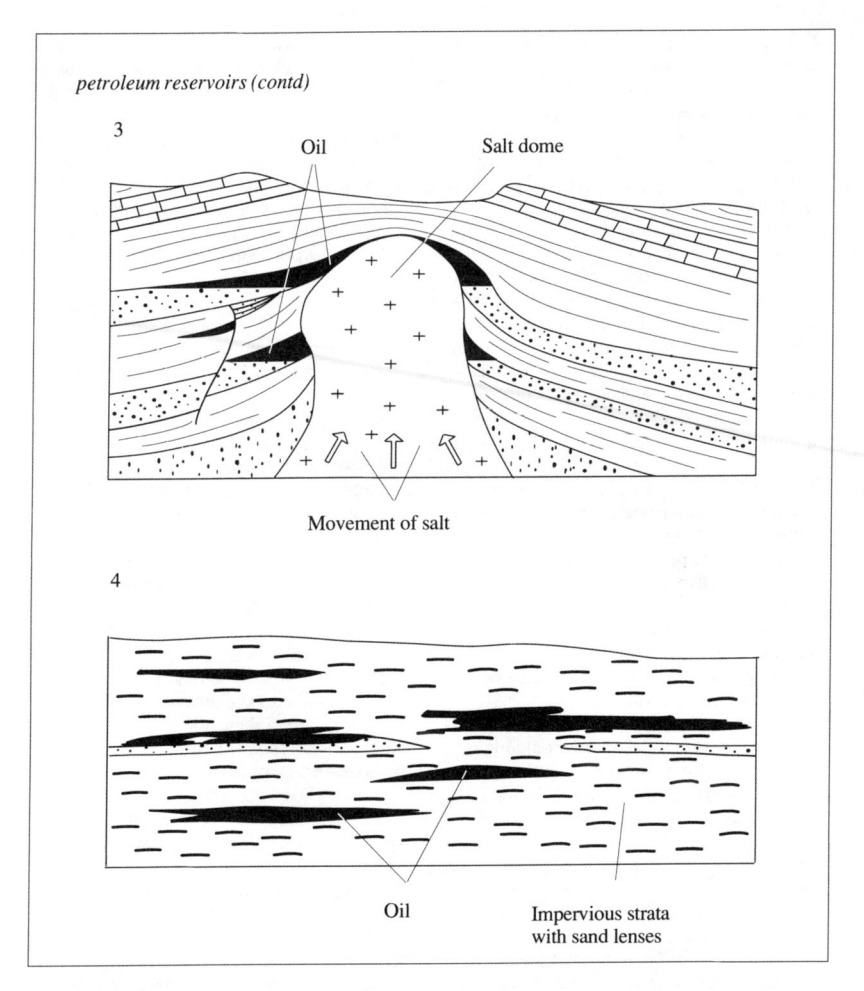

3

Oil — Salt dome — Movement of salt

4

Oil — Impervious strata with sand lenses

petzite A telluride of silver and gold. It is steel-grey to black and often shows tarnish.

pH See **pH value**.

phacoidal structure A rock structure in which mineral or rock-fragments of lens-like form are included. The term is applicable to igneous rocks containing softened and drawnout inclusions; also to metamorphic rocks such as crush-breccias and crush-conglomerates; and to certain gneisses. (Gk. *phakos*, lentil.)

phacolith A minor intrusion of igneous rock occupying the crest of an anticlinal fold. Its form is due to the folding, hence it is not the cause of the uparching of the roof. Cf. **laccolith**.

phanerocrystalline Said of an igneous rock in which the crystals of all the essential minerals can be discerned by the naked eye.

Phanerozoic The span of *obvious life*. More precisely the unit of geological time that comprises the Palaeozoic, Mesozoic and Cenozoic eras. See **geological column**.

pharmacolite A hydrated arsenate of calcium which crystallizes in the monoclinic system. It is a product of the late alteration of mineral deposits which carry arsenopyrite, and the arsenical ores of cobalt and silver.

pharmacosiderite Hydrated arsenate of iron. It crystallizes in the cubic system, and is a product of the alteration of arsenical ores.

phase The sum of all those portions of a material system which are identical in chemical composition and physical state, and are separated from the rest of the system by a distinct interface called the *phase boundary*.

phase reversal An interchange of the components of an emulsion; e.g. under certain conditions an emulsion of an oil in water may become an emulsion of water in the oil.

phase rule A generalization of great value in the study of equilibria between phases. In any system, $P+F = C+2$, where P is the number of phases, F the number of **degrees of freedom** (1), C the number of components.

phenakite A silicate of beryllium, crystallizing in the rhombohedral system. It is commonly found as a product of pneumatolysis. Sometimes cut as a gemstone, having brilliance of lustre but lacking fire. The name (Gk. 'the deceiver') refers to the frequency with which it has been confused with quartz.

phengite An end-member variety of muscovite mica, in which Si:A > 13:1 and some aluminium is replaced by magnesium or iron.

phenocrysts Large (megascopic) crystals, usually of perfect crystalline shape, found in a fine-grained matrix in igneous rocks. See **porphyritic texture**.

phi grade scale *Phi scale*. A logarithmic scale used for the mechanical analysis of sediments. It is expressed as the negative logarithm to base 2 from +10 for $1/1024$ mm to −5 for 32 mm diameter of grain on the Wentworth scale. Symbol, ϕ.

phillipsite A fibrous zeolite; hydrated silicate of potassium, calcium and aluminium, usually grouped in the orthorhombic system.

phlogopite Hydrous silicate of potassium, magnesium, iron and aluminium, crystallizing in the monoclinic system. It is a magnesium mica, and is usually found in metamorphosed limestones or in ultrabasic igneous rocks. Not so good an electrical insulator as **muscovite** at low temperatures, but keeps its water of composition until 950°C. Also *amber mica*.

pH meter Specialized millivolt meter which measures potential difference between reference electrodes in terms of pH value of the solution in which they are immersed.

phonochemistry The study of the effect of sound and ultrasonic waves on chemical reactions.

phonolite A fine-grained igneous rock of intermediate composition, consisting essentially of nepheline, subordinate alkali feldspar (sanidine) and sodium-rich coloured silicates. Colloq. *clinkstone*, because it rings under the hammer when struck.

phonon A quantum of lattice vibrational energy in a crystal. The thermal vibrations of the atoms can be described in terms of *normal modes* of oscillation, each mode specifying a correlated displacement of the atoms. The energy of a mode is quantized and can only exchange energy with other modes in units of $h\nu$, where ν is the mode frequency and h is Planck's constant. This quantum is the *phonon*. Lattice vibrations can be described in terms of waves in the lattice and the phonon is the particle of the field of the mechanical energy of a crystal. (Cf. **photon** as a particle of the electromagnetic field.) Phonons have been identified by neutron inelastic scattering experiments.

phonon dispersion curve A curve showing *frequency* as a function of *wavevector* for the modes of lattice vibrations in a crystal. Determined by neutron spectroscopy, their interpretation provides a powerful method of testing various models of interatomic forces in crystals.

-phore A suffix which denotes a group of atoms responsible for the corresponding property, e.g. *chromophore*.

phosgenite A chlorocarbonate of lead, crystallizing in the tetragonal system. It is found in association with cerussite.

phosphates (V) Salts of phosphoric acid. There are three series of orthophosphates: MH_2PO_4, M_2HPO_4 and M_3PO_4; the first yield acid, the second are practically neutral, and the third alkaline, aqueous solutions. Metaphosphates, MPO_3 and pyrophosphates, $M_4P_2O_7$, are also known. All phosphates give a yellow precipitate on heating with ammonium molybdate in nitric acid.

phosphatic deposits Beds containing calcium phosphate which are formed esp. in areas of low rainfall, and which may be exploited as sources of phosphate. See **phosphatic nodules**.

phosphatic nodules Rounded masses containing calcium phosphate, which are formed in the sea floor.

phosphorescence Luminescence which persists for more than 0.1 nanoseconds after excitation. See **fluorescence**.

phosphorite, rock-phosphate The fibrous concretionary variety of **apatite**.

phosphorus A non-metallic element in the fifth group of the periodic system. Symbol P, at.no. 15, r.a.m. 30.9738, valencies 3,5. White phosphorus is a waxy, poisonous, spontaneously flammable solid, mp 44°C, bp 282°C, rel.d. 1.8–2.3. Phosphorus occurs widely and abundantly in minerals (as phosphates) and in all living matter. There are numerous independent phosphate minerals in pegmatites and ores, and among their alteration products, but mainly the phosphate is *apatite*, a widespread accessory mineral of igneous rocks, which also occurs in sedimentary phosphate deposits, guano and bones. Phosphorous is manufactured by heating calcium phosphate with sand and carbon in an electric furnace. It is used mainly in the manufacture of phosphoric acid for phosphate fertilizers; also used in matches and organic synthesis.

phot The CGS unit of illumination. 1 phot = 1 lumen cm^{-2}, = 10^4 lumens m^{-2}.

photochemistry The study of the chemical effects of radiation, chiefly visible and ultraviolet, and of the direct production of radiation by chemical change.

photodisintegration The ejection of a neutron, proton or other particles from an atomic nucleus following the absorption of a *photon*. A γ-ray photon with energy 2.23 MeV can cause a deuteron nucleus to emit both a neutron and a proton.

photodissociation Dissociation produced by the absorption of radiant energy.

photoelasticity Phenomenon whereby strain in certain materials causes the material to become **birefringent**. Coloured fringes are observed when the transmitted light is viewed through crossed Nicol prisms.

photoelectron spectroscopy When visible light, ultraviolet light or X-rays are used as the excitation source, the energy of *photoelectrons* emitted from the material can be analysed to give information on surfaces, interfaces and bulk materials. Also used to deduce binding energies for deep core levels in atoms with a high degree of precision. Also *electron spectroscopy, photoionization spectroscopy.*

photographic borehole survey Check on the orientation and angle of a long borehole by insertion of a special camera which photographs a magnetic needle and a clinometer at a known distance down.

photolysis Molecular decomposition or dissociation as the result of the absorption of light.

photometry Volumetric analysis in which the end-point of a reaction is determined from colour changes detected by photoelectric means.

photon Quantum of light of electromagnetic radiation of energy $E = h\nu$, where h is Planck's constant and ν is the frequency. The photon has zero rest mass, but carries momentum $h\nu/c$, where c is the velocity of light. The introduction of this 'particle' is necessary to explain the photoelectron effect, the Compton effect, atomic line spectra, and other properties of electromagnetic radiation.

photophoresis The migration of suspended particles under the influence of light.

phreatic gases Those gases which are of atmospheric or oceanic origin and which are generated by contact with ascending magma.

pH value A logarithmic index for the hydrogen ion concentration in an aqueous solution. Used as a measure of acidity of a solution; given by pH = $\log_{10}(1/(H^+))$, where (H^+) is the hydrogen-ion concentration. A pH below 7 indicates acidity, and one above 7 alkalinity, at 25°C.

phyllite A name which has been used in different senses: (1) for the pseudohexagonal platy minerals (*phyllosilicates*) including mica, chlorite and talc (by some French authors); (2) for argillaceous rocks in a condition of metamorphism between slate and mica-schist (by most English authors). Phyllite in the latter (usual) sense is characterized by a silky lustre due to the minute flakes of white mica which, however, are individually too small to be seen with the naked eye.

phyllosilicates Those silicate minerals having an atomic structure in which SiO_4 groups are linked to each other to form continuous sheets, e.g. *talc*. See **silicates**.

physical chemistry The study of the dependence of physical properties on chemical composition, and of the physical changes accompanying chemical reactions.

physical geology The processes involved in the inorganic evolution of the Earth and esp. its morphology.

physiography The science of the surface of the Earth and the interrelations of air, water and land.

Piacenzian The highest stage of the Neogene (Pliocene). See **Tertiary**.

pickeringite Magnesia alum. Hydrated sulphate of aluminium and magnesium, crystallizing in the monoclinic system. It usually occurs in fibrous masses, and is formed by the weathering of pyrite-bearing schists.

picking belt Sorting belt, on which run-of-mine ore is displayed so that pickers can remove waste rock, debris or a special mineral constituent, which is not to be sent to the mill for treatment.

picotite A dark-coloured spinel containing iron, magnesium, aluminium; a chromium-bearing hercynite.

picrite An ultramafic coarse-grained igneous rock, consisting essentially of olivine and other ferromagnesian minerals, together with a small amount of plagioclase. Also used for volcanic rocks.

piedmont glacier A glacier of the 'expanded foot' type; one which, after being restricted within a valley, spreads out on reaching the flat ground into which the latter opens.

piedmont gravels Accumulations of coarse breccia, gravel and pebbles brought down from high ground by mountain torrents and spread out on the flat ground where the velocity of the water is checked. Literally, mountain foot gravels, typical of the outer zone of arid areas of inland drainage such as the Lop Nor Basin in Chinese Turkestan.

piedmontite A hydrated silicate of calcium, aluminium, manganese and iron, crystallizing in the monoclinic system; a member of the zoisite group. Also *manganepidote, piemontite.*

piezochemistry The study of the effect of high pressures in chemical reactions.

pig See **go-devil.**

pigeonite One of the monoclinic pyroxenes, intermediate in composition between clinoenstatite and diopside. It is poor in calcium, has a small optic axial angle and occurs in quickly chilled lavas and minor intrusions.

pillar Column of unserved ore left as roof support in stope.

pillar-and-stall See **bord-and-pillar.**

pillow lava A lava flow exhibiting pillow structure, generally formed in a subaqueous environment.

pillow structure A term applied to lavas consisting of ellipsoidal and pillow-like masses which have cooled under subaqueous conditions. The spaces between the pillows consist, in different cases, of chert, limestones, or volcanic ash.

pilotaxitic texture The term applied to the groundmass of certain holocrystalline igneous rocks in which there is a felt-like interweaving of feldspar microlites. Cf. **hyalopilitic texture.**

pinacoid An open crystal form which includes two precisely parallel faces.

pine oil Commercial frothing agent widely used in flotation of ores. Distillate of wood, varying somewhat in chemical composition according to timber used and scale of heating.

pingo A raised area in permafrost due to the local expansion of an ice mass.

pinite Hydrated silicate of aluminium and potassium which is usually amorphous. It is an alteration product of cordierite, spodumene, feldspar etc., close to muscovite in composition. See **pagodite.**

pipe A vertical conduit into the crust through which volcanic materials have passed. It may be filled with volcanic breccia and is often mineralized. See **chimney.**

pipeclay A white clay, nearly pure and free from iron, used in the pottery industry.

pipe factor Compensating factor used when samples are taken from casings which go into or through running sands or gravels, so that the amount of material raised from the section traversed by the drill does not correspond with the volume enclosed by the pipe.

pipe sample Sample obtained by driving an open-ended pipe into a heap of material and withdrawing the core it collects.

pisolite A type of limestone built of rounded bodies (*pisoliths*) similar to oöliths, but of less regular form and 2 mm or more in diameter.

pisolitic A term descriptive of the structure of certain sedimentary rocks containing pisoliths (see **pisolite**). Calcite-limestones, dolomitic limestones, laterites, iron ores and bauxites may be pisolitic.

pistacite See **epidote.**

pit (1) A place whence minerals are dug. (2) The shaft of a mine. The *pit eye* is the bottom, whence daylight is visible; the *pit frame* is the superstructure carrying poppet head and sheaves. The *pit head* is the surface landing-stage.

pitch (1) A dark-coloured, fusible, more or less solid material containing bituminous or resinous substances, insoluble in water, soluble in several organic solvents. Usually obtained as the distillation residue of tars. (2) Orientation of a linear element, e.g. mine tunnel, mineral lineation, *slickensides*, on an inclined surface, whereby the angle of pitch is measured between the inclined element and the horizontal. See **folding.**

pitchblende The massive variety of uraninite. Radium was first discovered in this mineral. This and helium are due to the disintegration of uranium.

pitchstone A volcanic glass which has a pitch-like (resinous) lustre and contains crystallites and microlites. It is usually of acid to subacid composition, contains a notable amount of water (4% or more), and is usually intrusive.

pitmen Men employed in shaft inspection and repair.

pixel The smallest element with controllable colour and intensity in a video display or in *computer graphics*; from *picture element.*

placers, placer deposits Superficial deposits, chiefly of fluviatile origin, rich in heavy ore minerals such as cassiterite, native gold, platinum, which have become concentrated in the course of time by long-continued disintegration and removal from the neighbourhood of the lighter associated minerals. See **auriferous deposit.**

plagioclase feldspars An isomorphous series of triclinic silicate minerals which consist of **albite** and **anorthite** combined in all proportions. They are essential constituents of the majority of igneous rocks. See **oligoclase, andesine, bytownite, labradorite, anorthite.**

plane of polarization The plane containing the incident and reflected light rays and the normal to the reflecting surface. The magnetic vector of plane-polarized light lies in this plane. The electric vector lies in the *plane of vibration* which is that containing the plane-polarized reflected ray and the normal to the plane of polarization. The description of plane-polarized light in terms of the plane of vibration is to be preferred as this specifies the plane of the electric vector.

plane of symmetry In a crystal, an imaginary plane on opposite sides of which faces, edges, or solid angles are found in similar positions. One half of the crystal is hence a mirror image of the other.

plane polarization When the vibrations of a transverse wave are confined to one direction, the wave is said to be plane-polarized. For electromagnetic waves the direction of the electric vector of a plane-polarized wave is the *plane of vibration*; the magnetic vector lies in a plane at right angles to this. Light reflected at the *Brewster angle* is plane-polarized. Polarization of radio waves and microwaves occurs as a result of the way these waves are transmitted from aerials.

planetology The study of the composition, origin and distribution of matter in the planets of the solar system.

plankton Animals and plants floating in the waters of seas, rivers, ponds and lakes, as distinct from animals which are attached to, or crawl upon, the bottom; esp. minute organisms and forms, possessing weak locomotor powers or unable to swim actively. See **nekton**.

plaque White-enamelled saucer-shaped disk used in spot checking of products made during ore treatment. It has taken the place of the old vanning shovel. A sample is gently manipulated on it with added water, to separate the light from the heavy constituents.

plasma A bright-green translucent variety of cryptocrystalline silica (*chalcedony*). It is used as a semi-precious gem.

plastic deformation The permanent deformation of a rock or mineral following the application of stress.

plate The rigid structures of the lithosphere, of about continental size, consisting of the crust and the upper mantle, floating on the viscous lower mantle. See **plate tectonics**.

plate amalgamation Trapping of metallic gold on an inclined plate made of copper or an alloy, which has been coated with a pasty film of mercury. Method largely superseded by use of **strake**.

plateau basalts Basic lavas of basaltic composition resulting from fissure eruptions and occurring as thin, widespread flows, forming extensive plateaux (e.g. the Deccan in India). See **dykes and sills**.

plateau eruptions Volcanic eruptions by which extensive lava flows are spread in successive sheets over a wide area and eventually build a plateau, as in Idaho. See **fissure eruption**.

plateau gravel Deposits of sandy gravel occurring on hilltops and plateaus at heights above those normally occupied by river-terrace gravels. Originally deposited as continuous sheets, plateau gravel has been raised by earth movements to its present level and deeply dissected. Of Pliocene or early Pleistocene age in the main.

plate tectonics See panels on pp. 180–1.

platforming Catalytic process for reforming low-grade into high-grade petroleum components using platinum; hence the name. See **catalytic reforming**.

platform tree *Christmas tree* with all the necessary valves for controlling the flow of oil from a producing platform.

platinum A metallic element. Symbol Pt, at.no. 78, r.a.m. 195.09, rel.d. at 20°C 21.45, electrical resistivity at 20°C 9.97×10^{-8} ohm metres, mp 1773.5°C, bp 3910°C, Brinell hardness 47. Platinum is the most important of a group of six closely related rare metals, the others being osmium, iridium, palladium, rhodium and ruthenium. It is heavy, soft and ductile, immune to attack by most chemical reagents and to oxidation at high temperatures. Used for making jewellery, special scientific apparatus, electrical contacts for high temperatures and for electrodes subjected to possible chemical attack. Also used as a basic metal for resistance thermometry over a wide temperature range. Native platinum is usually alloyed with iron, iridium, rhodium, palladium or osmium, and crystallizes in the cubic system. Its abundance in the Earth's crust is only 0.01 ppm, occurring as the metallic element and in the arsenide *sperrylite*, ($PtAs_2$), sulphides and a few other binary compounds.

P lattice Abbrev. for *primitive crystal lattice*. See **unit cell**.

playa lake Shallow lake formed in a flat arid or semi-arid region in the wet season and drying out in the summer.

Pleistocene The epoch of geological time following the Tertiary, and covering a time span of approx. the last 2 million years. It was during this period that ice covered a large part of the northern hemisphere; hence it has been called the *Great Ice Age*.

Pleistogene See **Quaternary**.

pleochroic haloes Dark-coloured zones around small inclusions of radioactive

plate tectonics

Plate tectonics is the concept of global tectonics in which huge plates of the lithosphere are considered to act as relatively rigid slabs floating on the relatively plastic asthenosphere. Major structures and processes in the crust are associated with the movement of these plates, including mid-oceanic ridges, mountain building, oceanic trenches, major tear faults, earthquake zones and volcanic belts.

The major plates of the Earth and some of these associated features are shown in Fig. 1 (opposite). The plates move away from the mid-oceanic ridges at *divergent plate boundaries*. The space between the separating plates is filled by volcanic rock, welling up from below to form new crust (new sea floor), a process called *sea-floor spreading* (Fig. 2). While the sea-floor crust has been moving away from the mid-oceanic ridges, the polarity of the Earth's magnetic field has reversed many times resulting in **magnetic anomaly patterns** in the new crust. There are also many transform faults across the axes of spreading.

At the leading edge, as the plate moves, there is a *convergent plate boundary*. The collision of the plates results in many earthquakes and in continental regions may cause the development of mountain chains, e.g. the Himalayas between the Indian and Eurasian plates (Fig. 3). When plates bearing oceanic crust collide, one may plunge beneath the other to produce a *subduction zone*. Island arcs with volcanoes and deep oceanic trenches may then develop (Fig. 4).

If the plates slide past each other, they may form *shear boundaries*, e.g. along the San Andreas Fault in California, where the Pacific Plate moves against the North American Plate.

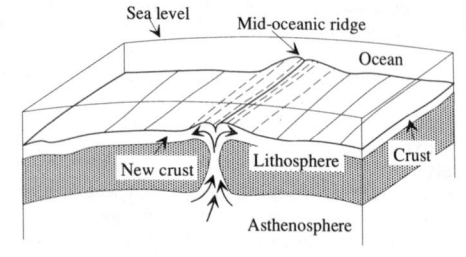

Fig. 2. Sea-floor spreading at a divergent plate boundary

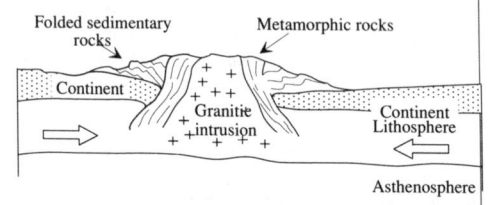

Fig. 3. Mountain range at the convergence of two continental plates

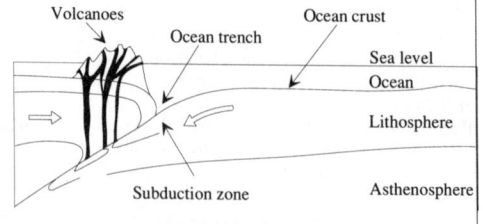

Fig. 4. Collision of oceanic plates

minerals which are found in certain crystals, notably biotite. The colour and pleochroism of the zones are stronger than those of the surrounding mineral, and result from radioactive emanations during the conversion of uranium or thorium into lead.

pleochroism The property of a mineral by which it exhibits different colours in different crystallographic direction on account of the selective absorption of transmitted light.

pleonaste Oxide of magnesium, iron and aluminium, with Mg:Fe from 3 to 1, crystallizing in the cubic system. It is a member of the

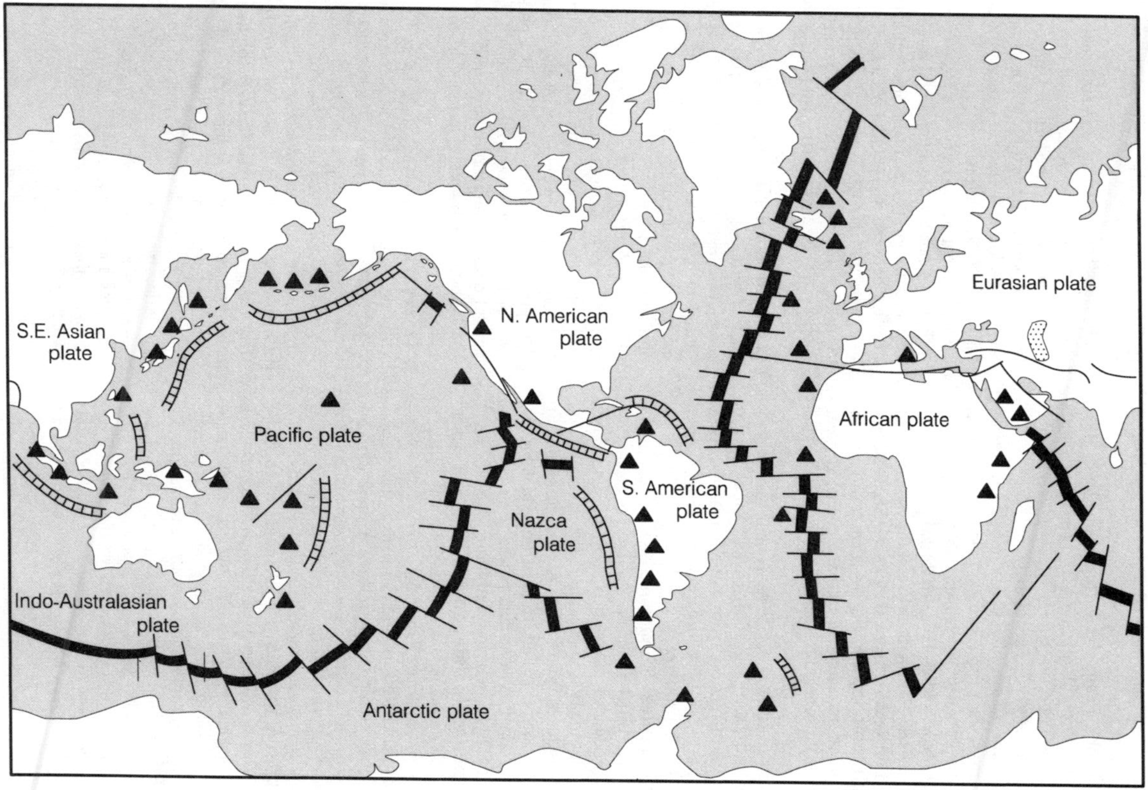

plate tectonics (contd)

Fig. 1.

The major plates are named in the map

⊢▮ Mid-oceanic ridges, offset by transform faults

≈≈≈ Oceanic trenches

— Collision zones and other plate boundaries

▲ Volcanoes

Pliensbachian A stage in the Lower Jurassic. See **Mesozoic**.

Plinian eruption A type of volcanic eruption characterized by repeated explosions.

Pliocene The epoch which followed the Miocene and preceded the Pleistocene. See **Tertiary**.

plug A vertical cylinder of solidified magma or pyroclastic material which represents the feeder pipe of a former volcano.

plumb- From the L. *plumbum*, lead; e.g. plumbic chloride ($PbCl_4$), plumbous chloride ($PbCl_2$).

plumbago See **graphite**.

plume Ascending partly molten material from the mantle believed to be responsible for intraplate volcanism.

plunging fold Fold whose axis is not horizontal. The angle between the axis and the horizontal is called the *plunge*. See **folding**.

plutonic intrusions A term applied to large intrusions which have cooled at great depth beneath the surface of the Earth. Also *major intrusions*.

plutonic rocks See panel on p.183.

plutonism The formation of rocks by the solidification from molten magma. This theory was put forward in the 18th century. Cf. **neptunism**.

plutonites See **plutonic rocks**.

plutonium Element. Symbol Pu, at. no. 94, product of radioactive decay of *neptunium*. Has many isotopes, the fissile isotope ^{239}Pu, produced from ^{238}U by neutron absorption in a reactor, being the most important for the production of nuclear power.

Pm Symbol for promethium.

pneumatic flotation cell Cell in which low-pressure air is blown in and diffused upward through the cell.

pneumatic lighting Use of a small compressed-air motor to drive a small dynamo which is connected to an electric lamp, thus avoiding wiring extensions underground where a compressed air service exists.

pneumatolysis The alteration of rocks by the concentrated volatile constituents of a magma, effected after the consolidation of the main body of magma. See **greisenization, kaolinization, tourmalinization**.

Po Symbol for **polonium**.

pockmark Concave cone-shaped depressions that occur in profusion in unconsolidated fine-grained seabed sediments off Canada and in the North Sea. They are typically 15–45 m across and 5–10 m deep, and were first recognized in 1970. They may be due to the escape of biogenic gas and their presence may be a potential hazard for engineering structures.

Poetsch process Freezing of waterlogged strata by circulation of refrigerated brine through the boreholes surrounding the section through which a shaft or tunnel is to be driven.

poikilitic texture Texture in igneous rocks in which small crystals of one mineral are irregularly scattered in larger crystals of another, e.g. small olivines embedded in larger pyroxenes, as in some peridotites. Also *poecilitic texture*.

poikiloblastic A textural term applicable to metamorphic rocks in which small crystals of one mineral are embedded in large crystals of another. The texture is comparable with the **poikilitic texture** of igneous rocks.

point defect See **defect**.

poise CGS unit of viscosity. The viscosity of a fluid is 1 poise when a tangential force of 1 dyne per unit area maintains unit velocity gradient between two parallel planes. Equal to 1 dyne cm^{-2} s, or in SI units 10^{-1} N m^{-2} s. Named after the physicist Poiseuille. Abbrev. *P*. See **viscosity, coefficient of viscosity**.

poisoning Loading of resin sites in ion-exchange with ions of unwanted species which therefore prevent the capture of those required by the process. In liquid-liquid ion exchange, fouling of the organic solvent in similar manner.

polar axis A crystal or symmetry axis to which no two- or four-fold axes are normal; thus the arrangements of faces at the two ends of such an axis may be dissimilar. The principal axis of tourmaline is a polar axis of three-fold symmetry, the top of the crystal being terminated by pyramid faces, the bottom end by a single plane in some cases.

polar crystal A crystal, such as sodium chloride, with ionic bonding between atoms.

polarimeter An instrument in which the optical activity of a liquid is determined by inserting Nicol prisms in the path of a ray of light before and after traversing the liquid.

polarimetry The measurement of optical activity, esp. in the analysis of sugar solutions.

polariscope Instrument for studying the effect of a medium on polarized light. Interference patterns enable elastic strains in doubly refracting materials to be analysed. It may consist of a polarizer and an analyser, with facilities for placing transparent specimens between them. The analyser and polarizer were formerly Nicol prisms. Modern polariscopes use light-polarizing films, e.g. Polaroid film, instead. See **microscope, photoelasticity**.

plutonic rocks

Named after Pluto the Greek god of the infernal regions, plutonic rocks are formed at considerable depth in the Earth. They are almost always coarse-grained and do not include the volcanic or metamorphic rocks. The great granitic masses of the Alps, Andes, Rocky Mountains, Himalayas and other mountain ranges as well as those of former geological ages, e.g. Devon and Cornwall, and in the Highlands of Scotland, are examples of plutonic rocks.

There is considerable controversy about the mechanism by which granitic rocks form. It may be crystallization from a **magma**, or by chemical alteration of existing rocks, or by processes involving both mechanisms. The chemical and hence mineralogical composition shows considerable variation, and has given rise to a considerable number of rock names. There is widespread agreement to use an internationally based (IUGS) system, shown in the figure, in which the principal names are given. It is based on the *modal* composition, i.e. the actual minerals present.

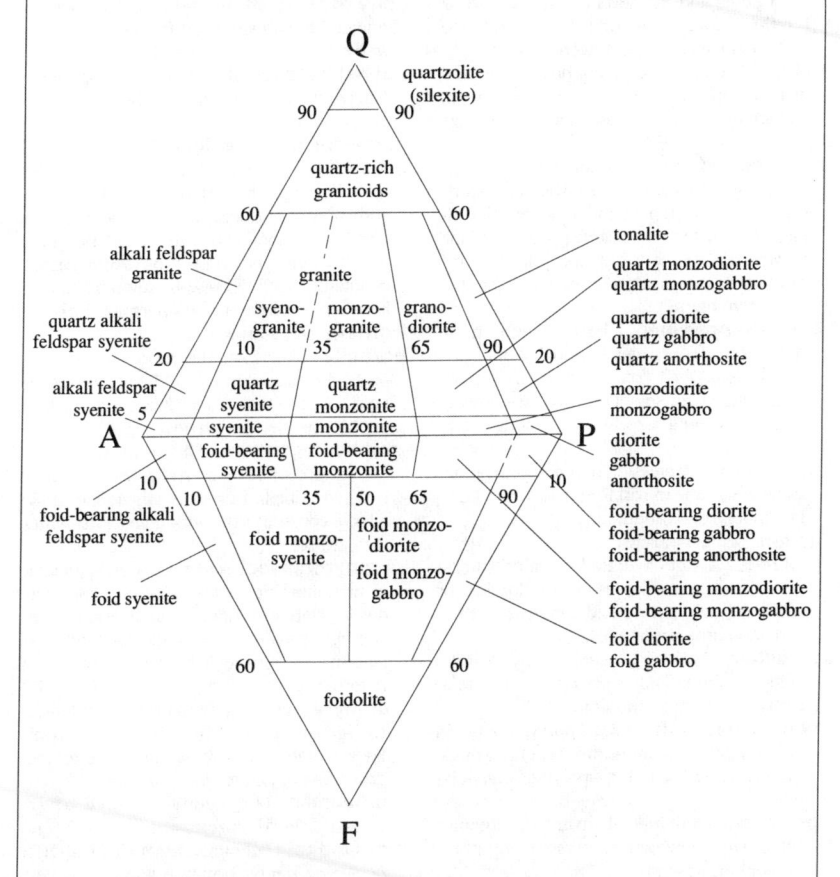

Classification and nomenclature of plutonic rocks with < 90% mafic (ferromagnesian and related) minerals

Q = quartz; A = alkali feldspar; P = plagioclase feldspar; F = feldspathoids ('foids'). For either triangle, the quantities Q+A+P, or F+A+P = 100%.

polarity A permanent property of a molecule which has an unsymmetrical electron distribution. All heteronuclear diatomic molecules are polar. See **dipole moment**.

polarity chron The time span of a **polarity chronozone** during which the Earth's magnetic polarity was predominantly of, or remained of one polarity.

polarity chronozone The fundamental polarity chronostratigraphical unit; the rocks of a polarity chron.

polarizer Prism of double refracting material, or Polaroid plate, which passes only plane-polarized light, or produces it through reflection. See **Nicol prism**.

polar wandering Movement of the magnetic poles and of the poles of the Earth's rotation through geological time. Can be partly explained by **plate tectonics**.

poling boards Fore-poling boards are used in tunnelling through loose (running) rock, and are driven horizontally ahead to support roof.

polished rod In an oil pumping station, the actuating rod which passes through the stuffing box at the top of the well; usually attached at the top to the cable going round the **horsehead**; at the lower end joined to the **pony rods** which are connected to the displacement pump.

polished specimen Characteristic hand specimen of ore, metal, alloy, compacted powder etc., one face of which is ground plane and mirror smooth by abrasive powder and/or polishing laps, or electrolytic methods.

polje A large depression found in some limestone areas, due in part to subsidence following underground solution.

pollen See **acritarchs**.

pollucite A rare hydrated alumino-silicate of caesium, occurring as clear colourless or white crystals with cubic symmetry. It occurs in granite pegmatites.

polonium Radioactive element. Symbol Po, at.no. 84. Important as an α-ray source relatively free from γ-emission.

poly- A prefix from Gk. *polys*, many. In chemistry (1) containing several atoms, groups etc; (2) denoting a polymer, e.g. polyethene.

polybasite Sulphide of silver and antimony, often with some copper, crystallizing in the monoclinic system.

polyhalite A hydrated sulphate of potassium, magnesium and calcium, $K_2MgCa_2(SO_4)_4.2H_2O$. Found in salt deposits.

polymerization The combination of several molecules to form a more complex molecule, usually by an *addition* or a *condensation* process. It is sometimes a reversible process.

polymorphism The property possessed by certain chemical compounds of crystallizing in several forms which are structurally distinct; thus TiO_2 (titanium dioxide) occurs as the mineral species *anatase, brookite* and *rutile*.

polyzoa See **bryozoa**.

pondermotive force In high-tension separation, the electrostatic force exerted on a particle as it passes through the field of a corona-type separator. In magnetic separation, the flux field intensity together with density of particle determines its deflection from a straight path.

pony rods In an oil well the rods which connect the **polished rod** to the pump. Also *drill rods*.

poppet head Headframe of hoisting shaft. The poppet is the bearing in which the winding pulley is set.

porcelain clay See **china clay**.

pore Cavity within or between particles in rock or aggregate. If these communicate with a free surface they are open, if not, closed. Number and size of cavities in a given volume determines **porosity**; degree of interconnection (ease of communication) between pores determines **permeability**.

Porifera See **sponges**.

porosity Of rocks, the ratio, usually expressed as a percentage, of the volume of the void space to the total volume of the rock.

porphyrite See **porphyry**.

porphyritic texture The term applied to the texture of igneous rocks which contain isolated euhedral crystals larger than those which constitute the groundmass in which they are set.

porphyroblastic A textural term applicable to metamorphic rocks containing conspicuous crystals in a finer groundmass, the former being analogous with the phenocrysts in a normal igneous rock, but having developed in the solid.

porphyry A general term used rather loosely for igneous rocks which contain relatively large isolated crystals set in a fine-grained groundmass, e.g. *granite porphyry*.

Portlandian The youngest stage of the Jurassic. See **Mesozoic**.

portlandite Calcium hydroxide, $Ca(OH)_2$, occurring rarely in nature but also in Portland cement, hence the name.

positive mineral A mineral in which the ordinary ray velocity is greater than that of the extraordinary ray, i.e. ω is less than ε. Quartz is a positive mineral for which ω = 1.544 and ε = 1.553. See **optic sign**.

potassium

A very reactive alkali metal, soft and silvery white. Symbol K, at.no. 19, r.a.m. 39.102, mp 63°C, bp 762°C, rel.d. 0.87. In the form of the element, it has little practical use, although its salts are used extensively. It is the eighth commonest element in the Earth's crust, with an abundance of 1.84%, and 380 ppm in seawater. It occurs as rock minerals particularly in *orthoclase feldspar*, $KAlSi_3O_8$, a major constituent of granitic and syenitic rocks, the micas especially *muscovite*, $KAl_3Si_3O_{10}(OH)_2$, *phlogopite*, $KMg_3AlSi_3O_{10}(OH)_2$ and *biotite*, $K(MgFe)_3(AlFe)Si_3O_{10}(OH)_2$, *leucite*, $KAlSi_2O_6$, a mineral of the undersaturated rocks, and zeolites, *apophyllite* and *phillipsite*.

Commercially potassium is obtained from the minerals of the evaporite deposits. These are mixed salts including *sylvine*, *carnallite*, *kainite*, *polyhalite*, *alunite* and *nitre*. *Sylvinite* is a name for mixtures of sylvite (KCl) and halite (NaCl). *Glauconite* is a sedimentary mineral containing potassium.

There are three naturally occurring isotopes of potassium, ^{39}K and ^{41}K, both stable isotopes, and ^{40}K with a half-life of 1.28×10^9 y. It breaks down to ^{40}Ca and 11% to ^{40}Ar, producing most of the argon of the Earth's atmosphere, and providing a significant amount of natural radioactivity and of radiogenic heat. The ratio, ^{40}Ar to ^{40}K, is used for dating rocks and minerals containing these elements. Potassium is an essential element of all organisms.

positive ore Ore blocked out on four sides in panels sufficiently small to warrant assumption that the exposed mineral continues right through the block as a calculable tonnage.

possible ore Ore probably existing, as indicated by the apparent extension of proved deposits, but not yet entered and sampled.

Post-Tertiary Name assigned to geological events which occurred after the close of the Tertiary era, i.e. during Pleistocene and Recent times.

potassium See panel on this page.

potassium alum Synonym for **alum**.

potassium-argon dating A method of determining the age in years of geological material, based on the known decay rate of ^{40}K to ^{40}A. See **potassium**.

potassium feldspar Silicate of aluminium and potassium, $KAlSi_3O_8$, occurring in two principal crystalline forms — *orthoclase* (monoclinic) and *microcline* (triclinic). Both are widely distributed in acid and intermediate rocks, esp. in granites and syenites, and the fine-grained equivalents. See **adularia**, **feldspar**, **sanidine**.

potassium mica See **muscovite**, **sericite**.

potential-determining ions Ions which leave the surface of a solid immersed in an aqueous liquid before saturation point (equilibrium) has been reached.

potentiometric titration A titration in which the end-point is indicated by a change in potential of an electrode immersed in the solution. This change in potential occurs as the solution changes from having excess substance to be determined to having excess titrant.

potstone A massive variety of **steatite**, more or less impure.

potters' clay See **ball clay**.

poughite A hydrated sulphate and telluride of iron, crystallizing in the orthorhombic system.

poundal Unit of force in the foot-pound-second system. The force that produces an acceleration of 1 ft s^{-2} on a mass of 1 pound. Abbrev. *pdl*. 32.2 pdl = 1lbf (lb wt); 1 pdl = 0.138 255 N.

pour point The temperature at which oil will no longer flow; determined by a laboratory test.

powder method See **powder photography**.

powder photography Method of identification of minerals or crystals, in which the powdered preparation, mounted vertically in a special camera in which it rotates, is subjected to a suitably modified beam of X-rays. The pattern, characterized by a set of concentric rings produced by rays diffracted at the Bragg angle relative to the incident beam, is diffracted on to a surrounding strip of film to give positive identification. Also *powder method*.

powered supports Pit props held to the roof of a coal seam by hydraulic pressure. In fully mechanized collieries they form part of a mechanically operated system.

pozzuolana, pozzolana A volcanic dust, first discovered at Pozzuoli in Italy, which has the effect, when mixed with mortar, of

ppm Abbrev. for *Parts Per Million*.
ppt, ppte Abbrevs. for *precipitate*.
Pr (1) Symbol for **praseodymium**. (2) The propyl radical C_3H_7—.
prase A translucent and dull leek-green variety of **chalcedony**.
praseodymium A metallic element, a member of the rare earth group. Symbol Pr, at.no. 59, r.a.m. 140.9077, rel.d. 6.48, mp 940°C. It closely resembles neodymium and occurs in the same minerals.
Precambrian See panel on p. 187.
precious stones See **gems and gemstones**.
precipitator (1) Device, usually in flue stacks, to remove solid particles from the effluent gas. Can be electrostatic, mechanical and/or chemical in action. (2) Device for purifying boiler feed water by adding chemicals to precipitate dissolved material.
pregnant solution In the **cyaniding** process for recovering gold, the gold-bearing solvent prior to precipitation and recovery. Also *pregs*.
pregs See **pregnant solution**.
prehnite Pale green and usually fibrous hydrated silicate of calcium and aluminium, crystallizing in the orthorhombic system. It occurs in altered igneous rocks.
préparation mécanique See **mineral processing**.
pressed amber See **ambroid**.
pressure leaching Chemical extraction of values from ore pulp in autoclaves, perhaps followed by precipitation as metal or refined salt.
Priabonian A stage of the Eocene. See **Tertiary**.
Pridoli The youngest epoch of the Silurian period.
primacord fuse Fuse based on pentaerythritol tetranitrate, with detonating effect. Speed of detonation, 7000 m/s.
primary A substance which is obtained directly, by extraction and purification, from natural raw material; e.g. benzene, phenol, anthracene are coal-tar *primaries*.
primary crushing The reduction of run-of-mine ore as severed to somewhere below 6 in. (15 cm) diameter lump, performed in jaw or gyratory breakers.
primary dispersion In geochemical prospecting, the diffusion of metals or other elements through the bedrock surrounding an ore body. Also *halo*.
primary gneissic banding Exhibited by certain igneous rocks, possibly due to the admixture of two magmas only partly miscible, injection of magma along bedding or foliation planes in the country rocks or selective mobilization of a rock under metamorphosis.
primary production Production which occurs when oil and gas flow naturally to the well bore without assistance. Also *primary recovery*. Cf. **secondary production, tertiary production**.
primary wave See **earthquakes**.
Primates An order of Mammals; dentition complete, but unspecialized; brain, esp. the neopallium, large and complex; pentadactyl; eyes well developed and directed forwards, the orbit being closed behind by the union of the frontal and jugal bones; basically arboreal animals; uterus a single chamber; few young produced, parental care lasting a long time after birth. Lemurs, Tarsiers, Monkeys, Apes and Man.
primer Cartridge in which detonator is placed in order to initiate explosion of string of high-explosive charges in borehole.
primitive crystal lattice See **unit cell**.
prism A hollow (open) crystal form consisting of three or more faces parallel to a crystal axis.
prism Triangular prisms made of glass and other transparent materials are used in a number of optical instruments. Equilateral prisms are used at minimum deviation in spectroscopes for forming spectra, 90° prisms are used for totally reflecting a ray through a right angle in binoculars, periscopes and range-finders.
prismatic system See **orthorhombic system**.
probable ore Inferred ore, partly exposed and sampled but not fully *blocked-out* in panels. See **blocking-out**.
processing routes Term used in considering alternative methods of treating a specified ore or concentrate. Main 'routes' are physical, chemical and pyro-metallurgical.
prochlorite See **ripidolite**.
production choke Aperture at the well head which limits the flow of oil from a well to the most economical or best allowable rate.
production platform Offshore platform from which the flow of oil from many wells is controlled and stored before onward transmission to the refinery.
production string The smallest casing in an oil well which reaches from the producing zone to the well head, up which the oil passes.
products pipeline Running from a refinery to distributors, a pipe often carrying many different products separated by **batching spheres**.

Precambrian

All rocks which were formed before the Cambrian. They consist of two divisions, the older a series of highly metamorphosed rocks, crystalline schists and gneisses with intrusive rocks, largely of Archaean age, e.g. the Lewisian Complex. Unconformably overlying this basement complex are Proterozoic sediments, e.g. the Torridonian (of Riphean age). There is little agreement on how Precambrian rocks should be divided, and almost no formal divisions. The table below shows a number of the more widely used terms.

In the highest Proterozoic there are impressions of soft-bodied animals and trace fossils (burrows and tracks) indicating a long period of earlier evolution. Primitive plant life existed well back into the the Archaean, and bacteria may have existed 3800 Ma ago. The algae are the only fossil group to have had a widespread development in the Precambrian. See **geological column**.

Eon			Era	Age, Ma
Phanerozoic				570
Precambrian	Proterozoic		Vendian	610
			Riphean	1650
			Aphebian	2500
	Archaean			4600

progradation Extension of the shoreline seawards by wave or current action.

progressive metamorphism Those progressive changes in mineral composition and texture which are observed in rocks within the aureole of contact metamorphism round igneous intrusions; also in rocks which have experienced regional metamorphism of varying degrees of intensity. The particular degree of metamorphism in the latter case is indicated by the 'metamorphic grade' of the rock.

promethium A radioactive element of the rare earth series. Symbol Pm, at.no. 61, having no known stable isotopes in nature. Its most stable isotope, ^{145}Pm, has a half-life of about 20 years.

promoter See **collector agent**.

prop Sturdy supporting post set across a lode or seam underground to hold up roof after excavation. See **powered supports**.

propagation of light Consists of transverse electromagnetic waves propagated through free space with a velocity of 2.9979×10^8 m/s and wavelengths about 400 to 800 nm. The ratio (velocity of light in free space/velocity in medium) is the refractive index of the medium. According to the special theory of relativity, the velocity of light is absolute and no body can move at a greater speed.

propane C_3H_8, an alkane hydrocarbon, a colourless gas at atmospheric pressure and temperature. Bp –45°C; found in crude petroleum. In liquid form it constitutes *liquefied petroleum gas* (LPG), a clean-burning fuel, used by some as a petrol substitute.

proppants Material like sand or special formulations used to keep open fissures in an oil-bearing sediment. See **fracking**.

prospect Area which shows sufficient promise of mineral wealth to warrant exploration. Methods of search include aerial survey, magnetometry, geophysical and geochemical tests, seismic probe, electroresistivity measurement, pitting, trenching and drilling.

protactinium A radioactive element. Symbol Pa, at.no. 91, half-life of 3.26×10^4 years. One radioactive isotope is ^{233}Pa which is an intermediate in the preparation of the fissionable ^{233}U from thorium.

Proterozoic A division of the Precambrian comprising the less ancient rocks of that system, and lying above the Archaean. See **Precambrian**.

protogenic Capable of supplying a hydrogen ion (proton).

proton The nucleus of the hydrogen atom; of positive charge and atomic mass number of unity. With neutrons, protons form the nucleus of all atoms, the number of protons being equal to the atomic number. It is the lightest *baryon*, 1.007 276 a.m.u. and the most stable. Beams of high energy protons produced in particle accelerators are used to study elementary particles.

protonic solvent Solvent that yields a proton, H^+, as the cation in self-dissociation.

protophilic Able to combine with a hydrogen ion.

Protozoa See panel on p. 189.

proustite Sulphide of silver and arsenic which crystallizes in the trigonal system. It is commonly associated with other silver-bearing minerals. Also *light red silver ore*, *ruby silver ore*. Cf. **pyrargyrite**.

proved reserves Tonnages of economically valuable ore which have been tested adequately by being blocked out into panels and sampled at close intervals.

psammitic gneiss A gneissose rock which has been produced by the metamorphism of arenaceous sediments.

psammitic schists Schists formed from arenaceous sedimentary rocks. Cf. **pelitic gneiss**, **pelitic schist**.

pseudo- A prefix which is sometimes used to indicate a tautomeric, isomeric, or closely related compound. Symbol ψ.

pseudocubic, pseudotetragonal, pseudohexagonal See **pseudosymmetry**.

pseudoleucite An aggregate showing the crystal shape of **leucite** but consisting mainly of potassium feldspar and nepheline.

pseudomalachite Phosphate and hydroxide of copper which resembles malachite and crystallizes in the monoclinic system.

pseudomorph A mineral whose external form is not the one usually assumed by its particular species, the original mineral having been replaced by another substance or substances.

pseudosymmetry A term applied to minerals whose symmetry elements place them on the borderline between two crystal systems, e.g. a mineral with the *c*-axis very nearly equal to the *b*- and *a*-axes might, on casual inspection, appear cubic, though actually tetragonal. It would be described as possessing *pseudocubic symmetry*. The phenomenon is due to slight displacement of the atoms from the positions which they would occupy in the class of higher symmetry. Also applied when the pseudosymmetry is due to twinning.

pseudotachylite Flinty crush-rock, resulting from the vitrification of rock fragments produced during faulting under conditions involving the development of considerable heat by friction, as in the Glencoe **cauldron subsidence**.

psilomelane A massive hydrated oxide of manganese which contains varying amounts of barium, potassium and sodium. It is a secondary mineral formed by alteration of manganese carbonates and silicates, and is used as an ore of manganese.

Pt Symbol for **platinum**.

pteropod ooze A calcareous deep-sea deposit which contains a large number of pteropod remains.

Pu Symbol for **plutonium**.

puddingstone A popular term for *conglomerate*. Hertfordshire Puddingstone, consisting of rounded flint pebbles set in a siliceous sandy matrix, is a good example.

puddling Concentration of diamond from 'blue ground' clays (weathered kimberlite) by forming an aqueous slurry in a mechanized stirring pan in which the heavier fraction is retained from periodic retrieval, while the lighter fraction overflows.

pug In metalliferous mining, the parting of soft clay which sometimes occurs between the walls of a vein and the country rock.

pulaskite A light-coloured alkali syenite consisting largely of alkali feldspar with subordinate ferromagnesian minerals and often a small amount of nepheline.

pulling by crystal Growing both metallic and non-metallic crystals by slowly withdrawing the crystal from a molten surface.

pulling tools The procedure whereby the drill string and bits are removed from the bore and stacked in the derrick before reuse. See **round trip**.

pulp Finely ground ore freely suspended in water at a consistency which permits flow, pumping or settlement in quiet conditions.

pulsator Mineral jig of Harz type. See **Harz jig**.

pulse-height analyser (1) A single or multichannel pulse height selector followed by equipment to count or record the pulses received in each channel. The multichannel units are known as *kicksorters*. (2) One

pulverized fuel   **pump jack**

Protozoa

The Protozoa (Gk. 'first animals') are the most primitive forms of animal life; they are usually very small and contain only one cell. This phylum ranges from the Cambrian or Precambrian to the Recent and includes two important fossil groups, the Foraminifera and the Radiolaria.

Foraminifera (often abbreviated to '*Forams*') are complex chambered marine organisms showing great diversity of form that allows them to be used as zone fossils, especially for the Tertiary. The shell is usually calcareous and may occur in very large numbers to become a major rock-forming constituent. *Nummulites* is a large disk-like genus which occurs in south-eastern England, and is the main constituent of the *Nummulite Limestone*, a thick Eocene formation that extends from the Mediterranean region to eastern Asia.

In the **Radiolaria** each individual generally has a siliceous skeleton surrounding a small mass of protoplasm. The skeleton may consist of isolated spicules or form a lattice-like array, assuming forms of great complexity. The radiolaria are entirely marine organisms of the open sea and range from Cambrian or earlier to Recent. At great depths in the oceans, calcium carbonate is more soluble than silica, and the sediments deposited there are siliceous. *Radiolarian ooze* is formed by the accumulation of vast numbers of radiolarians and occurs over large areas of the Central Pacific and Indian Oceans. *Radiolarian chert*, or *radiolarite*, is cryptocrystalline siliceous rock containing the remains of Radiolaria.

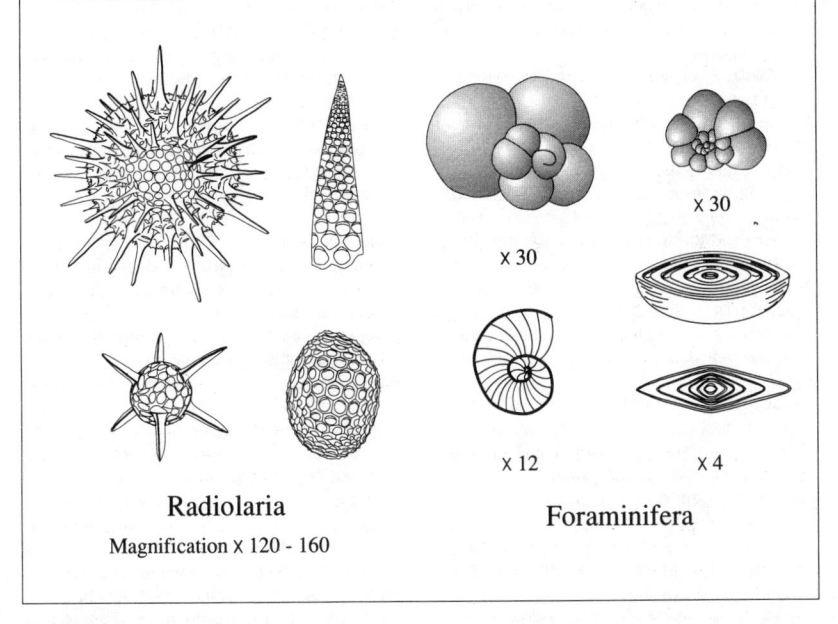

Radiolaria — Magnification × 120 - 160

Foraminifera — × 30, × 30, × 12, × 4

---

which analyses statistically the magnitudes of pulses in a signal.

pulverized fuel Finely-ground solid fuels which can be fed by airblast into the combustion chamber of large furnaces.

pumice An acid vesicular glass, formed from the froth on the surface of some particularly gaseous lavas. The sharp edges of the disrupted gas vesicles enable pumice to be used as an abrasive. It floats on water.

pumpellyite A complex greenish hydrated silicate of calcium, magnesium, iron and aluminium crystallizing in the monoclinic system. Found in low grade metamorphic rocks and in amygdales in some basaltic rocks.

pump jack Motor operated well-head pump

in which the reciprocating motion of the **horsehead** is transmitted to a displacement pump downhole. See **polished rod**.

pump rod Small diameter rods screwed together and used to connect a downhole pump to the surface. Also *pony rod*.

punched screens Suitably perforated robust plates which act as industrial screens.

punctuated equilibrium A concept of the process of evolution in which the fossil record is interpreted as long periods of stasis interrupted by relatively short periods of rapid change and speciation.

purple copper ore Bornite.

puy The name given to a small volcanic cone, esp. in the Auvergne, France.

P-wave See **earthquakes**.

pyralspite A group name for the *py*rope, *al*mandine and *s*pessartine garnets.

pyramid A crystal form with three or more inclined faces which cut all three axes of a crystal. See **bipyramid**.

pyramidal system See **tetragonal system**.

pyrargyrite Sulphide of silver and antimony which crystallizes in the trigonal system. It is commonly associated with other silver-bearing minerals. Cf. **proustite**. Also *dark red silver ore*.

pyribole A group name for pyroxene and amphibole.

pyrite, pyrites Sulphide of iron (FeS_2) crystallizing in the cubic system. It is brassy yellow and is the commonest sulphide mineral of widespread occurrence. Also *fool's gold*, *iron pyrites*, *mundic*. Pyrite(s) is sometimes used to include **copper pyrites, magnetic pyrites** etc.

pyro- A prefix used to denote an acid (and the corresponding salts) which is obtained by heating a normal acid, and thus contains relatively less water, e.g. *pyrosulphuric acid*, $H_2S_2O_7$.

pyrochlore A complex niobate of sodium, calcium and other bases, with iron, uranium, zirconium, titanium, thorium and fluorine; crystallizes in the cubic system. It is found in nepheline-syenites and in alkaline pegmatites.

pyroclast Crystal, glass or rock fragment generated by a disruptive volcanic eruption and not subjective to any secondary redeposition processes.

pyroclastic rocks A name given to fragmental deposits of volcanic origin.

pyroelectric effects In high-tension (electrostatic) separation, the electrical charging of particles by heating.

pyroelectricity Polarization developed in some hemihedral crystals by an inequality of temperature.

pyrogenic Resulting from the application of a high temperature.

pyrolusite Manganese dioxide crystallizing in the tetragonal system. It typically occurs massive and as a pseudomorph after manganite, and is used as an ore of manganese, as an oxidizer and as a decolorizer.

pyrolysis The decomposition of a substance by heat.

pyrolytic mining See **underground gasification**.

pyromeride An anglicized French term for nodular rhyolite. It is a quartz-felsite or devitrified rhyolite containing spherulites up to several centimetres in diameter which impart a nodular appearance to the rock.

pyromorphite Phosphate and chloride of lead, crystallizing in the hexagonal system. It is a mineral of secondary origin, frequently found in lead deposits; a minor ore of lead.

pyrope The fiery-red garnet; magnesium aluminium silicate crystallizing in the cubic system. It is often perfectly transparent and then prized as a gem, being ruby-red in colour. It occurs in some ultrabasic rocks and in eclogites.

pyrophyllite A soft hydrated aluminium silicate crystallizing in the monoclinic system. It occurs in metamorphic rocks; often resembles talc.

pyrostibnite See **kermesite**.

pyroxene group A most important group of rock-forming ferromagnesian silicates which, although falling into different systems (orthorhombic, monoclinic), are closely related in form, composition and structure. They are silicates of calcium, magnesium and iron, sometimes with manganese, titanium, sodium, or lithium. See **aegirine, augite, diallage, diopside, enstatite, hypersthene, orthopyroxene**.

pyroxenite A coarse-grained, holocrystalline igneous rock, consisting chiefly of pyroxenes. It may contain biotite, hornblende, or olivine as accessories. See **hypersthenite**.

pyrrhotite, pyrrhotine Iron sulphide, ca Fe_7S_8, with variable amount of sulphur. Hexagonal. Ni sulphide may be associated with it, as at Sudbury, Ontario, a major source of the world's nickel. Also *magnetic pyrites*.

pyrrole A heterocyclic compound having a ring of four carbon atoms and one nitrogen. A colourless liquid of chloroform-like odour, bp 131°C, rel.d. 0.984. Pyrrole is a secondary base, and is found in coal tar and in bone-oil. Numerous natural colouring matters are derivatives of pyrrole, e.g. chlorophyll and haemoglobin.

Quaternary

The Quaternary is completely different from any previous period. It has had a much shorter time span than any earlier period, less than two million years, but the period has exerted a profound influence on mankind: its processes and deposits mould the modern landscape and geography. The British Isles, north of a line from the Bristol Channel to the mouth of the Thames, and much of Europe were affected by glaciation which also markedly influenced sea levels. The record of past climatic fluctuations and the history of modern faunas, floras and the human race during and since the last glaciation lie in the Quaternary deposits.

In all previous periods it is possible to establish correlations based on the evolution and disappearance of species, but this method is of limited use in the Quaternary where climatic fluctuations are predominantly used in its chronology. The enormous variations between different regions at any one time, involving latitude and altitude, frequently render precise correlations of deposits and events difficult, and impossible over longer distances. The table shows the stage names that are widely used for the British Quaternary although there is no complete agreement on their use or validity.

Era	Period	Series	British stages	Climate	Age
Cenozoic	Quaternary	Holocene or Recent	Flandrian	temperate	10 000 years
		Pleistocene	Devensian	last glacial	
			Ipswichian	temperate	
			Wolstonian	cold, glacial	
			Hoxnian	temperate	
			Anglian	glacial	
			Cromerian	temperate	
			Beestonian	glacial	
			Pastonian	temperate	
			Baventian	cold	
			Antian	temperate	
			Thurnian	cold	
			Ludhamian	temperate	
			Waltonian	variable	1.64 Ma

Throughout the world, differently named, locally based, stages have been established, with which the British stages cannot be firmly correlated. The most recent stage, the Flandrian, is correlated by general agreement with the Holocene of the Continent, and the Devensian cold stage with the Weichselian of north-west continental Europe. In the Alps, successive glaciations are named as Günst (oldest), Mindel, Riss and Würm (youngest). The Würm glaciation can be correlated with the Devensian.

q Symbol for the quantity of heat which enters a system.
Q Symbol for throughput.
quadratic system See **tetragonal system**.
qualitative analysis Identification of the constituents of a sample without regard to their relative amounts. It often refers to elemental analysis, but it may also refer to the detection of acid-base or redox properties in a sample. See **quantitative analysis**.
quantitative analysis Identification of the relative amounts of substances making up a sample. It usually refers to elemental analysis, but may refer to any constituent of the sample. In addition to chemical methods, virtually every physical property can be a basis for some analytical method, and spectroscopic and electrochemical techniques are particularly often employed.
quantum (1) General term for the indivisible unit of any form of physical energy; in particular the *photon*, the discrete amount of electromagnetic radiation energy, its magnitude being $h\nu$ where ν is the frequency and h is Planck's constant. See **phonon**. (2) An interval on a measuring scale, fractions of which are considered insignificant.
quaquaversal fold A dome-like structure of folded sedimentary rocks which dip uniformly outwards from a central point. See **dome**.
quarry (1) An open working or pit for granite, building-stone, slate or other rock. (2) An underground working in a coalmine for stone to fill the goaf. Distinction between quarry and mine somewhat blurred in law, but usage implies surface workings.
quartering A method of obtaining a representative sample for analysis or test of an aggregate with occasional shovelfuls, of which a heap or cone is formed. This is flattened out and two opposite quarter parts are rejected. Another cone is formed from the remainder which is again quartered, the process being repeated until a sample of the required size is left.
quarter-wave plate A plate of quartz, cut parallel to the optic axis, of such thickness that a retardation of a quarter of a period is produced between ordinary and extraordinary rays travelling normally through the plate. By using a quarter-wave plate, with its axis at 45° to the principal plane of a Nicol prism, circularly polarized light is obtained.
quartz Crystalline silica, SiO_2, occurring either in prisms capped by rhombohedra (low-temperature quartz, stable up to 573°C) or in hexagonal bipyramidal crystals (high-temperature quartz, stable above 573°C). Widely distributed in rocks of all kinds; igneous, metamorphic and sedimentary; usually colourless and transparent (rock crystal), but often coloured by minute quantities of impurities as in citrine, cairngorm etc.; also finely crystalline in the several forms of chalcedony, jasper etc. See **cristobalite, silica, tridymite, twinning**.
quartz diorite A coarse-grained holocrystalline igneous rock of intermediate composition, composed of quartz, plagioclase feldspar, hornblende and biotite, and thus intermediate in mineral composition between typical diorite and granite. Also *tonalite*.
quartz dolerite A variety of dolerite which contains interstitial quartz usually intergrown graphically with feldspar, forming patches of micropegmatite. A dyke-rock of worldwide distribution, well represented by the Whin Sill rock in N. England.
quartzite The characteristic product of the metamorphism of a siliceous sandstone or grit. The term is also used to denote sandstones and grits which have been cemented by silica.
quartz keratophyre A type of soda-trachyte carrying accessory quartz.
quartz porphyrite A porphyrite carrying quartz as an accessory constituent; the representative in the medium grain-size group of the fine-grained dacite.
quartz porphyry A medium-grained igneous rock of granitic composition occurring normally as minor intrusions, and carrying prominent phenocrysts of quartz.
quartz topaz See **citrine**.
quartz wedge A thin wedge of quartz which provides a means of superposing any required thickness of quartz on a mineral section being viewed under a polarizing microscope, the wedge being cut parallel to the optic axis of a prism of quartz crystal. It enables the sign of the birefringence of biaxial minerals to be determined from their interference figure in convergent light.
Quaternary See panel on p. 191
quintal Unit of mass in the metric system, equal to 100 kg. Abbrev. *q*.
quitclaim A deed of relinquishment of a claim or portion of mining ground.

R

r Symbol (with subscript) for specific refraction.
r- Abbrev. for *racemic*.
R (1) General symbol for an organic hydrocarbon radical, esp. an alkyl radical. (2) Symbol used to indicate Rankine scale of temperature.
R- A prefix denoting right handed.
R Symbol for (1) the gas constant; (2) the **Rydberg constant**.
[*R*] Symbol, with subscript, for molecular refraction.
Ra Symbol for **radium**.
ra- Symbol for **radio-**, i.e. a radioactive isotope of an element, e.g. *ra*Na, *radiosodium*.
rabbit Apparatus for checking the internal diameter of pipes etc. to ensure that tools can pass. See **go-devil**. Also *pig*.
rabble Mechanized rake used to loosen sluice bed or move ore or concentrate through a kiln or furnace.
race Fragments of limestone sometimes found in certain brick earths of a hard marly character.
rack Reck. See **ragging frame**.
racking The operation of separating ore by washing on an inclined plane.
rad Former unit of radiation dose which is absorbed, equal to 0.01 J/kg of the absorbing (often tissue) medium. See **ionizing radiation**.
radiation The dissemination of energy from a source. The energy falls off as the inverse square of the distance from the source in the absence of absorption. The term is applied to electromagnetic waves (radio waves, infrared, light, X-rays, γ-rays etc.) and to acoustic waves. It is also applied to emitted particles (α, β, protons, neutrons etc.). See types of radiation: **black-body radiation, infrared radiation, ultraviolet radiation, visible radiation** etc. See **ionizing radiation**.
radiation chemistry Study of radiation-induced chemical effects (e.g. decomposition, polymerization etc.). Cf. **radiochemistry**.
radiation prospecting Form of geophysical prospecting which utilizes the radioactivity of uranium-, thorium- or radium-bearing minerals to identify potentially economic concentrations. Gamma-ray detectors (**Geiger counters** and **scintillation counters**) may be hand-held or airborne.
radical (1) A molecule or atom which possesses an odd number of electrons, e.g. Br.CH_3. It is often very short-lived, reacting rapidly with other radicals or other molecules. (2) A group of atoms which passes unchanged through a series of reactions, but is normally incapable of separate existence. This is now more usually called a *group*.
radio- A prefix denoting an artificially prepared radioactive isotope of an element.
radioactive dating See **radiometric dating**.
radioactive isotope Naturally occurring or artificially produced isotope exhibiting radioactivity; used as a source for medical or industrial purposes. Also *radioisotope*.
radioactive series Most naturally occurring radioactive isotopes belong to one of three series that show how they are related through radiation and decay. Each series involves the emission of an α-particle, which decreases the *mass number* by 4, and β- and γ-decay which do not change the mass number. The natural series have members having mass number (1) $4n$ (thorium series); (2) $4n + 2$ (*uranium-radium series*); (3) $4n + 3$ (*actinium series*). Members of the $4n + 1$ (*plutonium series*) can be produced artificially. Also *radioactive chain*.
radioactivity Spontaneous disintegration of certain natural heavy elements (e.g. radium, actinium, uranium, thorium) accompanied by the emission of α-rays, which are positively charged helium nuclei; β-rays, which are fast electrons; and γ-rays, which are short X-rays. The ultimate end-product of radioactive disintegration is an isotope of lead. See **artificial radioactivity, induced radioactivity**.
radiocarbon ^{14}C, a weakly radioactive isotope undergoing beta-decay with a half-life of 5770 years. It is present in the atmosphere in roughly constant amount, as it is produced from ^{14}N by cosmic rays. It is used in some tracer studies. It can also be used to date the time of death of once-living material (and hence the likely time of manufacture of an artifact). This is because living material has the same ratio of ^{14}C to ^{12}C as the atmosphere. After death, however, the ^{14}C decays and is not replaced.
radiocarbon dating A method of determining the age in years of fossil organic material or water bicarbonate, based on the known decay rate of ^{14}C to ^{14}N. See **radiocarbon**.
radiochemical purity The proportion of a given radioactive compound in the stated chemical form. Cf. **radioisotopic purity**.
radiochemistry Study of science and tech-

niques of producing and using radioactive isotopes or their compounds to study chemical compounds. Cf. **radiation chemistry**.

radiogenic Said of stable or radioactive products arising from radioactive disintegration.

radio-isotopic purity The proportion of the activity of a given compound which is due to material in the stated chemical form. Cf. **radiochemical purity**.

radiolaria See **protozoa**.

radiolarian chert, radiolarite A cryptocrystalline siliceous rock in part composed of the remains of Radiolaria. Most described examples seem to be of shallow-water origin, such as that which reaches a thickness of 3000 m in New South Wales and contains 20 million Radiolaria per cubic centimetre. See **Protozoa**.

radiolarian ooze A variety of non-calcareous deep-sea ooze, deposited at such depth that the minute calcareous skeletons of such organisms as Foraminifera pass into solution, causing a preponderance of the less soluble siliceous skeletons of Radiolaria. Confined to the Indian and Pacific Oceans, and passes laterally into red clay. See **Protozoa**.

radiolarite See **radiolarian chert**.

radiolysis Chemical decomposition of materials induced by ionizing radiation.

radiometer Instrument devised for the detection and measurement of electromagnetic radiant energy and acoustic energy, e.g. thermopile, bolometer, microradiometer.

radiometric age The radiometrically determined age of a fossil, mineral, rock or event, generally given in years. See **radiometric dating**.

radiometric dating The method of obtaining a geological age by measuring the relative abundance of radioactive parent and daughter isotopes in geological materials. See **uranium-lead, potassium-argon, rubidium-strontium** and **radiocarbon dating**.

radionuclide Any nuclide (isotope of an element) which is unstable and undergoes natural radioactive decay.

radiothorium Symbol RdTh. A disintegration product and isotope of thorium, with a half-life of 1.90 years.

radium Radioactive metallic element, one of the alkaline earth metals. Symbol Ra, at.no. 88, r.a.m. 226, half-life of ^{226}Ra 1602 years. The metal is white and resembles barium in its chemical properties; mp 700°C. It occurs in bröggerite, cleveite, carnotite, pitchblende and in certain mineral springs. Pitchblende and carnotites are the chief sources of supply.

radium emanation See **radon**.

radius of atom See **atomic radii**.

radon A zero-valent, radioactive element, the heaviest of the noble gases. Symbol Rn, at.no. 86, r.a.m. 222, half-life of ^{222}Rn 3.82 days, bp −65°C, mp −150°C. It is formed by the disintegration of radium. Isotopes are actinon (at.no. 219, half-life 4 seconds, from actinium) and thoron (at.no. 220, half-life 54 seconds, from thorium). Also *radium emanation* (obsolete).

raffinate In an extraction process in which a solvent is passed through a solid mixture of a desired product and an undesired material, the *raffinate* is the solution of desired product. The *extract* is the undesired material. In solvent refining practice in the oil industry, raffinate is that portion of the oil being treated that remains undissolved, not being removed by the selective solvent.

ragging Rough concentration or washing, for a low ratio of concentration. In mineral processing, grooves cut on surface of roll to improve grip on feed. In jigging, the bed of heavy mineral or metal shot maintained on the jig screen.

ragging frame Tilting table, which may be worked automatically, on which finely-ground ore is treated by sluicing. Also *rack, reck*.

rainbow quartz See **iris**.

rain chamber Washing tower or other space where rising dust and fumes are brought into contact with descending sprays of water.

rain prints More or less circular, vertical, or slanting pits occurring on the bedding planes of certain strata; believed to be the impressions of heavy raindrops falling on silt or clay, hard enough to retain the impression before being covered by a further layer of sediment.

rain-wash The creep of soil and superficial rocks under the influence of gravity and the lubricating action of rain.

raised beach Beach deposits which are found above the present high-water mark; due to the relative uplift of the land or to a falling sea level. See **eustatic movements**.

rake (1) A forked tool for loading coal underground. (2) An irregular vein of ironstone. (3) Train or *journey* of mineral trucks. (4) Another name for **pitch**.

rake classifier Inclined tank into which ore pulp is fed continuously, the slow settling portion overflowing and the coarser material gravitating down, to be gathered and raked up to a top discharge.

r.a.m. Abbrev. for *Relative Atomic Mass*. Also *RAM*.

Raman spectroscopy A method making use of *Raman scattering* for chemical analysis. Like *infrared spectroscopy*, it investigates molecular vibrations and rotations.

rankinite A monoclinic calcium disilicate, $Ca_3Si_2O_7$, found in highly metamorphosed siliceous limestones.

rank of coal A classification related to the percentage of carbon in dry mineral-free coal. The original vegetation has been modified by heat, pressure and chemical change after burial. Rank increases from *peat*, through *lignite* to *bituminous coal* and finally *anthracite*.

Rapakivi granite A type described from a locality in Finland, characterized by the occurrence of rounded pink crystals of orthoclase surrounded by a mantle of whitish sodic plagioclase. A widely-used textural term. Often used as an ornamental stone for facing buildings.

rare earth elements A group of metallic elements possessing closely similar chemical properties. They are mainly trivalent, but otherwise similar to the alkaline earth elements. The group consists of the lanthanide elements 57 to 71, plus scandium (21) and yttrium (39). Extracted from monazite, and separated by repeated fractional crystallization, liquid extraction, or ion exchange.

rare earths The oxides (M_2O_3) of the rare earth elements.

rare gases See **noble gases**.

raster graphics Computer graphics based on television technology in which the screen display is produced by an electron beam scanning one raster line at a time to cover the screen from top to bottom 30 times per second. The image is produced by modifying the intensity of the electron beam to each **pixel** from a map of the pixels in the computer memory. Used in computer-aided cartography.

rate-determining step Where a process consists of a series of consecutive steps, the overall rate of the process is largely determined by the step with the slowest rate, so that efforts to speed up the process must chiefly be directed to this step.

ray A linear landform on the surface of the Moon radiating outwards from a crater. Probably caused by ejecta from volcanic activity or the impact of a meteorite.

Rb Symbol for **rubidium**.

Re Symbol for **rhenium**.

reactants The substances taking part in a chemical reaction, those on the left-hand side of a reaction as written.

reaction pair Two minerals of different composition which exhibit the reaction relationship (see **reaction principle**). Thus forsterite at high temperature is converted into enstatite at a lower temperature, by a change in the atomic structure involving the addition of silica from the magma containing it. Forsterite and enstatite form a *reaction pair*.

reaction principle The conversion of one mineral species stable at high temperature into a different one at lower temperatures, by reaction between the crystal phase and the liquid magma containing it. The change may be continuous over a wide temperature range (*continuous reaction*), or may occur at a fixed temperature only (*discontinuous reaction*).

reaction rim The peripheral zone of mineral aggregates formed round a mineral or rock fragment by reaction with magma during consolidation of the latter. Thus quartz caught up by basaltic magma is partially resorbed, at the same time being surrounded by a reaction rim of granular pyroxene.

reaction series Series in which the minerals of igneous rocks are arranged in the order of temperature at which they crystallize from magmas.

reagent feeder Appliance which dispenses chemicals in continuously moving flow of ore or pulp at a controlled rate.

realgar A bright red monosulphide of arsenic; monoclinic. Occurs associated with orpiment.

Réaumur scale A temperature scale ranging from 0°R to 80°R (freezing point and boiling point of pure water at normal pressure).

Recent *Holocene*. See **Quaternary**.

reciprocal lattice The direct crystal lattice can be defined in terms of three vectors a_1, a_2, a_3. A reciprocal lattice whose vectors are b_1, b_2, b_3, defined by $a_i.b_i = \gamma$ and $a_i.b_j = 0$ It follows that $b_1 = \frac{\gamma}{V}(a_1 \times a_3)$ *etc*, where V is the volume of the unit cell of the direct lattice. In crystallographic work γ is chosen as 1, but in solid state physics as 2π. The reciprocal lattice is extensively used to discusss diffraction and scattering effects by crystals and in the band theory of solids.

reciprocating pump Pump which uses the displacing action of a plunger, piston or diaphragm to move water in a pulsated stream. Also *pulsometer pump*, using steam or compressed air in a valved system for similar pumping.

reck See **ragging frame**.

reconcentration Additional treatment of a mineral product to raise its grade or to separate out one constituent.

recovery Percentage of schedule tonnage

actually mined.

recrystallization The process of reforming crystals, usually by dissolving them, concentrating the solution, and thus permitting the crystals to reform. Frequently performed in the process of purification of a substance.

recumbent fold An overturned fold with a more or less horizontal axial plane.

red beds Red sedimentary deposits, mainly sandstones, siltstones and shales, coloured by iron oxides (hematite) and resulting from the arid continental conditions of their formation. In western Europe found particularly in the Old Red Sandstone (Devonian) and New Red Sandstone (Permo-Trias).

red clay A widespread deep-sea deposit; essentially a soft, plastic clay consisting dominantly of insoluble substances which have settled from the surface waters; these substances are partly of volcanic, partly of cosmic origin, and include nodules of manganese and phosphorus, crystals of zeolites and rare organic remains such as shark's teeth.

reddle A red and earthy mixture of hematite, often with a certain admixture of clay.

redox Abbrev. for *oxidation-reduction*. Symbol, Eh. See **oxidation potential**.

red oxide of copper See **cuprite**.

red oxide of zinc See **zincite**.

redruthite A name frequently applied to the mineral *chalcocite* because of its occurrence, among other Cornish localities, at Redruth.

red silver ore For *dark red silver ore*, see **pyrargyrite**; for *light red silver ore*, see **proustite**.

reducing agent A substance which is capable of bringing about the chemical change known as *reduction* in another substance, itself being *oxidized*.

reduction Any process in which an electron is added to an atom or an ion. Three common types of reduction are removal of oxygen from a molecule, the liberation of a metal from its compounds, and diminution of positive valency of an atom or ion. Always occurs accompanied by oxidation.

reduction The extraction of gold from ore. The *reduction officer* is the official in charge of mill, extraction plant or reduction works. (South Africa.)

red zinc ore See **zincite**.

reef See **coral reef**.

reef Originally an Australian term for a **lode**. Now used for a gold-bearing tabular deposit or flattish lode.

reef knolls Large masses of limestone formed by reef-building organisms; found typically in the Craven district of Yorkshire where they have weathered out as rounded hills above the lower ground on the shales. These are of Carboniferous age.

reef picking On Rand, removal of gold-bearing blanket ore from barren waste rock, a reversal of the more usual hand sorting.

reel barge Vessel carrying a very large diameter reel on which long lengths of oil pipe are wound and payed out during laying.

re-entry Finding and connecting to a capped well on the sea-bottom.

reflectivity Proportion of incident energy returned by a surface of discontinuity.

reflectometer Instrument measuring ratio of energy of reflected wave to that of incident wave in any physical system.

reflector A device consisting of a bright metal surface shaped so that it reflects in a desired direction light or heat falling on it.

reflux Boiling a liquid in a flask, with a condenser attached so that the vapour condenses and flows back into the flask, thus providing a means of keeping the liquid at its bp without loss by evaporation.

refraction Phenomenon which occurs when a wave crosses a boundary between two media in which its phase velocity differs. This leads to a change in the direction of propagation of the wavefront in accordance with **Snell's law**.

refractive index The absolute refractive index of a transparent medium is the ratio of the phase velocity of electromagnetic waves in free space to that in the medium. It is given by the square root of the product of the complex relative permittivity and complex relative permeability. Symbol n. In anisotropic media, which include more minerals, there are two refractive indices (ε, ω) in uniaxial minerals and three (α, β, γ) in biaxial minerals. See **gems and gemstones**, **refraction**, **Snell's law**.

refractivity Specific **refraction**.

refractometer Instrument for measuring refractive indices. Refractometers used for liquids usually measure the critical angle at the surface between a liquid and a prism of known refractive index. See **gems and gemstones**.

refractor A device by which the direction of a beam of light is changed by causing it to pass through the boundary between two transparent materials of different refractive index.

refractory ore (1) Gold ore non-responsive to amalgamation process. (2) Ore of mineral or rock used in fabrication of *refractories*, i.e. materials used in lining furnaces etc., e.g. **chromite**, **kyanite**.

refringent Refractive.

regelation Process by which ice melts when

regeneration

subjected to pressure and freezes again when pressure is removed. Regelation operates when forming a snowball by pressure, in the flow of glaciers, and in the slow passage through a block of ice by a weighted loop of wire.

regeneration (1) Reconstitution of liquid used in chemical treatment of ores before returning it to head of attacking process (e.g. in cyanide process). (2) Freshening of 'poisoned' ion-exchange resins.

regional metamorphism All those changes in mineral composition and texture of rocks due to compressional and shearing stresses, and to rise in temperature occasioned by intense earth movements over a widespread area. The characteristic products are the crystalline schists and gneisses.

regolith The mantle of rock material that overlies bedrock. Also applied to lunar materials.

regression The retreat of the sea from the land (stratigraphical usage).

rejuvenation A term applied to the action of a river system which, following the uplift of the area drained by it, can resume down cutting in the manner of a younger stream.

relative atomic mass Mass of atoms of an element formerly in *atomic weight units* but now more correctly given on the *unified scale* where 1u is 1.660×10^{-27} kg, where u is the **atomic mass unit**. Abbrevs. *r.a.m.*, *RAM*. See **atomic weight**.

relative density The ratio of the mass of a given volume of a substance to the mass of an equal volume of water at a temperature of 4°C. Originally *specific gravity*.

relative molecular mass Preferred term for **molecular weight**.

release mesh That at, and below, which screen size mineral is released from a closed crushing or grinding circuit and passed to next stage of treatment.

relic Block of ore temporarily or permanently left close to a drive through solid rock, forming a wall between this and the stoped portion of the deposit. Cf. **remnant**.

relief well A well drilled into a reservoir to reduce the pressure in a burning or blown-out well, or to inject water to flood it. Also *killer well*.

rem Former unit of radiation *dose equivalent*, replaced by the sievert. 1 rem = 0.01 Sv. See **ionizing radiation**.

remanent magnetization That magnetization formed in a rock at the time of its formation, by the Earth's magnetic field or during some subsequent event. See **magnetic anomaly patterns**.

remnant Block of ore or stope pillar left *well*

retaining mesh

clear of the underground travelling ways on completion of stoping. Cf. **relic**.

replacement The process by which one type of rock occupies the space previously occupied by another rock; also applies to minerals.

representative sample Sample cut from the bulk of material or ore deposit in such a way as to make it reasonably representative of the whole body.

reptile See **dinosaurs, vertebrates**.

reserves Block of ore proved by development to warrant extraction. This is normally done by driving levels and winzes or raises so as to expose its four sides in a rectangular panel which is sufficiently sampled and tested.

reservoir See **petroleum reservoirs**.

reservoir pressure The pressure in a natural oil or gas reservoir which causes flow to the borehole. Maintenance of a proper pressure in the reservoir by not allowing too high an extraction rate is important in ensuring good recovery.

residence time The term given by dividing the volume of any reservoir or pool or processing unit by the rate of flow through it. The expression gives the average time spent in the pool by the substance or organism under consideration.

residual deposits Accumulation of rock waste resulting from weathering *in situ*. They cover the whole range of grain size, from residual boulder beds to residual clays.

resin-in-pulp Ion-exchange method of continuously treating ores by acid leaching. Baskets containing **ion-exchange resins** are jigged through tanks containing finely-ground ore pulp as it flows from vessel to vessel. Abbrev. *RIP*.

resin poisons In ion-exchange processes, substances which reduce efficiency of resin loading by masking activated resin sites.

resorption The partial or complete solution of a mineral or rock fragment by a magma, as a result of changes in temperature, pressure, or composition of the latter.

resources Mineral deposits that are at present known and counted as reserves plus those that may eventually become available, whether already known but not economically workable or inferred but not yet discovered.

resurgent gas Superheated steam and other volatiles which play an active role in volcanic action, and which are derived from the water included in sedimentary rocks at the time of accumulation.

retaining mesh In sizing ore before further treatment, the screen aperture above which

retentivity size the material is arrested.

retentivity Residual magnetic induction after the field producing saturation has been removed from a ferromagnetic material.

retinite A large group of resins, characterized by the absence of succinic acid.

retreating systems Systems in which the removal of ore or coal is commenced from the boundary of the deposit, which is then worked towards the entry through undisturbed rock.

retrogressive, retrograde metamorphism A term descriptive of those changes which are involved in the conversion of a rock of high metamorphic grade to one of lower grade, through the advent of metamorphic processes less intense than those which determined the original mineral content and texture of the rock.

return airway, aircourse Airway leading foul air away from the mine workings to the upcast shaft.

returning charge In custom smelting, that imposed by the smelter per unit of mineral treated. It may be modified by penalties or premiums if the ore or concentrate varies from a specified composition.

returns Oil rig term for the material carried back by the returning drilling mud; provides essential information about conditions downhole.

reversal of spectrum lines The appearance of a line as a broad, diffuse bright line with a narrow dark line down the centre. The effect is caused by cool vapour surrounding a hot source such as an electric arc, which produces a narrow absorption line on the short range of continuous spectrum given by the same vapour, at a high temperature, at the centre of the arc. Only certain lines are thus affected.

reversed drainage See **drainage patterns**.

reversed fault A type of **fault** in which compression has forced the strata on the side towards which the fracture is inclined to over-ride the strata on the downthrow side. Cf. **normal fault**.

reversed magnetization See **magnetic anomaly patterns**.

reversible reaction A chemical reaction which can occur in both directions, and which is therefore incomplete, a mixture of reactants and reaction products being obtained, unless the equilibrium is disturbed by removing one of the products as rapidly as it is formed. Examples of reversible reactions are the formation of an ester and water from an alcohol and an acid, the dissociation of vapours, e.g. ammonium chloride, and the ionic dissociation of electrolytes. Also *equilibrium reaction*.

Reynolds number The dimensionless group

$$\frac{\text{density} \times \text{velocity} \times \text{linear measure}}{\text{viscosity}},$$

all values being in the same system of units, e.g. FPS or SI. If used in pipe or tube flow the linear dimension is internal diameter and this has special application in heat exchange calculations. In other applications, e.g. mixers in vessels, the linear dimension may be, for example, the diameter of a moving part.

Rh Symbol for **rhodium**.

Rhaetian Uppermost stage of the Triassic. *Rhaetic* is an old lithostratigraphic term. See **Mesozoic**.

rhenium A metallic element in the sub-group manganese, technetium, rhenium. Symbol Re, at.no. 75, r.a.m. 186.2, rel.d. 21, mp 3000°C. Valencies 2,3,4,6,7. A very rare element, occurring in molybdenum ores. A small percentage increases the electrical resistance of tungsten. Used in high-temperature thermocouples.

rheolaveur Coal-cleaning plant where raw coal is sluiced through a series of troughs, the high-ash fraction gravitating down to a separate discharge.

rheology The science of the flow of matter. The critical study of elasticity, viscosity and plasticity.

rheomorphism Flowage of rocks resulting from severe deformation, esp. applied to those undergoing partial melting as a result of heating to a high temperature.

rhodium A metallic element of the platinum group. Symbol Rh, at.no. 45, r.a.m. 102.9055, rel.d. at 20°C 12.1, mp ca 2000°C; electrical resistivity approx. 5.1×10^{-8} ohm metres. A noble silvery-white metal, it resembles platinum and is alloyed with the latter to form positive wire of the platinum-rhodium-platinum thermocouple. Used for plating silver and silverplate to prevent tarnishing, in catalysts and in alloys for high temperature thermocouples.

rhodochrosite Manganese carbonate which crystallizes in the trigonal system, occurring as rose-pink rhombohedral crystals. It is a minor ore of manganese. Also *manganese spar*, *dialogite*.

rhodonite Manganese silicate, generally with some iron and calcium; crystallizes in the triclinic system. It is rose-coloured, and is sometimes used as an ornamental stone.

rhombic dodecahedron A crystal form of the cubic system, consisting of 12 exactly

rhombic system

similar faces, each of which is a regular rhombus. Does not occur in the orthorhombic system, despite its name.

rhombic system See **orthorhombic system**.

rhombohedral class A class of the trigonal system, a characteristic form being the **rhombohedron**, which is exhibited by crystals of quartz, calcite, dolomite etc. See **Bravais lattices**.

rhombohedron A crystal form of the trigonal system, bounded by six similar faces, each a rhombus or parallelogram.

rhomb-porphyry A medium-grained rock of intermediate composition, usually occurring in dykes and other minor intrusions; characterized by numerous phenocrysts of anorthoclase which are rhomb-shaped in cross-section, set in a finer-grained groundmass. Related to laurvikite among the coarse-grained, and to kenyte among the fine-grained rocks.

rhomb-spar An old-fashioned synonym for **dolomite**.

R_H, r_H, rH See **rH value**.

rH value Logarithm, to the base 10, of the reciprocal of hydrogen pressure which would produce same electrode potential as that of a given oxidation-reduction system, in a solution of same **pH value**. The greater the oxidizing power of a system, the greater the rH value.

Rhynie Chert A silicified peaty bed containing well-preserved plant remains as well as spiders, scorpions and insects; discovered at Rhynie, Aberdeenshire, in the Middle Old Red Sandstone (i.e. Devonian Age).

rhyolite General name for fine-grained igneous rocks having a similar chemical composition to granite, commonly occurring as lava flows, although occasionally as minor intrusions, and generally containing small phenocrysts of quartz and alkalifeldspar set in a glassy or cryptocrystalline groundmass. Sometimes *liparite*. See **obsidian, pitchstone, pumice, volcanic rocks**.

rhythmic crystallization A phenomenon exhibited by rocks of widely different composition but characterized by development of orbicular structure.

rhythmic sedimentation A more-or-less consistently repeated sequence of two or more rock units which can be recognized as forming a pattern, e.g. cyclothem, **Bouma cycle, varved clays**.

ria A normal valley drowned by a rise of sea level relative to the land. Cf. **fiords**, in the production of which glacial action plays an essential part. A good example of ria type of

ring stress

coastline is S.W. Ireland, the rias being long synclinal valleys lying between anticlinal ridges.

richterite Hydrated metasilicate of sodium, calcium and magnesium, occurring as monoclinic crystals in alkaline igneous rocks and in thermally metamorphosed limestones and skarns. A member of the amphibole group.

Richter scale See **earthquakes**.

riddle Strong coarse sieve used to size gravel, furnace clinker etc. Large pieces removed by hand in riddling are called *knockings*, remaining on-screen material *middlings*, and through passing particles *fells, undersize or smalls*.

rider (1) A *horse*, i.e. mass of country rock occurring in a mineral deposit. (2) A thin seam of coal above a thick one. (3) A guide for a *bowk*, in sinking.

riebeckite A dark blue hydrated silicate of sodium and iron found in alkaline igneous rocks as monoclinic prismatic crystals. A member of the amphibole group: the blue asbestos *crocidolite* is a variety occurring in metamorphosed ironstones.

riffler A device for dividing a stream of crushed material, e.g. coal, into truly representative samples.

rift and grain The two directions, approximately at right angles to one another, along which granite and other massive igneous rocks can be split; rift being the easier of the two.

rift valley See **graben**.

rig A well-boring plant, e.g. for oil.

rill stoping Overhand or upward stoping, in which ore is detached from above the miner so as to form an inverted stoped pyramid spreading from a winze at its apex, through which broken ore is withdrawn.

ring complex See **ring dyke**.

ring dyke An almost vertical intrusion of igneous rock which rose along a more or less cylindrical fault which had an approximately circular outcrop. In some Tertiary instances several successive ring dykes, separated by 'screens' of *country rock* and approximately concentric, form *ring complexes*. See **dykes and sills**.

rings and brushes Name applied to the patterns produced when convergent or divergent plane-polarized light, after passing through a doubly refracting crystal cut perpendicular to the optic axis, is examined by an analyser. See **interference figure**.

ring size Description of rock too large for handling by screening, in accordance with the diameter of a ring which can be slipped over it.

ring stress The inward stress on the walls of

an unsupported underground excavation.
RIP Abbrev. for **Resin-In-Pulp**.
rip-bit See **jackbit**.
Riphean The later part of the Proterozoic. See **Precambrian**.
ripidolite A species of the chlorite group of minerals, crystallizing in the monoclinic system. It is essentially a hydrated silicate of magnesium and aluminium with iron. Also *prochlorite*.
ripple marks Undulating ridges and furrows found on the bedding planes of certain sedimentary rocks, due to the action of waves or currents of air or water on the sediments before they were consolidated. Such ripple and rill marks can be seen in the process of formation today on most sandy beaches, on sand dunes, and in deserts.
rising shaft A shaft which is excavated from below upward. Cf. **sinking shaft**.
Riss A glacial stage in the Pleistocene epoch of the Alps. See **Quaternary**.
Rittinger's law Law stating that energy required in a crushing operation is directly proportional to the area of fresh surface produced, i.e. $E = k_r(1/d_2 - 1/d_1)$, where E is the energy used in crushing, k_r is a constant, depending on the characteristics of the material and on the type and method of operation of the crusher, and d_1 and d_2 are the average initial and final linear dimensions of the material crushed.
river capture The beheading of a stream by a neighbouring stream which has greater power of erosion. See **drainage patterns**.
rivers, geological work of The *corrasion* (wearing away) of river banks and beds, *corrosion* (solvent and chemical action of river water) and *hydraulic action* and *attrition* of the transported material.
river terrace A nearly flat surface along the side (or sides) of a valley marking the position of a former flood plain when the river was at a higher level. The terrace is usually built up of gravel, sand or alluvium, or may be a *bench* cut into solid rock, and it can occur on either side of a valley where the meandering river has not removed all of the earlier deposits.
Rn Symbol for **radon**.
roasting The operation of heating sulphide ores in air to convert to oxide. Sometimes the sulphur-bearing gases produced are used to make sulphuric acid.
roche moutonnée A mound of bare rock which is usually smoothed on the upstream side and roughened by plucking on the downstream side, as a result of a moving ice sheet.
rock Any mineral matter making up the Earth. As used by geologists, the term also includes unconsolidated material such as sand, mud, clay and peat, in addition to the harder materials described as rock in conventional usage.
rock burst Sudden failure of stope pillars, walls or other rock buttresses adjacent to underground works, with explosively violent disintegration.
rock crystal The name given to colourless quartz whether in distinct crystals or not; particularly applicable to quartz of the quality formerly used in making lenses.
rock cycle The cycle of rock change in which rocks are uplifted, eroded, transported, deposited, possibly metamorphosed and intruded, and then uplifted to start a new cycle. The concept was first developed by James Hutton.
rocker A short and easily-portable rocking trough or cradle for washing concentrates, gold-bearing sand etc.
rock flour A term used for finely comminuted rock material found at the base of glaciers and ice sheets. It is mud-like and is composed largely of unweathered mineral particles.
rock-forming minerals The minerals which occur as dominant constituents of igneous rocks, including quartz, feldspars, feldspathoids, micas, amphiboles, pyroxenes and olivine.
rock head See **stone head**.
rock meal A white and light variety of calcium carbonate, resembling cotton; it becomes a powder on the slightest pressure.
rock milk A very soft white variety of calcium carbonate which breaks easily in the fingers; it is sometimes deposited in caverns or about sources holding lime in solution.
rock phosphate See **phosphorite**.
rock roses See **desert rose**.
rock salt See **halite**.
rock wall failure Collapse due to one or more of the following: rock fall, simple dropping; rock flow, slope failure; plane shear, failure along weakness plane; rotational shear stress where soil has dropped leaving a rounded cavity.
rock wool Fibrous insulating material made by blowing steam through molten slag. Also *mineral wool*, *slag wool*.
roller bit Drilling bit with three or more conical rollers carrying teeth. The rollers have axes inclined to that of the drill pipe and rotate individually with it. Also *tricone bit*. See **drilling rig**.
rolls Crushing rolls are pairs of horizontally-mounted cylinders, faced with manganese steel, between which ore is crushed as they

rotate inward.

roméite Naturally occurring hydrated antimonite of calcium, sometimes with manganese and iron; crystallizes in the cubic system, often as brownish octahedra.

roof bolting Method of roof support in which steel bolts are inserted in drill holes so as to pin supporting steel beams under the roof of a stope.

roofing slate A term widely applied to rocks of fine grain in which regional metamorphism has developed in good slaty cleavage.

roof pendant A mass of country rock projecting downwards, below the general level of the roof, into an intrusive rock body.

room-and-pillar See **bord-and-pillar**.

ropy lava Ses **pahoehoe**.

roscoelite A mineral, essentially *muscovite* in which vanadium has partly replaced the aluminium. Its colour is clove-brown to greenish-brown.

rose opal A variety of opaque common opal having a fine red colour.

rose quartz Quartz of a pretty rose-pink colour, due probably to titanium in minute quantity. The colour is apt to be destroyed by exposure to strong sunlight. See **Bohemian gemstone**.

rose topaz The yellow-brown variety of topaz changed to rose-pink by heating. These crystals often contain inclusions of liquid carbon dioxide.

Rosiwal intercept method Particle-size analysis technique based on measurement of the intercepts made by a line drawn through a selection of particle images on a photomicrograph or in the field of view of a projection microscope.

Rossi-Forell scale A scale of apparent intensity of earthquake movements, now replaced by the **Mercalli scale**.

rotary drier Tubular furnace sloped gently from feed to discharge end, through which moist material is tumbled as it rotates slowly, while rising hot gases remove moisture.

rotary drill The drill *downhole* and connected by the drilling pipe to the rotary table in the oil derrick or drilling platform. It may consist of several drills, reamers and stabilizing **collars**, themselves joined by special tool joints. It is usually lubricated and cleared by **drilling mud** forced down the drilling tube and out through nozzles in the drill head. The oil and debris passes up through the annular space between the drill tube and casing. The most important component fixed to the rotary table is the **kelly**. See **drilling rig**.

rotary table Heavy circular component mounted just above the derrick floor which carries the **kelly bushing**. The table is rotated by the draw-works machinery and thus rotates the **kelly** which slides in the bushing. The table thus turns the drill string. See **drilling rig**.

rotating crystal method A widely used method of X-ray analysis of the atomic structure of crystals. A small crystal, less than 1 mm in maximum size, is rotated about an axis at right angles to a narrow incident beam of X-rays. The diffraction of the beam by the crystal is recorded photographically or with a detector.

rotating, rotary disk contactor A liquid-liquid extraction device which consists of a column with annular stators fitted to it and a central shaft carrying rotors of diameter nearly equal to the holes in the stators. Liquids flow counter-current through the column and, due to the effect of the high speed of rotation of the shaft and rotors, improved contact is obtained.

rotation axes of symmetry Symmetrically placed lines, rotation about which causes every atom in a crystal structure, as revealed by X-ray analysis, to occupy identical positions a given number (2, 3, 4, 6) of times. Cf. **screw axes**.

rotation of the plane of polarization A property possessed by optically active substances. See **optical activity**.

rotatory dispersion Variation of rotation of the plane of polarized light with wavelength for an optically active substance.

rotatory evaporator A device for facilitating the evaporation of a liquid, generally under reduced pressure, by continuously rotating the flask in which it is contained.

rotatory power See **optical rotation**.

Rotliegendes The name of the lower series of the Permian. See **Palaeozoic**.

rottenstone A material used commercially for polishing metals; formed by the weathering of impure siliceous limestones, the calcareous material being removed in solution by percolating waters.

roughing Production of an impure concentrate as an early stage in ore processing, thus reducing bulk for more thorough treatment.

round trip Removing and dismantling the drill string to replace the drill bit, and then re-assembling and placing it back *downhole*. Also *trip*.

royals In the **cyaniding** process for recovering gold, a term for the sludge, rich in gold, formed when the **pregnant solution** is precipitated on zinc dust or on to resin or carbon.

Ru Symbol for **ruthenium**.

rubellite The pink or red variety of tourmaline, sometimes used as a semi-precious gemstone.

rubicelle A yellow or orange-red variety of spinel; an aluminate of magnesium.

rubidium A metallic element in the first group of the periodic system, one of the alkali metals. Symbol Rb, at.no. 37, r.a.m. 85.47, mp 38.5°C, bp 690°C, rel.d. 1.532. The element is widely distributed in nature, but occurs only in small amounts; the chief source is carnallite. The metal is slightly radioactive.

rubidium-strontium dating A method of determining the age in years of geological material, based on the known decay rate of ^{87}Rb to ^{87}Sr.

ruby The blood-red variety of the mineral corundum, the oxide of aluminium (Al_2O_3), which crystallizes in the trigonal system. Also *true ruby* (to distinguish it from the various types of **false ruby**) and *Oriental ruby*, though the adjective *Oriental* is quite unnecessary, since it merely stresses the fact that rubies come from the East (Burma, Thailand, Sri Lanka, Afghanistan). See **balas ruby, ruby spinel**.

ruby silver ore See **proustite, pyrargyrite**.

ruby spinel That variety of magnesian spinel, $MgAl_2O_4$, which has the colour, but none of the other attributes, of true ruby. Also *almandine spinel. Spinel ruby* is a deceptive misnomer.

rudaceous Used of a sedimentary rock which is coarser in grain size than sand.

Rudistes A group of heavily-built lamellibranchs of coral-like form which are characteristic of the Cretaceous rocks formed in the southern ocean (the Tethys) of the period; Rudistids also occur in the Cretaceous Trinity Series of Texas and Mexico.

runite See **graphic granite**.

runoff The resultant discharge of a river from a catchment area; surface water as distinct from that rising from deep-seated springs.

run-of-mine coal Coal raised from the mine before screening or other treatment.

Rupelian A stage in the Oligocene. See **Tertiary**.

rusty gold Native gold which has become surface-filmed by adherent staining substances and is non-amalgamable and non-treatable by cyanide process in consequence.

ruthenium A metallic element. Symbol Ru, at.no. 44, r.a.m. 101.07, mp 2400°C, rel.d. 12.26. The metal is silvery-white, hard and brittle. It occurs with the platinum metals in osmiridium, and is used in certain platinum alloys.

rutilated quartz See **needle stone**.

rutile Titanium dioxide which crystallizes as reddish-brown prismatic crystals in the tetragonal system. It is found in igneous and metamorphic rocks, and in sediments derived from these, also in quartz (see **flèches d'amour**), and it is a source of titanium.

Ryazanian The oldest stage of the Cretaceous. See **Mesozoic**.

Rydberg constant The frequency of atomic spectrum lines in a given series that are related by the Rydberg formula. The Rydberg constant R involved was first deduced from spectroscopic data but has since been shown to be a universal constant.

$$R = \frac{2\pi^2 e^4}{ch^3} M_r ,$$

where M_r is the reduced mass of the electrons, e is the electronic charge, c is the velocity of light and h is Planck's constant.

S

s Symbol for (1) distance along a path; (2) solubility; (3) specific entropy.

σ Symbol for (1) conductivity; (2) normal stress; (3) Stefan-Boltzmann constant; (4) surface charge density; (5) **surface tension**; (6) wave number.

S Symbol for **siemens**.

S Symbol for (1) area; (2) **entropy**; (3) Poynting vector.

sabkha A flat salt-encrusted coastal plain, common in Arabia.

saccharoidal textures Granular textures which resemble sugar; found esp. in limestones and marbles.

safety cage A cage fitted with a 'safety catch' to prevent it from falling if the hoisting rope breaks.

safety lamp Oil-burning miners' lamp which will not immediately ignite firedamp or gas in a coalmine, e.g. a *Davy lamp*. Also used for detecting gas.

sahlite, salite A mineral of the clinopyroxene group, intermediate in composition between diopside and hedenbergite.

Sakmarian The lowest stage of the Permian system in eastern Europe and USSR.

sal-ammoniac Chloride of ammonia, which crystallizes in the cubic system. It is found as a white encrustation around volcanoes, as at Etna and Vesuvius.

salic minerals Those minerals of the *norm* which are rich in silicon and aluminium, including quartz, feldspars and feldspathoids.

salina See **salt lakes**.

salite See **sahlite**.

salt dome A diapiric salt plug which has arched up, or broken through, the sediments into which it has been intruded. Such rock salt deposits have become plastic under pressure and the plugs or domes so formed are impervious to oil. Differences in gravity between the dome and the intruded rocks can be measured at the surface and may indicate the presence of an oil reservoir. See **petroleum reservoirs**.

salting Fraudulent enrichment of ore samples, made to increase apparent value of a mine. Originally, the sprinkling of salt in dry mines to allay dust.

saltpetre See **potassium nitrate**.

salt lakes, saline lakes Enclosed bodies of water, e.g. lake, lagoon, marsh, spring etc., in areas of inland drainage, whose concentration of salts in solution is much higher than in ordinary river water. Also *salina*. See **soda lakes**.

samarium A hard and brittle metallic element. Symbol Sm, at.no. 62, r.a.m. 150.35, mp 1350°C, bp 1600°C, rel.d. 7.7. Found in allanite, cerite, gadolinite and samarskite. Feebly, naturally radioactive, can be produced by decay of fission fragments, and forms a reactor poison.

sand A term popularly applied to loose, unconsolidated accumulations of detrital sediment, consisting essentially of rounded grains of quartz. In the mechanical analysis of soil, sand, according to international classification, has a size between 1/16 mm and 2 mm. See **silt**. In coral sand the term implies a grade of sediment the individual particles of which are fragments of coral, not quartz. See **particle size, Wentworth scale**.

sand dunes Rounded or crescentic mounds of loose sand which have been piled up by wind action on seacoasts or in deserts. See **barchan**.

sand fill Underground support of worked-out stopes by return of ore tailings from mill, usually by hydraulic flow.

sands Particles of crushed ore of such a size that they settle readily in water and may be leached by allowing the solution to percolate. See **slimes**.

sandstone Compacted and cemented sedimentary rock, which consists essentially of rounded grains of quartz, between the diameters of 0.06 and 2 mm, with a variable content of 'heavy mineral' grains. According to the nature of the cementing materials the varieties *calcareous sandstone*, *ferruginous sandstone*, *siliceous sandstone* may be distinguished; *glauconitic sandstone*, *micaceous sandstone* etc. are so termed from the presence in quantity of the mineral named.

sand volcano A structure formed by sand flowing upwards through an overlying bed of sediment and spilling out on to the surface; not related to any type of volcanic activity or volcanic rock.

sanidine A form of potassium feldspar similar in chemical composition to orthoclase, but physically different, formed under different conditions and occurring in different rock types. It is the high-temperature form of orthoclase, to which it inverts below 900°C. Occurs in lavas and dyke rocks.

Santonian A stage of the Upper Cretaceous. See **Mesozoic**.

saphir d'eau French 'water sapphire'. A misnomer for an intense blue variety of the mineral cordierite, occurring in water-worn masses in the river gravels of Ceylon; used

saponite Hydrated aluminosilicate of magnesium. A clay mineral of the smectite (montmorillonite) group, occurring as white soapy masses in serpentinite. Also called *bowlingite*.

sapphire The fine blue transparent variety of crystalline corundum, of gemstone quality; obtained chiefly from Sri Lanka, Kashmir, Thailand, Cambodia and Australia.

sapphirine A silicate of magnesium and aluminium with lesser iron, crystallizing in the monoclinic system. It occurs as blue grains in metamorphosed, aluminous, silica-poor rocks.

sapropel Slimy sediment laid down in stagnant water, largely consisting of decomposed algal material. A source material for oil and natural gas.

sapropelite A term applied to coals derived from algal materials. Cf. **humite**.

sardonyx A form of chalcedony in which the alternating bands are reddish-brown and white. Cf. **onyx**.

sarsen Irregular masses of hard sandstones which are found in the Reading and Bagshot Beds of the Tertiary System in S. England. They often persist as residual masses after the softer sands have been denuded away.

sassolite The mineral boric acid, H_3BO_3.

satin spar Name given to fine fibrous varieties of calcite, aragonite and gypsum, the gypsum variety being distinguished from the others by its softness. (It can be scratched by a finger-nail.)

saturated Igneous rocks that lack silica minerals and feldspathoids so that in the normal mode they are neither *oversaturated* nor *undersaturated*.

sauconite One of the montmorillonite group of clay minerals in which zinc has replaced magnesium. See **smectites**.

saussurite A fine-grained mixture of zoisite and other minerals resulting from the more or less complete alteration of feldspar. Sometimes simulates **jade**.

saxonite A coarse-grained, ultrabasic rock, consisting essentially of olivine and orthopyroxene, usually hypersthene. A hypersthene-peridotite.

scandium A metallic element, classed with the rare earth metals. Symbol Sc, at.no. 21, r.a.m. 44.956. It has been found in cerite, orthite, thortveitite, wolframite and euxenite; discovered in the last named. Scandium is the least basic of the rare earth metals.

scanning electron microscope A form of *electron microscope* in which a very fine beam of electrons at 3–30 kV is made to scan a chosen area of specimen as a *raster* of parallel contiguous lines. Abbrev. *SEM*. Usually the specimen is a solid object and secondary electrons, which are emitted from the surface or from near the surface in numbers depending on its nature and topography, are collected and after analysis form a signal which modulates the beam of a cathode ray tube scanned in synchrony with the scanning beam. The images resemble those seen in a hand lens but have much finer resolution (say 5 nm), can be usefully magnified about 100 000 times and have, for comparable magnifications, much greater depth of focus than a light microscope. Many SEMs are also capable of *X-ray analysis*.

scanning heating Induction heating where the workpiece is moved continuously through the heating region, as in *zone refining* of germanium. Also *progressive heating*.

scanning microscope An instrument which enables the surface structure of a specimen to be examined. The specimen is scanned by a small-diameter electron probe which gives rise to secondary-electron emission from the specimen. The secondary-emission current is detected and causes a picture to be projected on a cathode-ray screen.

scanning transmission electron microscope Microscope which uses field emission from a very fine tungsten point as the source of electrons. The electrons transmitted through the sample are either unscattered, elastically scattered or inelastically scattered; they are collected, separated and analysed to produce an image. Abbrev. *STEM*.

scanning tunnelling electron microscope Microscope which uses a probe with an 'atomic micro-tip' floated, using *superconducting levitation* over the surface being scanned. The tip/surface is of the order of atomic diameters so that the electron current obtained is through *quantum mechanical tunnelling*. A horizontal resolution of ~0.2 nm and a vertical resolution of ~0.01 nm is obtained. Applications are principally to the study of surface effects.

scapolite A group of minerals forming an isomorphous series, varying from meionite, a silicate of aluminium and calcium with calcium carbonate, to marialite, a silicate of aluminium and sodium with sodium chloride. Common scapolite is intermediate in composition between these two end-member minerals. See **dipyre**, **mizzonite**. The scapolites crystallize in the tetragonal system and are associated with altered lime-rich igneous and metamorphic rocks. A transparent honey yellow variety is cut as a gemstone.

scarp face See **escarpment**.

scavenging Final stage of froth flotation in which a low-grade concentrate or middling is removed.

scheelite An ore of tungsten. It occurs in association with granites and pegmatites, has the composition calcium tungstate and crystallizes in the tetragonal system.

schillerization A play of colour (in some cases resembling iridescence due to tarnish) produced by the diffraction of light in the surface layers of certain minerals.

schist A metamorphic rock which has a tendency to split on account of the presence of folia of flaky and elongated minerals, such as mica, talc and chlorite; formed from original sedimentary or igneous rocks by the action of regional metamorphism.

schistosity The tendency in certain rocks to split easily along weak planes produced by regional metamorphism and due to the abundance of mica or other cleavable minerals lying with their cleavage planes parallel.

schorlomite A black variety of andradite garnet richer in titania (5%–20% TiO_2) than melanite.

schorl-rock A rock composed essentially of aggregates of black tourmaline (schorl) and quartz. A Cornish term for the end product of tourmalinization. See **tourmaline**.

scientific alexandrite Synthetic corundum coloured with vanadium oxide and resembling true alexandrite in some of its optical characters.

scientific emerald Beryl glass coloured with chromic oxide, resembling true emerald in colour.

scintillation counter Counter consisting of a *phosphor* or *scintillator*, e.g. NaI(Tl), which, when radiation falls on it, emits light which is detected and amplified by a photomultiplier, the height of the pulses from which are proportional to the energy of the event. These pulses are further amplified and passed to a single- or multi-channel pulse height analyser, to measure the energy and intensity of the radiation.

sclerometer An instrument used for measuring the hardness of minerals or metals by impressing a polished surface with a diamond point.

scolecite A member of the zeolite group of minerals; a hydrated silicate of calcium and aluminium, occurring usually in fibrous or acicular groups of crystals.

scoria A cavernous mass of volcanic rock which simulates a clinker.

scorodite An orthorhombic hydrated arsenate of iron and aluminium.

Scottish topaz A term applied in the gemstone trade to yellow transparent quartzes, resembling Brazilian topaz in colour, used for ornamental purposes. Not a true topaz. See **cairngorm, citrine**.

scree The accumulation of rock debris strewn on a hillside or at a mountain foot, resulting from the mechanical weathering of rocks.

screen, screening Perforated or woven cloths (metal, fibre, rods, bars), used to size ore or products as part of treatment required to regulate concentration. Include *grizzlys, trommels,* and mechanically- and electrically-vibrated screens.

screw axes Axes of symmetry about which the atoms in a mineral are symmetrically disposed. Rotation about a 4-fold screw axis, for example, will carry an atom 1 into the positions successively occupied by similar atoms 2, 3 and 4, after rotations of 90°, 180°, 270° and 360°. Cf. **rotation axes of symmetry**.

scrubber (1) Part of a process plant used to purify gas streams. Contaminated gas is passed upwards through a column, and mixes intimately with a counter-current of liquid solvent, reactant solution or slurry which is passing downwards. This removes the impurities by absorption. Desulphurization of flue gases to remove sulphur dioxide is carried out in this way before the scrubbed gases are discharged into the atmosphere. (2) A kind of *dampener* that uses baffles to change the velocity of a flowing gas stream by altering its direction and flow area.

Scythian The oldest epoch of the Triassic period.

sea-floor spreading The process by which new oceanic crust is generated at oceanic ridges by the convective upwelling of magma. The plates on either side of the *divergent junction* move very gradually apart. See **magnetic anomaly patterns, plate tectonics**.

seam (1) A tabular, generally flat deposit of coal or mineral; a stratum or bed. (2) A joint or fissure in a coal bed.

seamount An elevation from the ocean floor which may be flat topped (a *guyot*) or peaked (a *sea peak*).

seat earth A fossil soil which underlies a coal seam.

secondary dispersion In geochemical prospecting, the dispersion of elements or minerals from an ore body or **halo** by physical agents such as stream, river or groundwater flow, glacial ice, wind or wave action.

secondary enrichment The name given to the addition of minerals to, or the change in the composition of the original minerals in, an ore body, either by precipitation from downward-percolating waters or upward-moving gases and solutions. The net result of the changes is an increase in the amount of metal present in the ore at the level of secondary enrichment.

secondary gneissic banding A prominent mineral banding exhibited by coarse-grained crystalline rocks which have been subjected to intense regional metamorphism, involving rock-flowage. Often it is difficult to distinguish from **primary gneissic banding**.

secondary mineral (1) Mineral formed after the formation of the rock enclosing it. (2) Mineral of minor interest in an ore body undergoing exploitation.

secondary production Oil production in which means like pumping gas or water into the reservoir are needed to assist the flow to the well bore. Also *secondary recovery*. Cf. **primary production, tertiary production**.

secondary wave See **earthquakes**.

secular changes Changes which are extremely slow and take many centuries to accomplish; they may apply to climate, levels of land and sea, or, as in geomagnetism, to long period changes in the magnetic fields at any place.

sedimentary rocks All those rocks which result from the wastage of pre-existing rocks. They include the fragmental rocks deposited as sheets of sediment on the floors of seas, lakes and rivers and on land; also deposits formed of the hard parts of organisms, and salts deposited from solution, in some cases by organic activity. Igneous and metamorphic rocks are excluded.

sedimentary structure Any physical structure in a sedimentary rock that was formed at the time of its deposition, e.g. **cross bedding, sole mark**.

Segrè chart A chart on which all known nuclides are represented by plotting the number of protons vertically against the number of neutrons horizontally. Stable nuclides lie close to a line which rises from the origin at 45° and gradually flattens at high atomic masses. Nuclides below this line tend to be β-emitters whilst those above tend to decay by positron emission or electron capture. Data for half-life, cross-section, disintegration energy etc. are frequently added.

seif dune A longitudinal sand dune developed parallel to the dominant wind direction.

seismic prospecting See panels on pp. 207–8.

seismology The study of earthquakes, particularly their shock waves. Studies of the velocity and refraction of seismic waves enable the deeper structure of the Earth to be investigated. See **earthquakes**.

selective absorption Absorption of light, limited to certain definite wavelengths, which produces so-called absorption lines or bands in the spectrum of an incandescent source, seen through the absorbing medium.

selenite The name given to the colourless and transparent variety of **gypsum** which occurs as distinct monoclinic crystals, esp. in clay rocks.

selenium A non-metallic element. Symbol Se, at.no. 34, r.a.m. 78.96, valencies 2,4,6. A number of allotropic forms are known. *Red selenium* is monoclinic; mp 180°C, rel.d. 4.45. *Grey (metallic) selenium*, formed when the other varieties are heated at 200°C, is a conductor of electricity when illuminated; mp 220°C, bp 688°C, rel.d 4.80, electrical resistivity 12×10^{-8} ohm metres. Selenium is widely distributed in small quantities, usually as selenides of heavy metals. It is obtained from the flue dusts of processes in which sulphide ores are used, and from the anode slimes in copper refining. It is used as a decolorizer for glass, in red glass and enamels, and in photoelectric cells and rectifiers. Selenium is similar to sulphur in chemical properties, but resembles tellurium more closely still.

SEM Abbrev. for **Scanning Electron Microscope**.

semi-submersible A seagoing vessel with adjustable buoyancy. A drilling rig of this type is towed to its location floating high in the water for passage, but once on site is flooded and partially submerged to provide a stable platform no longer riding on the waves. Emergency vessels of this type may have fire fighting facilities for attending well or platform accidents in offshore oilfields.

Senonian The youngest epoch of the Cretaceous period. See **Mesozoic**.

sensitive tint plate A thin, optically-orientated plate of a crystal, usually gypsum, used to measure the optical properties of minerals and other crystalline substances with a polarizing microscope.

separator (1) Concentrating machine, used to separate constituent minerals of mixed ore from one another. (2) An item of process plant for separating one substance from a mixture with another, the two substances usually being in two different **phases**, e.g. oil from water, oil from gas, gas from oil or ash from flue gases.

sepiolite See **meerschaum**.

septaria Concretionary nodules containing irregular cracks which have been filled with

seismic prospecting

A standard method of applied geophysical exploration is the determination of the underground structure by *seismic methods* which are used esp. in petroleum exploration. In *seismic reflection surveying*, explosives or other energy sources produce sudden pulses of short duration that are *reflected* and detected by small detectors or *geophones* (Fig. 1). The signals from each geophone are amplified, fed into sophisticated data processing equipment and arranged to produce a *seismic reflection record*. This method is often known as *reflection shooting*.

Among the many energy sources that can be used are explosives, compressed gas exploders, electrical energy sources moving a piston or plate, gas exploders, mechanical hammers, implosive sources and, increasingly, hydraulic or electromagnetic vibrators.

If the waves from the energy source reach a boundary and are then transmitted along it, the method is **refraction shooting**. Snell's law (of optics) can be extended to cover these mechanisms. The lower bed must have a higher velocity of transmission to obtain refraction shooting. Detection at the geophones is related to their spacing and distance from the shot-point. Because of its faster travel, the refracted wave may arrive before the direct or reflected waves.

Offshore, similar techniques are used from *survey ships*. In **continuous seismic profiling** the energy source and receiver are both towed at the sea surface and a facsimile picture obtained of the sub-seabed geology.

Air guns, discharging air under high pressure into the sea, are extensively used as the seismic energy source. *Sparker* methods use high-voltage electrical discharge under water. The *two-way time* or *two-way travel time*, (abbrevs. *TWT, TWTT*) is the time for the seismic wave to travel from an energy source close to the surface to the reflecting horizon and back (Fig. 2). It is usually measured in seconds or milliseconds and subject to some corrections is used as the ordinate in plotting results on seismograms or reflection profiles. If the velocity through the rocks is accurately known the TWT values can be converted to an equivalent depth.

figures on next page

calcite or other minerals. Also *septarian nodules*.

septechlorites A group of sheet silicates closely related chemically to the chlorites, and structurally to the serpentines and kandites. Includes **chamosite** and **greenalite**.

sericite A fine-grained white potassium mica, like muscovite in chemical composition and general characters but occurring as a secondary mineral, often as a decomposition product of orthoclase.

series A time-stratigraphic unit intermediate between **system** and **stage**, and corresponding to an **epoch** of geological time.

serpentine Hydrated magnesium silicate which crystallizes in the monoclinic system. The three chief polymorphic forms are **antigorite**, **chrysotile** and **lizardite**. The serpentine minerals occur mainly in altered ultrabasic rocks, where they are derived from olivine or from enstatite. Usually dark green, streaked and blotched with red iron oxide, whitish talc etc. The translucent varieties are used for ornamental purposes; those with a fibrous habit form one type of asbestos.

serpentine jade A variety of serpentine, resembling bowenite, occurring in China.

serpentinization A type of metamorphism effected by water, which results in the replacement of the original mafic silicates in peridotites by the mineral serpentine and secondary fibrous amphibole.

Serpukhovian The youngest epoch of the Mississippian period.

Serravallian A stage of the Miocene. See **Tertiary**.

set A frame of timber used in a shaft or tunnel.

shadow zone See **earthquakes**.

seismic prospecting (contd)

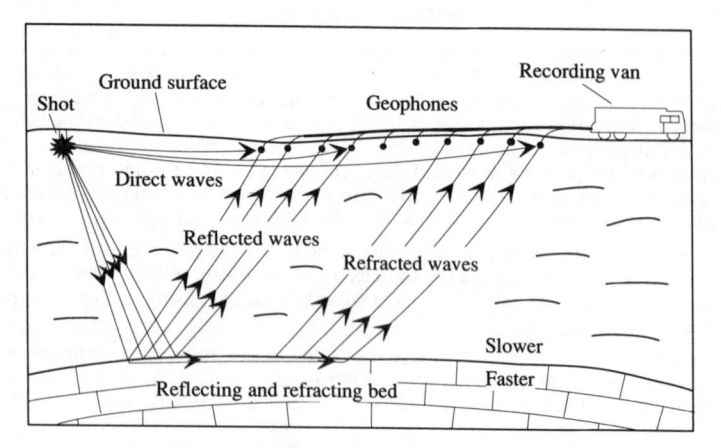

Fig. 1. Seismic surveying on land

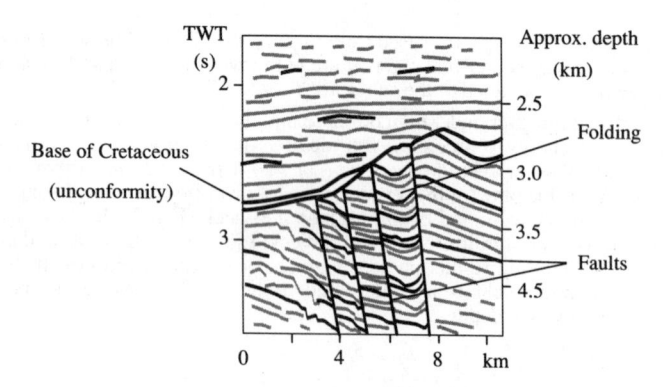

Fig. 2. Seismic reflection line showing faulting, folding and unconformity

shaft A passage, usually vertical, leading from ground level into an underground excavation, for purposes of ventilation, access etc.

shaft pillar Solid block of coal or ore left unworked round the bottom of a shaft or pit for support.

shaft station Room excavated underground adjacent to shaft, to accommodate special equipment such as pumps, crushing machine, truck tipples, ore sorting equipment and surge storage bins.

shaking table See **Wilfley table**.

shale A consolidated clay rock which possesses closely-spaced well-defined laminae. Cf. **mudstone**. See **oil shale**.

shale oils Oils obtained by the pyrolysis of oil shale at ca 550°C and characterized by a large proportion of unsaturated hydrocarbons, e.g. alkenes and di-alkenes.

shard A fragment of volcanic glass, often with curved edges. Glass shards are important constituents of some pyroclastic rocks.

sharp gas Mine air so contaminated with methane as to burn inside the Davy-type lamp and therefore to be dangerous.

shear A type of deformation in which parallel planes in a body remain parallel but are relatively displaced in a direction parallel to themselves; in fact, there is a tendency for adjacent planes to slide over each other. A rectangle, if it is subjected to a

shearer loader

shearing force parallel to one side, becomes a parallelogram. See **elasticity of shear, strain**.

shearer loader Machine which cuts coal from seam and loads it in the same operation to a conveyor belt working parallel to the face. In fully mechanized mining the assembly, together with roof support props, is moved hydraulically and can be remotely controlled.

shear ram Hydraulically-operated sliding jaws designed to cut off flow near the **blowout preventer**. It compresses the pipe and cuts it. The well is then *shut in*, but the gear above can be removed.

shear wave See **earthquakes**.

shear zones Bands in metamorphic rocks consisting of crushed and brecciated material and many parallel fractures. See **strain-slip cleavage**.

sheridanite A mineral in the chlorite group poor in iron and relatively low in silica.

shield A large stable area of the Earth's crust consisting of Precambrian rocks. Effectively synonymous with **craton**, e.g. Canadian Shield.

shingle Loose detritus, generally of coarser grade than gravel though finer than boulder beds, occurring typically on the higher parts of beaches on rocky coasts.

shoad, shode Water-worn fragments of vein minerals found on the surface away from the outcrop. Also *float-ore*.

shoe The replaceable steel wearing part of the head of a stamp or muller of a grinding pan.

shonkinite A coarse-grained, feldspar-rich syenite, consisting largely of pyroxenes and some olivine. Named after the laccolith, Shonkin Sag, Montana.

short Brittle.

shoshonite A potassic variety of basaltic trachyandesite.

shot drilling Boring of deep holes by means of hard steel shot fed down rotating hollow cylinder.

shot firer Miner who tests for gas and then fires explosive charges in colliery.

SI See **SI units**.

Si Symbol for **silicon**.

sial The discontinuous shell of granitic composition which forms the foundation of the continental masses and which is in turn underlain by the *sima*. So called because it is essentially composed of *si*liceous and *al*uminous minerals.

siderite (1) Ferrous carbonate, crystallizing in the trigonal system and occurring in sedimentary iron ores and in mineralized veins. Sometimes *chalybite*. (2) A name for iron meteorites as a class.

silk

siderophile Descriptive of elements which have an affinity for iron, and whose geochemical distribution is influenced by this property.

siderophyllite The iron and aluminium-rich end-member of the biotite micas.

Siegennian A stage in the Lower Devonian. See **Palaeozoic**.

siemens SI unit of electrical conductance; reciprocal of *ohm*. Abbrev. *S*.

Siemens-Halske process Chemical extraction of sulphidic copper from its ores and concentrates with sulphuric acid and ferrous sulphate.

sievert A unit of radiation dose, being that delivered in 1 hr at a distance of 1 cm from a point source of 1 mg of radium element enclosed in platinum 0.5 mm in thickness. Numerically equal to ca 8.4 röntgens or 21.6 Ci/kg.

Silesian The Upper Carboniferous of Western Europe. See **Palaeozoic**.

silex Silica brick used to line grinding mills when contamination by abraded steel must be avoided.

silica See panels on pp. 210–211.

silica glass Fused quartz, occurring in shapeless masses on the surface of the Libyan Desert, in Moravia, in parts of Australia and elsewhere; believed to be of meteoritic origin. See **tektites**.

silica poisoning Loading of the resins used in ion-exchange with silica, thus reducing the efficiency of reaction with desired ions.

silicates See panels on pp. 210–211.

siliceous deposits Those sediments, incrustations, or deposits which contain a large percentage of silica in one or more of its modes of occurrence. They may be chemically or mechanically formed, or may consist of the siliceous skeletons of organisms such as diatoms and Radiolaria. See **silicification**.

siliceous sinter Cellular quartz or translucent to opaque opal, found as incrustations or fibrous growths and deposited from thermal waters containing silica or silicates in solution.

silicification The process by which silica is introduced as a cement into rocks after their deposition, or as an infiltration or replacement of organic tissues or of other minerals such as calcite. See **novaculite**.

silicon See panels on pp. 210–211.

silk A sheen resembling that of silk, exhibited by some corundums, including ruby, and due to minute tubular cavities, or to rutile needles, in parallel orientation. The colour of such stones is paler than normal by reason of the inclusions.

silicon, silica, silicates

Silicon is the non-metallic element, symbol Si, obtained from the dioxide of silica, SiO_2. The element, silicon, is the second commonest in the Earth's crust, after oxygen, amounting to 27%, but it is never found in the elemental form. It has at. no. 14, r.a.m. 28.086, valence 4, three stable isotopes, rel.d. 2.33, mp 1410°C, bp 2355°C. It has a blue-grey metallic appearance and is produced by reduction of quartz sand, SiO_2, with carbon. It is used as an alloying agent for steels and other metals, and in semi-conductors.

Silica, SiO_2, forms some of the commonest minerals of the crust, and as silicates in combination with aluminium and other elements makes up the bulk of all the rocks of the crust, over a thousand silicate minerals being known. Silica itself occurs in six principal polymorphs, of which by far the most common is quartz which is widely distributed (as α–quartz) in acid igneous, metamorphic and sedimentary rocks. β–quartz is a high temperature form, as are *tridymite* and *cristobalite* which are found as scarce minerals in volcanic rocks. *Coesite* and *stishovite* are high pressure varieties that occur in siliceous rocks subjected to the impact of meteorites. In cryptocrystalline form, silica occurs as *chalcedony, flint, chert* and *agate* and their varieties. Some contain minor components of combined water. *Opal* is hydrated silica much prized in its precious varieties as a gemstone. Some varieties of natural highly siliceous glasses occur. Coloured varieties of quartz, some used as gemstones, include *amethyst* (purple), *cairngorm* (brownish-yellow), *citrine* (yellow), *rose quartz* (rose pink), *milky quartz* (translucent white) and *smoky quartz* (grey-brown), as well as *rock crystal* (clear).

Silica and the **silicates** are based on the silica tetrahedron which contains an ion of silicon linked to four oxygen atoms at the corners of the tetrahedron, as in Fig. 1, opposite. Each tetrahedron is linked to neighbouring tetrahedra in all three dimensions, directly in *silica* ('tectosilicate', Fig. 2), and with other ions (Mg, Fe, Ca, K, Na, (OH) etc.) intervening in the silicates (Figs. 3–6).

Among the principal rock groups, the *olivines*, which are common constituents of basic and ultrabasic rocks, have SiO_4 tetrahedra linked by divalent atoms (mainly Mg, Fe^{2+}) in six-fold coordination (Fig. 5). In *pyroxenes*, which occur in all types of igneous rock, two of the four corners of each tetrahedron are linked to the next tetrahedron, forming continuous chains with the composition $(SiO_3)_n$, which are linked laterally by cations (Fig. 3a). *Amphiboles* have two chains of silica tetrahedra, linked together, with the composition $(Si_4O_{11})_n$. The double chains are joined by cations and hydroxyl ions (Fig. 3b). *Micas* have a sheet structure of tetrahedra linked together in a hexagonal pattern, with cations linking sheets above and below and additional (OH) ions occurring in the structure (Fig. 4). Sheet structures also form the chlorites, clay minerals, e.g. *kaolinite, illite, montmorillonite* (smectite) and *vermiculite* groups. In the *sorosilicates* (Fig. 6) pairs of tetrahedra share one oxygen, e.g. in the *melilites*, with an Si_2O_7 structure. The total number of structures within the silicates is very large.

Silica is important in biological processes, especially in some plants. In diatoms the cell walls are siliceous as are the skeletons of radiolaria. There are huge quantities of siliceous minerals used, largely for structural purposes, and very substantial amounts for other purposes, e.g. in the chemical industry, manufacture of glass, abrasives, fillers, insulation etc.

figures on next page

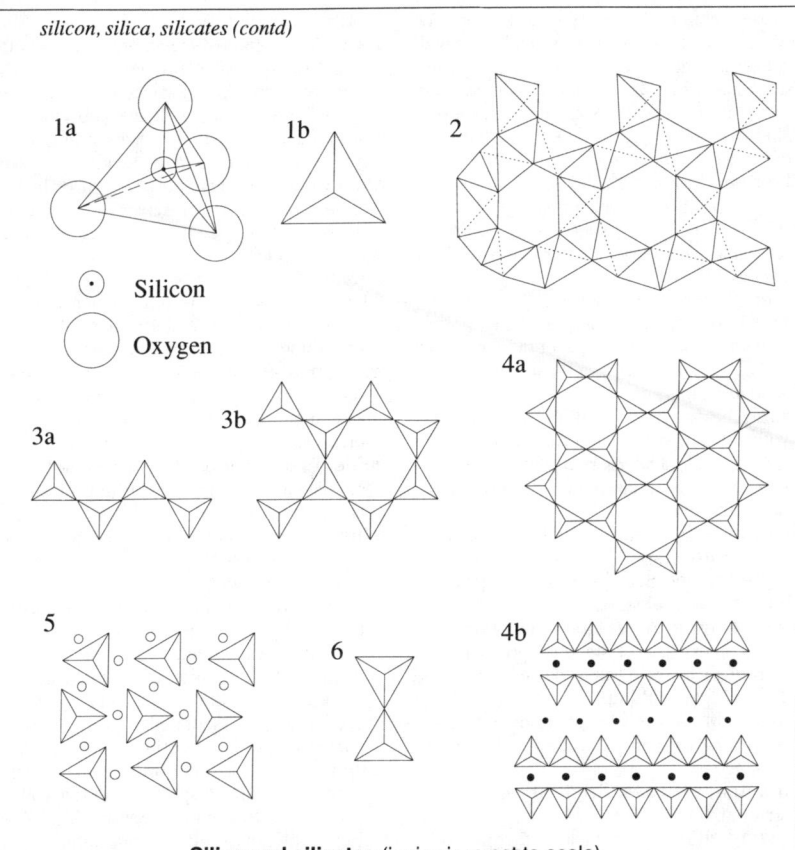

Silica and silicates (ionic sizes not to scale)

1a, the SiO$_4$ tetrahedron. 1b, conventional representation of the SiO$_4$ tetrahedron. 2, tectosilicate. 3a, single chain inosilicate. 3b, double chain inosilicate. 4a, plan view of one tetrahedral silica layer of a phyllosilicate. 4b, elevation of four superimposed tetrahedral layers of a phyllosilicate, which are linked by cations shown as solid circles. 5, nesosilicate, the open circles represent Mg or Fe ions. 6, sorosilicate.

sill A minor intrusion of igneous rock injected as a tabular sheet between, and more or less parallel to, the bedding planes of rocks. See **dykes and sills**.

sillénite Cubic bismuth trioxide. Cf. **bismite**.

sillimanite An orthorhombic aluminium silicate. The high-temperature polymorph of Al$_2$SiO$_5$, occurring in high-grade metamorphosed argillaceous rocks. *Fibrolite* is a fibrous variety, used as a gemstone. *Kyanite* and *andalusite* are other aluminium silicates.

silt Material of an earthy character intermediate in grain size between sand and clay. See **particle size**, **Wentworth scale**.

Silurian The youngest period of the Lower Palaeozoic, covering a time span between approx. 440–408 million years. Named after the *Silures*, an ancient Welsh tribe. The corresponding system of rocks. See **Palaeozoic**.

silver A pure white metallic element. Symbol Ag. at.no. 47, r.a.m. 107.868, rel.d. at 20°C 10.5, mp 960°C, bp 1955°C, casting temp. 1030–1090°C, Brinell hardness 37, electrical resistivity approx. 1.62×10^{-8} ohm metres. The metal is not oxidized in air. Occurs massive, or assumes arborescent or filiform shapes. The best electrical conductor and the main constituent of photographic emulsions. Native silver often has variable admixture of other metals; gold, copper or sometimes platinum. Used for ornaments, mirrors, cutlery, jewellery etc. and for certain components in the food and chemical industries where cheaper metals fail to withstand corrosion. The principal minerals are native silver, *argentite* (AgS) and complex sulphides, but most production is from argentiferous lead, copper, zinc, tin and gold ores.

silver amalgam A solid solution of mercury and silver, which crystallizes in the cubic system. Of rare occurrence, it is found scattered in mercury or silver deposits. Also *arquerite*.

silver glance See **argentite**.

silver lead ore The name given to galena containing silver. When 1% or more of silver is present it becomes a valuable ore of silver. Also *argentiferous galena*.

sima The lower layer of the Earth's crust with the composition of a basic or ultrabasic igneous rock. Such rocks contain silicon and magnesium as their principal constituents, hence the name.

Sinemurian A stage in the Lower Jurassic. See **Mesozoic**.

single crystal See **X-ray crystallography**.

singles See **coal sizes**.

sinistral fault A tear fault in which the rocks on one side of the fault appear to have moved to the left when viewed across the fault. Cf. **dextral fault**.

sinker bar A heavy bar attached to the cable above the drilling tools used in percussive drilling.

sink-float process See **heavy-media separation**.

sinking shaft Shaft excavated from above downwards. Cf. **rising shaft**.

sinter A concretionary deposit of opaline silica which is porous, incrusting, or stalactitic in habit; found near geysers, as at Yellowstone National Park, US. Also *geyserite*.

SI units See panel on p. 213.

skarn A rock containing calcium silicate minerals produced by metasomatic alteration of limestone close to the contact of an igneous intrusion. Skarns are sometimes enriched in ore minerals such as magnetite or scheelite. US *tactite*.

skirting board At delivery on to belt conveyor, boards which direct falling rock toward centre of belt, or prevent spillover.

skutterudite Grey or whitish arsenide of cobalt, which crystallizes in the cubic system. Often contains appreciable nickel and iron substituting for cobalt.

slack Small coal dirt, as in *slack heap*, a tip or dump.

slant rig A rig with special facilities for drilling at an angle from the vertical. The head-string components are hauled up rails set at the angle.

slate A fine-grained metamorphic rock with good fissility along cleavage planes.

slate, spotted An argillaceous rock altered by low or moderate grade metamorphism to produce porphyroblasts which impart a speckled appearance to the rock.

slaty cleavage The property of splitting easily along regular, closely spaced planes of fissility, produced by pressure in fine-grained rocks.

slicing Removal of a layer from a massive ore body. In top slicing this is horizontal, a mat of timber separating it from the overburden. Side and bottom slicing are also practised.

slickensides Smooth, grooved, polished surfaces produced by friction on fault planes and joint faces of rocks which have been involved in faulting.

slide Crack or plane along which movement has taken place; the clay filling of such a crack; a fault.

slimes Particles of crushed ore which are of such a size that they settle very slowly in water and through a bed which water does not readily percolate. Such particles must be leached by agitation. By convention these particles are regarded as less than 1/400 in (0.0635 mm) in diameter (mesh number 200). *Primary slimes* are naturally weathered ore, or associated clays. *Secondary slimes* are produced during comminution.

slip joint Special **coupling** used on floating platforms, which is splined or keyed to allow relative vertical movement in the drilling pipe while transmitting rotary torque.

slip planes The particular set or sets of crystallographic planes along which slip takes place in metal and other crystals. These are usually the most widely spaced set or sets of planes in the crystals concerned. See **gliding planes**.

sluice A long trough for washing gold-bearing sand, clay, or gravel. Also *sluice box*, *launder*.

slump Downslope gravity movement of unconsolidated sediments, esp. in a subaqueous environment.

slurry A thin paste produced by mixing some materials, esp. Portland cement, with water, sufficiently fluid to flow viscously. Used, e.g. to repair (*fettle*) slag-eroded brickwork in smelting furnace etc.

smalls See **riddle**.

smaltite Cobalt arsenide, crystallizing in the cubic system and usually associated with *chloanthite*, nickel arsenide.

smaragdite A fibrous green amphibole, pseudomorphous after pyroxene in such rocks as eclogite.

SI units

The *international system of units*, the SI (Système International d'Unités), is a coherent system of metric units, with seven *base* and two *supplementary* units from which other units are derived.

	physical quantity	SI unit	Symbol	Definition of SI unit
Base units	length	metre	m	
	mass	kilogram	kg	
	time	second	s	
	electric current	ampere	A	
	thermodynamic temperature	kelvin	K	
	amount of substance	mole	mol	
	luminous intensity	candela	cd	
Supplementary units	plane angle	radian	rad	
	solid angle	steradian	sr	
Some derived units	energy	joule	J	$m^2 kg s^{-2}$
	force	newton	N	$m kg s^{-2}$
	pressure	pascal	Pa	$m^{-1} kg s^{-2}$
	power	watt	W	$m^2 kg s^{-3}$
	electric charge	coulomb	C	sA
	electric potential difference	volt	V	$m^2 kg s^{-3} A^{-1}$
	electric resistance	ohm	Ω	$m^2 kg s^{-3} A^{-2}$
	electric capacitance	farad	F	$m^{-2} kg^{-1} s^4 A^2$
	magnetic flux	weber	Wb	$m^2 kg s^{-2} A^{-1}$
	magnetic flux density	tesla	T	$kg s^{-2} A^{-1}$
	frequency	hertz	Hz	s^{-1}
Other units with special names	length	ångstrom	Å	10^{-10} m
	length	micrometre	μm	10^{-6} m
	acceleration of free fall	gal	Gal	10^{-2} ms^{-2}
	pressure	bar	bar	10^5 Pa
	magnetic flux density	gamma	γ	10^{-9} T

The *micron* is sometimes used for micrometre and *milligal* (mGal, $10^{-5} ms^{-2}$) is also commonly used. Decimal multiples can be used with the following prefixes and symbols.

multiple	prefix	symbol	multiple	prefix	symbol
10^{-1}	deci	d	10	deca	da
10^{-2}	centi	c	10^2	hecto	h
10^{-3}	milli	m	10^3	kilo	k
10^{-6}	micro	μ	10^6	mega	M
10^{-9}	nano	n	10^9	giga	G
10^{-12}	pico	p	10^{12}	tera	T

The SI system is an *MKSA* (metre-kilogram-second-ampere) system, for which earlier equivalents in other systems are shown in **fundamental dynamical units**.

smectites A group of clay minerals including montmorillonite, beidellite, nontronite, saponite, sauconite and hectorite. They are 'swelling' clay minerals and can take up water or organic liquids between their layers, and they show cation exchange properties.

smithsonite Carbonate of zinc, crystallizing in the trigonal system. It occurs in veins and beds, and in calcareous rocks, and is commonly associated with hemimorphite. The honeycombed variety is known as *dry bone ore*. In Britain sometimes *calamine*.

smoker See **black smoker, white smoker, chimney**.

smoky quartz Dark greyish-brown transparent quartz, used as a gemstone. See **cairngorm**.

smut (1) Bad soft coal containing earthy matter. (2) Worthless outcrop material of a coal seam.

Snell's law A wave refracted at a surface makes angles relative to the normal to the surface, which are related by the law $n_1 \sin\theta_1 = n_2 \sin\theta_2$. n_1 and n_2 are the refractive indices on each side of the surface, and θ_1 and θ_2 are the corresponding angles, The two rays and the normal at the point of incidence on the boundary lie in the same plane. See **gems and gemstones**.

snubbing In high pressure oil wells the process by which drill pipe is removed through the **blowout preventer** stack while retaining pressure at all times.

soapstone See **steatite**.

socket Portion of drill hole left undisturbed after blasting, and liable to contain unexploded charge.

soda-ash Impure (commercial) sodium carbonate. Widely used in pH control of flotation process.

soda lakes Salt lakes the water of which has a high content of sodium salts (chiefly chloride, sulphate and acid carbonate). These salts also occur as an efflorescence around the lakes.

sodalite A cubic feldspathoid mineral, essentially silicate of sodium and aluminium with sodium chloride, occurring in certain alkali-rich syenitic rocks.

soda nitre Sodium nitrate, crystallizing in the trigonal system. It is found in great quantities in northern Chile, where beds of it are exposed at the surface and are known as *caliche*. Also *Chile saltpetre, nitratine.*

sodium See panel on p. 215.

sodium nitrate See **soda nitre**.

soft-rock geology An informal term for the geology of sedimentary rocks.

soil The material, normally unconsolidated and directly below the ground surface, composed of rock material, weathered to a greater or lesser extent, including organic matter, and able to support plant life. There is a wide range of compositions and textures. In civil engineering soil is taken to include a broader definition of soft unconsolidated materials, e.g. some geological clays and sands. See **soil mechanics, seat earth**.

soil mechanics The process of determining the properties of any soil, e.g. water content, bulk, density, permeability, shear strength etc.

soil sampling Taking of samples of soil or overburden as part of a **geochemical prospecting** exercise to identify anomalous concentrations of metals, or the presence of **tracers**.

solar constant The total electromagnetic energy radiated by the Sun at all wavelengths per unit time through a given area, normal to the solar beam, at the mean distance of the Earth and after correction for loss by absorption in the Earth's atmosphere. Its current value is 1.37 kW per m^2. It is not, in fact, truly constant, and variations of the order of 0.1% are detectable.

sole mark A physical structure found on the underside of a bed of sandstone or siltstone, that is the *mould* of the top surface of the bed on which it lies. The mould may represent a sedimentary structure (e.g. a groove or a ripple mark) or the remains of a **trace fossil**.

Solenhofen stone An exceedingly fine-grained and even-bedded limestone, thinly stratified, of Upper Jurassic age, occurring in S.E. Bavaria; formerly widely used in lithography.

solfatara The name applied to a volcanic orifice which is in a dormant or decadent stage and from which gases (esp. sulphur dioxide) and volatile substances are emitted.

solifluction, solifluxion Soil-creep on sloping ground, characteristic of, though not restricted to, regions subjected to periods of alternating freezing and thawing.

solution mining Winning of soluble salts by use of percolating liquor introduced through shafts, drives and/or bores. Resulting saturated solution is pumped to surface for further treatment.

solvent extraction In chemical extraction of values from ores or concentrates, selective transfer of desired metal salt from aqueous liquor into an immiscible organic liquid after intimate stirring together followed by phase separation.

sommaite An alkaline igneous rock, similar to essexite but with leucite in place of nepheline.

sodium

Metallic element, one of the alkali metals. Symbol Na, at.no. 11, r.a.m. 22.9898, rel.d. 0.978, mp 97.5°C, bp 883°C, valence 1. Sodium does not occur in nature in the free state and is a soft silvery-white metal, very reactive, with a moderate thermal neutron cross-section. Sodium is the seventh commonest element in the Earth's crust, with an abundance of 2.27%, and 1.06% in seawater. Its principal mode of occurrence is in complex silicates, especially in igneous and metamorphic rocks. It is also present as various salts in evaporite deposits that are commercial sources of sodium.

In the silicates, sodium is a constituent of the *plagioclase feldspars*, sodium-calcium aluminosilicate, ($NaAlSi_3O_8 - CaAlSi_2O_8$), which are major components of igneous rocks. The pure sodium end-member is *albite*. There is a high sodium content in many alkaline igneous rocks, which may contain alkali pyroxenes (e.g. *aegirine*) and amphiboles (e.g. *riebeckite*), and feldspathoids (*nepheline, sodalite, hauyne* etc). Among alteration products, there are several sodium-bearing zeolites (*analcite, natrolite, stilbite* etc.) which are found in amygdales in basalts, and in related occurrences. *Jadeite* is a sodium aluminium pyroxene of high-pressure metamorphic rocks.

Halite, rock salt, (NaCl), is the main mineral of the non-silicate occurrences and there are numerous other salts: nitrates, suphates, carbonates and borates. They occur in sedimentary evaporite deposits. Halite is the main component of salt domes which may form **petroleum reservoirs** in adjacent sediments. Metallic sodium is made by electrolysis of fused caustic soda or salt and is used in some nuclear reactors as a liquid metal heat-transfer fluid. There are many chemical products from sodium salts, including the commonest type of glass, soda-lime. Yellow 'sodium lighting' for street lighting makes use of the intense sodium lines of its atomic spectrum. Sodium is also an essential element of all living organisms.

sorosilicate A silicate mineral whose atomic structure contains paired silicon-oxygen tetrahedra (Si_2O_7 groups), e.g. *melilite*. See **silicates**.

source In radioactivity, the origin of α-, β-, or γ-rays.

sources of neutrons Fast neutrons are obtained by (1) nuclear transformations and (2) fission in nuclear reactors. To obtain slow neutrons, a moderator such as paraffin wax is used.

sour crude Crude oil containing significant amounts of sulphur compounds, e.g. hydrogen sulphide and mercaptans, giving it an unpleasant odour. Hydrogen sulphide is normally removed from the crude oil before shipment. Cf. **sweet crude**.

sour gas Natural gas containing gaseous impurities such as hydrogen sulphide (H_2S), hydrogen cyanide (HCN) or carbon dioxide (CO_2). Sour gas will normally be treated to remove these impurities before use. See **scrubber**.

South African jade See **Transvaal jade**.

sövite A coarse-grained calcite carbonatite, commonly containing biotite and apatite.

sp. Abbrev. for *species* (sing.); pl. *spp*.

space group Classification of crystal lattice structures into groups with corresponding symmetry elements.

space lattice Three-dimensional regular arrangement of atoms characteristic of a particular crystal structure. There are 14 such simple symmetrical arrangements, known as **Bravais lattices**. See **symmetry class**.

Spanish topaz Not a true topaz but an orange-brown quartz, the colour resembling that of the honey-brown Brazilian topaz. It is often amethyst which has been heat treated. See **citrine**.

spar Transparent to translucent crystalline mineral with vitreous lustre and clean cleavage planes, e.g. *fluorspar*. See **Iceland spar**.

Sparagmite A comprehensive term which includes the late Precambrian rocks of Scandinavia. These, like the Torridonian Sandstone of N. Scotland, consist of conglomerates and red feldspathic grits and arkoses.

sparker Marine seismic method employing a high-voltage electrical discharge under water as the energy source. See **seismic surveying**.

spathic iron See **siderite** (1).

spear pyrites The name given to twin crystals of marcasite which show re-entrant angles, in form somewhat like the head of a spear. Cf. **cockscomb pyrite**.

specific charge Charge/mass ratio of elementary particle, e.g. the ratio e/m_e of the electronic charge to the rest mass of the electron = 1.759×10^{11} coulomb per kilogram.

specific heat capacity The quantity of heat which unit mass of a substance requires to raise its temperature by one degree. Abbrev. *s.h.c.* This definition is true for any system of units, including SI, but whereas in all earlier systems a unit of heat was defined by putting the s.h.c. of water equal to unity, SI employs a single unit, the joule, for all forms of energy including heat, which makes the s.h.c. of water 4.1868 kJ/kg K.

specific latent heat See **latent heat**.

spectral distribution curve The curve showing the relation between the radiant energy and the wavelength of the radiation from a light source.

spectral line Component consisting of a very narrow band of frequencies isolated in a spectrum. These are due to similar quanta produced by corresponding electron transitions in atoms. The lines are broadened into bands when the equivalent process takes place in molecules.

spectral series Group of related spectrum lines produced by electron transitions from different initial energy levels to the same final one. The recognition and measurement of series has been of great importance in atomic and quantum theories.

spectrograph Normally used of spectroscope designed for use over wide range of frequencies (well beyond visible spectrum) and recording the spectrum photographically. The **mass spectrograph** separates particles of different specific charge in an analogous manner to the separation of spectrum lines of an optical spectrum.

spectrometer Instrument used for measurements of wavelength or energy distribution in a heterogeneous beam of radiation.

spectrophotometer An instrument for measuring photometric intensity of each colour or wavelength present in an optical spectrum.

spectroradiometer A spectrometer for measurements in the infrared.

spectroscope General term for instrument (spectrograph, spectrometer etc.) used in spectroscopy. The basic features are a slit and collimator for producing a parallel beam of radiation, a prism or grating for 'dispersing' different wavelengths through differing angles of deviation, and a telescope, camera or counter tube for observing the dispersed radiation.

spectroscopy The practical side of the study of spectra, including the excitation of the spectrum, its visual or photographic observation, and the precise determination of wavelengths.

spectrum Arrangement of components of a complex colour or sound in order of frequency or energy, thereby showing distribution of energy or stimulus among the components. A mass spectrum is one showing the distribution in mass, or in mass-to-charge ratio of ionized atoms or molecules. The mass spectrum of an element will show the relative abundances of the isotopes of the element.

specular iron The name given to a crystalline rhombohedral variety of hematite which possesses a splendent metallic lustre often showing iridescence.

specular reflection General conception of wave motion in which the wavefront is diverted from a polished surface, so that the angle of the incident wave to the normal at the point of reflection is the same as that of the reflected wave. Applicable to heat, light, radio and acoustic waves. See **reflection laws**.

speed of light The constancy and universality of the speed of light *in vacuo* is recognized by *defining* it (1983) to be exactly $2.997\,924\,58 \times 10^8$ ms^{-1}. This enables the SI fundamental unit of *length*, the metre, to be defined in terms of this value.

speise, speiss Metallic arsenides and antimonides produced in the smelting of cobalt and lead ores.

speleothems Secondary calcium carbonate encrustations deposited in caves by running water.

sperrylite Platinum diarsenide, crystallizing in the cubic system; has a brilliant metallic lustre and is tin-white in colour.

spessartine, spessartite Manganese garnet; silicate of manganese and aluminium, crystallizing in the cubic system. Usually contains a certain amount of either ferrous or ferric iron. The colour is dark, orange-red, sometimes having a tinge of violet or brown, Strictly, *spessartite* is the name for a lamprophyre rock (below). See **garnet**.

spessartite A lamprophyre composed of hornblende and plagioclase feldspar, with other mafic minerals and subordinate alkali feldspar.

sphalerite Zinc sulphide which crystallizes in the cubic system as black or brown crystals with resinous to adamantine lustre. The commonest zinc mineral and ore, deposits with fluorite, galena etc. Also *blende*, *zinc blende*.

sphene Calcium titanium silicate, with varying amounts of iron, manganese and the rare earths, it crystallizes in the monoclinic system as lozenge-shaped black or brown crystals and occurs as an accessory mineral in many igneous and metamorphic rocks. Also *titanite*.

sphenoid A wedge-shaped crystal form consisting of four triangular faces. The tetragonal and orthorhombic analogue of the cubic tetrahedron.

spheroidal jointing Spheroidal cracks found in both igneous and sedimentary rocks. Some are due to cooling and resultant contraction in the igneous rock body; others are due to a shell-like type of weathering.

spheroidal structure A structure exhibited by certain igneous rocks, which appear to consist of large rounded masses, surrounded by concentric shells of the same material. Presumably a cooling phenomenon, comparable with perlitic structure, but on a much bigger scale, and exhibited by crystalline, not glassy, rocks.

spherulite A crystalline spherical body built of exceedingly thin fibres radiating outwards from a centre and terminating on the surface of the sphere, which may vary in diameter in different cases from a fraction of a millimetre to that of a large apple.

spherulitic texture A type of rock fabric consisting of spherulites, which may be closely packed or embedded in an originally glassy groundmass. Commonly exhibited by rhyolitic rocks.

spilite A fine-grained igneous rock of basaltic composition, generally highly vesicular and containing the sodium feldspar, albite. The pyroxenes or amphiboles are usually altered. These rocks are frequently developed as submarine lava flows and exhibit pillow structure.

spinel A group of closely related oxide minerals crystallizing in the cubic system, usually in octahedra. Chemically, spinels are aluminates, chromates, or ferrates of magnesium, iron, zinc etc., and are distinguished as *iron spinel* (hercynite), *zinc spinel* (gahnite), *chrome spinel* (picotite) and *magnesian spinel*. See **ruby-**, **synthetic-**, **balas ruby**, **chromite**, **franklinite**, **magnetite**.

spinel ruby See **ruby spinel**.

spit A long bank of sediment formed by longshore drift. It is attached to the land at the upstream end.

spitzkasten Crude **classifiers**, consisting of one or more pyramid-shaped boxes with regulated holes in down-pointing apexes. Ore pulp streaming across either settles (*coarse particles*) to bottom discharge or overflows (*fires*). In the *spitzlutten*, efficiency is improved by adding upflow of low-pressure hydraulic water.

splitting limits Divergence between assay of ore, concentrate or metal made by vendor and purchaser, inside which mutual adjustment will be made without going to arbitration.

spodumene A silicate of aluminium and lithium which crystallizes in the monoclinic system; a pyroxene. It usually occurs in granite-pegmatites, often in very large crystals. The rare emerald-green variety *hiddenite* and the clear lilac coloured variety *kunzite* are used as gems.

sponge beds See **sponges**.

sponges See panel on p. 218.

spontaneous emission Process involving the emission of energy in an atomic system without external stimulation in contrast to *laser* and *stimulated emission*s. Spontaneous emission is a strictly quantum effect.

spores See **acritarchs**.

spot price The price agreed for an immediate shipment or parcel. For crude oil and products in Europe, the price is set in the Rotterdam market and differs from the price negotiated for a long-term contract or supply, or from one set by an organization such as OPEC.

spotted slate See **slate, spotted**.

sprag (1) Timber prop; short piece of wood, used to prevent the wheels of a train from revolving. (2) Slanting prop used to support coalface. Also *gib*.

spray column Tower packed with coarse material, e.g. coke, or set with grids or trays down which a liquid trickles counter-current to rising gas so as to facilitate interaction. See **Glover tower**.

spray discharge In high-tension separation the corona radiated from the electric discharge wire (18 000 volts or more) on to a passing stream of finely divided mineral particles. In classification, a spray-shaped discharge from the hydrocyclone.

spud To begin actual well-drilling operations. 'Spudding in'.

spudding bit Large drill bit for making the initial *top hole* which takes the **anchor string** or **top casing**. Very deep, high pressure wells may require over a 1000 ft of top hole into which the casing is cemented.

spur A hilly projection extending from the flanks of a valley.

stabbing The process of locating one tube above and in line with another so that they can be screwed together.

---

sponges

Members of the phylum Porifera, the sponges are a very varied group of sessile aquatic, generally marine, animals. Each individual has an internal skeleton composed of spicules of calcite or silica which may remain free or be fused together. Nearly all sponges are attached to a foreign object or are rooted in the mud of the seabed. Sponges range from the Cambrian, or earlier, to the Recent and were most common during the Cretaceous Period. *Sponge beds* are deposits, either calcareous or siliceous, which contain a large number of sponge spicules.

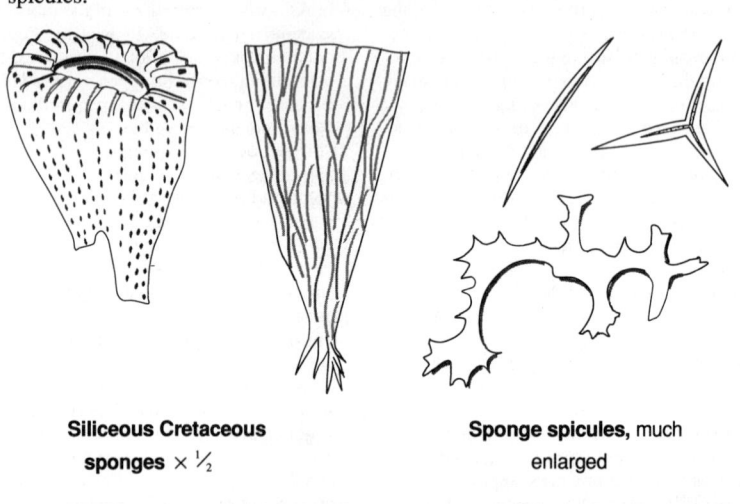

Siliceous Cretaceous sponges × ½

Sponge spicules, much enlarged

---

stacking faults Some crystal structures may be thought of as the stacking of planes of atoms in a definite sequence. Under some circumstances, e.g. deformation, the stacking order can be disturbed and such a sequence is said to have a stacking fault, a type of planar defect.

stage A chronostratigraphic succession of rocks which were deposited during an *age* of geological time. A subdivision of a geological series.

stalactite A concretionary deposit of calcium carbonate which is formed by percolating solutions and hangs icicle-like from the roofs of limestone caverns.

stalagmite A concretionary deposit of calcium carbonate, precipitated from dripping solutions on the floors and walls of limestone caverns. Stalagmites are often complementary to stalactites, and may grow so that they eventually join with them.

stall The working compartment or room in the **bord-and-pillar** method of working coal; a coalminer's working place.

stamp (1) To crush. (2) A freely falling weight, attached to a long rod and lifted by means of a cam; once widely used for crushing ores.

standard atmosphere Unit of pressure, defined as 101 325 N/m^2, equivalent to that exerted by a column of mercury 760 mm high at 0°C. Abbrev. *atm*.

standard temperature and pressure A temperature of 0°C and a pressure of 101 325 N/m^2. Abbrevs. *stp*, *STP*. See **standard atmosphere**.

stannite Sulphide of copper, iron and tin, which crystallizes in the tetragonal system and usually occurs in tin-bearing veins. Also *bell-metal ore*, *tin pyrites*.

starlite A name suggested (from a fancied resemblance to starlight) for the blue zircons which are heat-treated and used as gemstones.

star ruby, -sapphire, -quartz The prefix 'star' has reference to the narrow-rayed star of light exhibited by varieties of the minerals named. The star is seen to best advantage when they are cut *en cabochon*. It is caused by reflections from exceedingly fine inclusions lying in certain planes. See **asterism**.

stassfurtite A massive variety of boracite which sometimes has a subcolumnar structure and resembles a fine-grained white marble or granular limestone. From Stassfurt, Germany. See **boracite**.

staurolite Silicate of aluminium, iron and magnesium, with chemically combined water, sometimes occurring as brown cruciform twins, and crystallizing in the orthorhombic system. It is typically found in medium grade regionally metamorphosed argillaceous sediments.

St. David's The middle epoch of the Cambrian period.

steady state State in dynamic equilibrium, with entropy at its maximum.

steam Water in the vapour state; formed when specific latent heat of vaporization is supplied to water at boiling point. The specific latent heat varies with the pressure of formation, being approximately 2257 kJ/kg at atmospheric pressure.

steam coal Two varieties classified by the National Coal Board are *dry steam coal* (also *semi-anthracite*) and *coking steam coal* (rank 202, 203, 204).

steatite, soapstone A coarse, massive, or granular variety of talc, greasy to the touch. On account of its softness it is readily carved into ornamental objects.

Stefan-Boltzmann law Total radiated energy from a black body per unit area per unit time is proportional to the fourth power of its absolute temperature, i.e. $E = \sigma T^4$ where σ (Stefan-Boltzmann constant) is equal to 5.6696×10^{-8} W m^{-2} K^{-4}.

Steinmann trinity An association of cherts, spilites and serpentines, characteristic of a former ocean-floor environment.

STEM Abbrev. for **Scanning Transmission Electron Microscope**.

stempipe See **drilling pipe**.

step-faults A series of tensional or normal faults which have a parallel arrangement, throw in the same direction, and hence progressively 'step down' a particular bed.

Stephanian The uppermost series of the Carboniferous system, corresponding in Britain to the beds above the Coal Measures. See **Palaeozoic**.

stephanite A sulphide of silver and antimony which crystallizes in the orthorhombic system. It is usually associated with other silver-bearing minerals. Also *brittle silver ore*.

step-out well See **appraisal well**.

stibnite Antimony sulphide, which crystallizes in grey metallic prisms in the orthorhombic system. It is sometimes auriferous and also argentiferous. It is widely distributed but not in large quantity, and is the chief source of antimony. Formerly *antimony glance*. Also *antimonite*.

stichtite A lilac or pink trigonal hydrated and hydrous carbonate of magnesium and chromium.

stilb Unit of luminance, equal to 1 cd/cm^2 or 10^4 cd/m^2 of a surface.

stilbite A zeolite; silicate of sodium, calcium and aluminium with chemically combined water; crystallizes in the monoclinic system, the crystals frequently being grouped in sheaf-like aggregates. Found both in igneous rock cavities and in fissures in metamorphic rocks. Also *desmine*.

stilpnomelane Monoclinic hydrated iron magnesium potassium aluminium silicate resembling biotite. It occurs in metamorphosed sediments and in iron ores.

stink damp Underground ventilation tainted by sulphuretted hydrogen.

stipe One of the branches of a fossil **graptolite**.

stishovite A high-density form of silica. Synthesized at 1.6×10^{10} N/m^2 and 1200°C, and also found occurring naturally in Meteor Crater, Arizona, in shock-loaded sandstone.

stock Similar to **boss**.

stock pile (1) Temporarily stored tonnage of ore, middlings, concentrates or saleable products. (2) A country's holdings of strategic minerals.

stockwork An irregular mass of interlacing veins of ore; good examples occur among the tin ores of Cornwall and in the Erzgebirge. Ger. *Stockwerk*.

Stokes' law (1) The resisting force offered by a fluid of dynamic viscosity η to a sphere of radius r, moving through it at steady velocity v is given by

$$R = 6\pi\eta r v,$$

whence it can be shown that the **terminal velocity** of a sphere of density ρ, falling under gravitational acceleration g through fluid of density ρ_0 is given by

$$v = \frac{2gr^2}{9\eta}(\rho - \rho_0).$$

Applies only for viscous flow with Reynolds number less than 0.2. (2) Incident radiation is at a higher frequency and shorter wavelength than the re-radiation emitted by an absorber of that incident radiation.

stone head (1) First solid rock met while sinking a shaft or drill hole. Also *rock head*. (2) A heading or tunnel in stone.

stony meteorites Those meteorites which consist essentially of rock-forming silicates. See **achondrite, aerolites, chondrite**.

stope (1) To excavate ore from a reef, vein or lode. (2) Space formed during extraction of ore underground. Types are flat, open, overhand, rill, shrinkage and underhand, with variations to suit shape, geology and size of deposit.

stoping A mining term applied by R. A. Daly to a process in the emplacement of some igneous rock bodies, by which blocks of the overlying country rock are wedged off and sink into the advancing magma.

s.t.p., STP Abbrevs. for **Standard Temperature and Pressure**.

strain When a material is distorted by forces acting on it, it is said to be in a state of *strain*, or *strained*. Strain is the ratio

$$\frac{\text{deflection}}{\text{dimension of material}}$$

and thus has no units. The main types of strain are direct (tensile or compressive) strain:

$$\frac{\text{elongation or contraction}}{\text{original length}} ;$$

shear strain:

$$\frac{\text{deflection in direction of shear force}}{\text{distance between shear forces}} ;$$

volumetric (or bulk) strain:

$$\frac{\text{change of volume}}{\text{original volume}} .$$

See **elasticity of elongation, elasticity of shear, elasticity of bulk**.

strain-slip cleavage A cleavage in which the cleavage planes are parallel shear planes; between each pair the rocks are puckered into small sigmoidal folds.

strait work (1) Narrow headings in coal. (2) A method of working coal by driving parallel headings and then removing the coal between them.

strake Gently-sloped, flat table used for catching grains of heavy water-borne mineral. See **blanket strake, tye**.

strategic minerals Minerals considered essential for the security of a nation but not available in sufficient quantity from domestic sources in time of war.

stratification The layering in sedimentary rocks due to chemical, physical or biological changes in the sediment. See **lamination**.

stratigraphical break The geological record is incomplete, the succession of strata being broken by unconformities and non-consequences, these representing longer or shorter periods of time during which no sediment was deposited or erosion predominated.

stratigraphical level See **horizon**.

stratigraphic column See **geological column**.

stratigraphic facies See **facies, stratigraphic**.

stratigraphy The definition and description of the stratified rocks of the Earth's crust, their relationships and structure, their arrangement into chronological groups, their lithology and the conditions of their formation, and their fossil contents. The subject does not exclude igneous and metamorphic rocks where these are part of the succession.

stratotype The type representative of a named stratigraphic unit or of a stratigraphic boundary.

stratum A single bed of rock bounded above and below by divisional planes of **stratification**. A stratum differs from a lamination only in thickness. Pl. *strata*; adj. *stratified*.

stratum contours Contours drawn on the surface of a bed of rock. The position of the outcrop of the bed can be predicted from the intersection of these contours with the surface of the ground.

streak The name given to the colour of the powder obtained by scratching a mineral with a knife or file, or by rubbing the mineral on paper or an unglazed porcelain surface (*streak plate*). For some minerals, this differs from the body colour.

stream order See **drainage patterns**.

stream sampling Sampling of stream or river water to identify anomalous concentrations of dissolved metals; in gravels to find chemical or mineral concentrations, e.g. tracers. See **geochemical prospecting**.

stream tin Cassiterite occurring as derived grains in sands and gravels in the beds of rivers.

stress The force per unit area acting on a material and tending to change its dimensions, i.e. cause a *strain*. The *stress* in the material is the ratio of force applied, to the area of material resisting the force, i.e.

$$\frac{\text{force}}{\text{area}} .$$

The two main types of stress are *direct* or *normal* (i.e. *tensile* or *compressive*) *stress* (symbols σ or f) and *shear stress* (symbols τ, f_s or q). The usual units are kPa, MPa, lbf/in^2, tonf/in^2, kN/m^2, MN/m^2, bar or hbar. See **bar**.

stress zone Depth of rock surrounding an underground excavation, e.g. a stope, which is now bearing the transferred stress originally supported by the removed ore.

striae Parallel lines or grooves occurring on

glaciated pavements, roches moutonnées etc.; produced by rock material frozen into the base of a moving ice sheet; also seen on slickensided rock surfaces along which movement has taken place during faulting.

striae Parallel lines occurring on the faces of some crystals; caused by oscillation between two crystal forms. The striated cubes of pyrite are good examples.

strike The horizontal direction which is at right angles to the dip of a rock. See **drainage patterns**.

strike fault A fault aligned parallel to the strike of the strata which it cuts. Cf. **dip fault**.

strike lines See **stratum contours**.

strike-slip fault A fault whose movement is parallel to its strike.

string The succession of tubes and other drilling and well-top equipment joined together make a *string*. Also *drill string*. See **drilling rig**.

strip mining Form of opencast work, in which the overburden is removed (stripped) after which the valuable ore is excavated, the work usually being done in series of benches, steps, or terraces.

stripping Removal of barren overburden in opencast work.

stromatolite A layered or domed calcareous sediment formed by algal mats which both trap sediment and also precipitate lime.

stromatoporoid limestone A calcareous sediment, rich in remains of the reef builder *Stromatopora*, important from the Cambrian to Cretaceous.

Strombolian eruption A type of volcanic eruption characterized by frequent small explosions as trapped gases break through overlying viscous lava.

strontium Metallic element. Symbol Sr, at.no. 38, r.a.m. 87.62, rel.d. 2.54, mp 800°C, bp 1300°C. Silvery-white in colour, it is found naturally in *celestine* and in *strontianite*; it also occurs in mineral springs. Similar chemical qualities to calcium. Compounds give crimson colour to flame and are used in fireworks. The radioactive isotope, ^{90}Sr, is produced in the fission of uranium and has a long life, hence its presence in 'fall-out' after a nuclear explosion. The radiogenic isotope ^{87}Sr is produced by radioactive decay of rubidium-87, ^{87}Rb, and is used (as the ratio ^{87}Sr/^{86}Sr) in *isotopic age determination*.

strontium titanate A colourless isotropic, highly refringent artificial gemstone that is a simulant for **diamond**.

struvite Magnesium ammonium phosphate hexahydrate, crystallizing in the orthorhombic system. Found in guano and dung, and common in human calculi.

stylolite An irregular suture-like boundary found in some limestones. In three dimensions it has a tooth-and-socket arrangement and appears to have been formed by pressure solution after deposition.

subduction zone The area where a plate moves under an overriding plate. Associated with regions of high seismic activity.

subglacial drainage The system of streams beneath a glacier or ice sheet; formed chiefly of meltwaters. Cf. **englacial** streams.

subhedral See **hypidiomorphic**.

sublevel caving Method of mining massive ore deposit in which ore is drawn down to a delivery road under-running the deposit, and overburden is allowed to cave in, the process being repeated.

sublevel stoping Method in which ore is blasted in stopes and drawn down to a sublevel in the footwall through ore passes.

submarginal ore Developed ore which, at current market price of extracted values, cannot be profitably treated.

submarine canyon A trench on the continental shelf. It sometimes has tributaries.

submarine fan A fan of terrigeneous material formed at the foot of **submarine canyons** and large rivers, often as **turbidite** deposits.

suboutcrop See **blind apex**.

subsequent drainage See **drainage patterns**.

subsoil Residual deposits lying between the soil above and the bedrock below, the three grading into one another.

succinite A variety of **amber**, separated mineralogically because it yields succinic acid.

sulphate of iron See **melanterite**.

sulphate of lead See **anglesite**.

sulphate of lime See **anhydrite, gypsum**.

sulphate of strontium See **celestine**.

sulphide zone Primary (unaltered) zone of sulphide-mineral lode, underlying leached (superficial) zone and that of secondary enrichment in which there has been redeposition of values oxidized from leached zone by penetrating water.

sulphur A non-metallic element occurring in many allotropic forms. Symbol S, at.no. 16, r.a.m. 32.06, valencies 2, 4, 6. Rhombic (β-) sulphur is a lemon yellow powder; mp 112.8°C, rel.d. 2.07. Monoclinic (β-) sulphur has a deeper colour than the rhombic form; mp 119°C, rel.d. 1.96, bp 444.6°C. Chemically, sulphur resembles oxygen, and can replace the latter in many compounds, organic and inorganic. It is abundantly and widely distributed in nature with an abundance in the Earth's crust of 340 ppm, and 900 ppm in seawater. It occurs as the native element in volcanic regions in fumaroles and hot springs, and in sediments, esp. the cap rocks of salt domes. Hydrogen sulphide (H_2S) and sulphur oxides (SO_2, SO_3) also occur in fumarolic or volcanic gases. H_2S is recovered from natural gas. Many ore bodies are of sulphides of metals (galena, PbS, sphalerite, ZnS etc.). In the evaporite deposits there are many sulphate minerals. Sulphur is used in the manufacture of sulphuric acid and carbon disulphide; in the preparation of gunpowder, matches, fireworks and dyes; as a fungicide, and in medicine; and for vulcanizing rubber.

sump The prolongation of a shaft or pit, to provide for the collection of water in a mine. The pump sump is that from which casual water or ore pulp is delivered to the mine or mill pumps.

sun cracks Polygonal cracks, usually in a fine-grained sedimentary rock, indicative of desiccation at the time of formation. Common in beds laid down under arid conditions.

sunstone See **aventurine feldspar**.

superficial deposits See **drift**.

supergene enrichment Part of a mineral vein, lode or massive deposit where material removed from the *leached zone* is reprecipitated. See **secondary enrichment**.

superimposed drainage A river system unrelated to the geological structure of the area, as it was established on a surface since removed. Cf. **consequent drainage**.

superposition, law of Strata which overlie other strata are always younger, except in strongly folded areas.

surface energy The free potential energy of a surface equal to the **surface tension** multiplied by the surface area.

surface pipe See **anchor string**. Also *surface casing*.

surface tension A property possessed by liquid surfaces whereby they appear to be covered by a thin elastic membrane in a state of tension, the surface tension being measured by the force acting normally across unit length in the surface. The phenomenon is due to unbalanced molecular cohesive forces near the surface. Units of measurement are dyne cm^{-1}, Nm^{-1}. See **capillarity**.

surfactant flooding Recovery enhancement process in oil wells in which surface-tension reducing compounds are forced into the surrounding strata and release oil held there.

surge bin, tank Hopper (dry material) or reservoir with means of agitation (ore pulps), used to minimize irregularities in process delivery and flow.

suspect terrane See **terrane**.

sutured A textural term descriptive of the sinuous interlocking grain boundaries of rocks which have undergone extensive recrystallization, e.g. quartzites.

S-wave Abbrev. for *Secondary wave*. See **earthquakes**.

sweet crude Oil containing no hydrogen sulphide, H_2S. Cf. **sour crude**.

sweetening A refining process that improves the smell of some light products without necessarily reducing their sulphur content; thus **mercaptans** may be oxidized to less offensive disulphides. Sodium plumbite, hypochlorite or an oxidizing copper salt are used.

sweet gas Natural gas that is free of malodorous sulphur compounds and therefore smells 'sweet'. Most natural gas is contaminated by such compounds, e.g. hydrogen sulphide (H_2S) and **mercaptans** in addition to carbon dioxide, and needs **sweetening** treatment before it is ready for use.

swell Volumetric increase due to crushing, which creates more void space in a given weight of rock.

Swiss lapis An imitation of lapis lazuli, obtained by staining pale-coloured jasper or ironstone with blue pigment. Also *German lapis*.

syenite A coarse-grained igneous rock of intermediate composition, composed essentially of alkali-feldspar to the extent of at least two-thirds of the total, with a variable content of mafic minerals, of which common hornblende is characteristic. See **plutonic rocks**.

syenite-porphyry An igneous rock of syenitic composition and medium grain size, commonly occurring in minor intrusions; it consists of phenocrysts of feldspar and/or coloured silicates set in a microcrystalline groundmass.

syenodiorite See **monzonite**.

syenogranite A variety of granite composed of alkali feldspar with subordinate plagioclase, quartz and biotite or hornblende. See **plutonic rocks**.

sylvanite Telluride of gold and silver, which crystallizes in the monoclinic system and is usually associated with igneous rocks and, in veins, with native gold. It is an ore of gold. Also *yellow tellurium*.

sylvine, sylvite Potassium chloride, which crystallizes in the cubic system. It occurs in bedded salt deposits (*evaporites*), and as a

sylvinite sublimation product near volcanoes; it is a source of potassium compounds, used as fertilizers. See **sylvinite**.

sylvinite A general name for mixtures of the two salts sylvine and halite, the latter predominating, occurring at Stassfurt, Germany and elsewhere. Also used as a commercial name for **sylvine**.

symmetry The quality possessed by crystalline substances by virtue of which they exhibit a repetitive arrangement of similar faces. This is a result of their peculiar internal atomic structure, and the feature is used as a basis of crystal classification.

symmetry class Crystal lattice structures can show 32 combinations of symmetry elements, each combination forming a possible symmetry class.

syncline A concave-upwards fold with the youngest rocks in the centre.

syngenetic A category of ore bodies comprising all those which were formed contemporaneously with the enclosing rock. Cf. **epigenetic**.

synthetic ruby, synthetic sapphire In chemical composition and in all their physical characters, including optical properties, these stones are true crystalline ruby or sapphire; but they are produced in quantity in the laboratory by fusing pure precipitated alumina with the predetermined amount of pigmentary material. They can be distinguished from natural stones only by the most careful expert examination.

synthetic spinel This is produced, in a wide variety of fine colours, by the **Verneuil process**; in chemical and optical characters identical with natural magnesian spinel, it is widely used as a gemstone.

Syrian garnet A name for **almandine** of gemstone quality.

system (1) The chronostratigraphical equivalent of a **period** of geological time. (2) Name given to the succession of rocks which were formed during a certain period of geological time, e.g. *Jurassic System*. (3) Term applied to the sum of the phases which can be formed from one or more components of minerals under different conditions of temperature, pressure and composition.

Système International See **SI system**.

systems of crystals The 7 large divisions into which all crystallizing substances can be placed, viz. cubic, tetragonal, hexagonal, trigonal, orthorhombic, monoclinic, triclinic. This classification is based on the degree of **symmetry** displayed by the crystals. See **Bravais lattices**.

T

T Symbol for **tesla**.

table of strata A column which depicts a series of rocks arranged in chronological order, the oldest being at the bottom. It is usual to draw this to scale so that the average thicknesses of the beds are also shown.

tachylite, tachylyte A black glassy igneous rock of basaltic composition, which occurs as a chilled margin of dykes and sills. In Hawaii it forms the bulk of certain lava flows.

tachyon A theoretical particle moving faster than the speed of light.

Taconic orogeny A period of intense folding which affected the eastern parts of North America at the end of the Ordovician period. The effects are best seen in the Taconic Mts. on the borders of New York State and Massachusetts.

tactite See **skarn**.

tactoid A rod-shaped droplet or flat particle appearing in colloidal solutions which exhibit double refraction.

tactosol Sol containing **tactoids**.

taenite A solid solution of iron and nickel occurring in iron meteorites; it appears as bright white areas on a polished surface. It crystallizes in the cubic system and has 27% to 65% Ni.

tagged atom   See **labelled atom**. Also *radioactive tracer*.

tailings (1) Rejected portion of an ore; waste, *gangue*. (2) Portion washed away in water concentration. May be impounded in a **tailings dam** or pond, or stacked dry on a dump. Also US *tails*.

tailings dam Dam used to hold mill residues after treatment. These arrive as fluent slurries. Dam may include arrangements for run-off or return of water after the slow-settling solids have been deposited.

tail race The launder or trough for the discharge of water-borne tailings.

tail(s) US term for **tailings**.

talbot Unit of luminous energy, such that 1 lumen is a flux of 1 talbot per second.

talc A monoclinic hydrated magnesium silicate, $MgSi_8O_{20}(OH)_4$. It is usually massive and foliated and is a common mineral of secondary origin associated with serpentine and schistose rocks; also found in metamorphosed siliceous dolomites. Purified talc is used medically, in toilet preparations and in many other ways. See **steatite**.

talus See **scree**.

tamarugite   A hydrated sulphate of sodium and aluminium, crystallizing in the monoclinic system.

Tamman's temperature The temperature at which the mobility and reactivity of the molecules in a solid become appreciable. It is approximately half the melting point in kelvins.

tantalum A metallic element. Symbol Ta, at.no. 73, r.a.m. 180.948, rel.d at 20°C 16.6, mp 2850°C, electrical resistivity 15.5×10^{-8} ohm metres, Brinell hardness 46. It occurs in crystals and grains (usually containing, in addition, small amounts of niobium) in the Ural and Altai Mts. It is used as a substitute for platinum for corrosion-resisting laboratory apparatus, as acid-resisting metal in chemical industry, and in the form of carbide in cemented carbides. Used in surgical insertions because of its lack of reaction to body fluids.

tantalite Tantalate and niobate of iron and manganese, crystallizing in the orthorhombic system. The principal ore of tantalum, occurring in pegmatites and granitic rocks and in alluvial deposits. When the Ta content exceeds that of Nb, the ore is called tantalate. See **columbite**.

taphrogenesis Vertical movements of the Earth's crust, resulting in the formation of major faults and rift valleys.

tapiolite A tantalate resembling **tantalite** but crystallizing in the tetragonal system.

tarbuttite Hydrated zinc phosphate, which crystallizes in the triclinic system. The crystals are often found in sheaf-like aggregates.

tar pit An outcrop where natural bitumen occurs. The tar frequently contains the skeletons of trapped animals.

tar sands   Sedimentary deposits of oil-bearing sands. The oil may be separated by steam heating or solvent extraction, and **reforming processes** are used to recover normal oil products. The Athabasca Tar Sands are extensive deposits in Canada and are estimated to contain recoverable oil equivalent to three quarters of the present world crude reserves, but extraction is expensive and difficult.

TAS Abbrev. for *Total Alkali Silica* diagram. A graph on which chemical analyses of volcanic rocks may be plotted and used as the basis for classification when modes cannot be determined; a recommended method of the *IUGS system*.

tasmanite A type of practically pure spore coal; a variety of **cannel coal**. See **boghead coal**.

tear fault A horizontal displacement of a

series of rocks along a more or less vertical plane. Cf. **normal fault, thrust plane**.

technetium Radioactive element not found in ores. First produced as a result of deuteron and neutron bombardments of molybdenum. Symbol Tc, at.no. 43, the most common isotope ^{99}Tc has half-life of 2.1×10^6 years. ^{99m}Tc has a half-life of 6.1 h and is used in nuclear medicine.

tectonic Said of rock structures which are directly attributable to earth movements involved in folding and faulting.

tectonics The study of the major structural features of the Earth's crust.

tectosilicates Those silicates having a structure in which atoms of silicon and oxygen are linked in a continuous framework.

tektites A group term which covers moldavites, billitonites and australites. They are natural glasses of non-volcanic origin and may be of extra-terrestrial origin.

tektosilicates See **tectosilicates**.

telluric bismuth An intermetallic compound, Bi_2Te_3, crystallizing in the trigonal system. The name has also been used as a synonym for **tetradymite**.

telluric current Current in, or put into, the Earth, which is used in exploration of strata.

tellurite Orthorhombic tellurium dioxide.

tellurium A semi-metallic element, tin-white in colour. Symbol Te, at.no. 52, r.a.m. 127.60, rel.d. at 20°C 6.24, mp 452°C, valencies 2, 4, 6, electrical resistivity 2×10^{-8} ohm metres. Used in the electrolytic refining of zinc in order to eliminate cobalt; alloyed with lead to increase the strength of pipes and cable sheaths. The chief sources are the slimes from copper and lead refineries and the fine dusts from telluride gold ores. There are numerous but rare silver, gold, bismuth and iron tellurides.

tellurobismuth See **telluric bismuth**.

TEM Abbrev. for *Transmission Electron Microscope*.

temperature A measure of whether two systems are relatively hot or cold with respect to one another. Two systems brought into contact will, after sufficient time, be in thermal equilibrium and will have the same *temperature*. A *thermometer* using a temperature scale established with respect to an arbitrary zero (e.g. **Celsius scale**) or to absolute zero (**Kelvin thermodynamic scale of temperature**) is required to establish the relative temperatures of two systems. See Zeroth law of **thermodynamics**.

temperature coefficient The fractional change in any particular physical quantity per degree rise of temperature.

tenebrescence The reversible bleaching observed in the hackmanite variety of sodalite. This mineral has a pink tinge when freshly fractured; the colour fades on exposure to light, but returns when the mineral is kept in the dark for a few weeks or is bombarded by X-rays.

tennantite Sulphide of copper and arsenic, which crystallizes in the cubic system. It is isomorphous with **tetrahedrite**. The crystals frequently contain antimony, and grade into tetrahedrite. Also *fahlerz*.

tenorite Copper oxide, crystallizing in the triclinic system. Occurs in minute black scales as a sublimation product in volcanic regions or associated with copper veins. *Melaconite* is a massive variety.

tepee buttes Conical hills of Cretaceous shale, with steep, smooth slopes of talus and a core of shell-limestone, formed *in situ* by the growth of successive generations of lamellibranchs (*Lucina*). Found in the Great Plains of the US.

tephra A general term for all fragmental volcanic products, e.g. *ash*, *bombs*, *pumice*.

tephrite A fine-grained igneous rock resembling basalt and normally occurring in lava flows; characterized by the presence of a feldspathoid mineral in addition to, or in place of, feldspar. See **volcanic rocks**.

tephroite An orthosilicate of manganese, which crystallizes in the orthorhombic system. It forms a member of the olivine isomorphous group, and occurs with zinc and manganese minerals.

terbium A metallic element, a member of the rare earth group. Symbol Tb, at.no. 65, r.a.m. 158.9254. It occurs in the same minerals as dysprosium, europium and gadolinium.

terrace See **river terrace**.

terrane A geologically consistent area, discontinuous with that of its neighbours. A *displaced* or *suspect* terrane is one whose distinctive stratigraphical or structural features and its geological history, indicate that it is *foreign* to the region.

terrestrial equator An imaginary circle on the surface of the Earth, the latter being regarded as being cut by the plane through the centre of the Earth perpendicular to the polar axis; it divides the Earth into the northern and southern hemispheres, and is the primary circle from which terrestrial latitudes are measured.

terrestrial magnetism The magnetic properties exhibited within, on and outside the Earth's surface. There is a nominal (magnetic) North pole in Canada and a nominal South pole opposite, the positions varying cyclically with time. The direction indicated by a compass needle at any one point is that of the horizontal component of the field at that point. Having the characteristics of flux from a permanent magnet, the Earth's magnetic field probably depends on currents within the Earth and also on those arising from ionization in the upper atmosphere, interaction being exhibited by the Aurora Borealis.

terrestrial poles The two diametrically opposite points in which the Earth's axis cuts the Earth's surface are the geographical poles. The magnetic poles, the positions to which the compass needle will point, are unstable and differ from the geographical. North magnetic pole: ca 76° N, 101° W; South magnetic pole: ca 66° S, 139° E. See **terrestrial magnetism**.

terrigenous sediments Sediments derived from the erosion of the land. They include sediments deposited on land and land-derived material deposited in the sea.

Tertiary The first period or sub-era of the Cenozoic era, covering an approx. time span from 65–2 million years ago. See panel on p. 227.

Tertiary igneous rocks The various types of igneous rocks which were intruded or extruded during early Tertiary times, esp. over a region stretching from Britain to Iceland, e.g. in the Inner Hebrides and northeast Ireland (the Thulean Province).

tertiary production Special methods of increasing oil flow after **primary production** (natural flow by gravity or intrinsic pressure) and **secondary production** (pressurizing the reservoir) are exhausted. Includes chemical treatment and water injection. Also *tertiary recovery*.

teschenite A coarse-grained basic (gabbroic) igneous rock consisting essentially of plagioclase, near labradorite in composition, titanaugite, ilmenite and olivine (or its decomposition products); primary analcite occurs in wedges between the plagioclase crystals, which it also veins. An analcime (foid) gabbro or dolerite. See **plutonic rocks**.

tesla SI unit of magnetic flux density or magnetic induction equal to 1 weber m^{-2}. Equivalent definition; the magnetic induction for which the maximum force it produces on a current of unit strength is 1 newton. Symbol T.

test The external shell of many invertebrates, e.g. echinoderms.

Tethys An east-west ocean lying between **Laurasia** (to the north) and **Gondwanaland** (to the south) during Palaeozoic and Mesozoic time, from which the Alpine and Himalayan mountains arose; a Mesozoic geosyncline.

tetradymite An ore of tellurium, of composition Bi_2Te_2S; crystallizes in the trigonal system. Bismuth tellurides are commonly found in gold-quartz veins.

tetragonal system The crystallographic system in which all the forms are referred to three axes at right angles; two are equal and are taken as the horizontal axes, whilst the vertical axis is either longer or shorter than these. It includes such minerals as zircon and cassiterite. Also *pyramidal system*. See **Bravais lattices**.

tetrahedrite A sulphide of copper and antimony, crystallizing in the tetrahedral division of the cubic system; frequently contains arsenic and other metals. It is used as an ore of copper and, in some cases, of other metals. Also *fahlerz, fahl ore, grey copper ore*. See **tennantite**.

tetratohedral Containing a quarter of the number of faces required for the full symmetry of the crystal system.

texture The physical quality of a rock which is determined by the relative sizes, disposition, and arrangement of the component minerals. The nomenclature and classification of rocks are governed by mineral composition and texture. See **graphic texture, ophitic texture, poikilitic texture**.

thallium White malleable metal like lead. Symbol Tl, at.no. 81, r.a.m. 204.37, rel.d. 11.85, mp 303.5°C, bp 1650°C. Several thallium isotopes are members of the uranium, actinium, neptunium and thorium radioactive series. Thallium isotopes are used in scintillation crystals. Its compounds are very poisonous and it has common valencies of one and three. There are a few rare independent minerals.

thalweg (German 'valley way'.) The name frequently used for the longitudinal profile of a river, i.e. from source to mouth. See **valleys**.

thanatocoenosis Assemblage of fossil remains of organisms which were not associated during their life but were brought together after death.

Thanetian A stage in the Palaeocene. See **Tertiary**.

thenardite Sodium sulphate, crystallizing in the orthorhombic system and occurring in saline residues of alkali lakes.

theralite A coarse-grained, holocrystalline igneous rock composed essentially of the minerals labradorite, nepheline, purple titanaugite, and often with soda-amphiboles, biotite, analcite or olivine. A nepheline (foid) gabbro. See **plutonic rocks**.

Tertiary

The Tertiary derives its name from an old and disused division of all geological time into three parts. It forms the lower part of the Cenozoic era (also *Cainozoic, Kainozoic*, meaning 'recent life') and consists of five epochs: Palaeocene, Eocene, Oligocene, Miocene and Pliocene.

There was a very marked change from the flora and fauna of the Cretaceous to plants and animals of more modern aspect. On land there was a reversal of roles of reptiles and mammals, the latter becoming predominant.

The Palaeogene had a temperate to warm climate and forests became widespread in the Tertiary. Lamellibranchs, gastropods and echinoderms were abundant. In the Neogene (Miocene and Pliocene), the climate was temperate and warm, cooling in the Pliocene of more northern latitudes.

Following earlier Palaeogene tremors, earth-movements during the Miocene built many of the mountain ranges of the world (Himalayas, Rockies, Alps etc.). Britain was on the edge of the area affected and folding occurred in southern England, including the anticline of the Wealden axis and the syncline of the London basin.

During the Eocene, lavas, mainly basaltic, were erupted from a series of fissures in many parts of the world including north-east Ireland and western Scotland, and plutonic centres developed at a number of localities including Skye, Rhum, Ardnamurchan, Mull, Arran, the Mourne Mountains and Rockall in the North Atlantic. There were also extensive **dyke swarms**.

Era	Sub-era, period	Series	Stage	Age, Ma
Cenozoic	Quaternary (see separate panel)	Holocene		
		Pleistocene		1.64
	Tertiary — Neogene	Pliocene	Piacenzian	3.40
			Zanclian	5.2
		Miocene	Messinian	6.7
			Tortonian	10.4
			Serravallian	14.2
			Langhian	16.3
			Burdigalian	21.5
			Aquitanian	23.3
	Tertiary — Palaeogene	Oligocene	Chattian	29.3
			Rupelian	35.4
		Eocene	Priabonian	38.6
			Bartonian	42.1
			Lutetian	50.0
			Ypresian	56.5
		Palaeocene	Thanetian	60.5
			Danian	65.0

thermal conductivity A measure of the rate of flow of thermal energy through a material in the presence of a temperature gradient. If (dQ/dt) is the rate at which heat is transmitted in a direction normal to a cross-sectional area A when a temperature gradient (dT/dx) is applied, then the thermal conductivity is

$$k = -\frac{dQ/dt}{A\,(dt/dx)}.$$

SI unit is W m^{-1} K^{-1}. Materials with high electrical conductivities tend to have high thermal conductivities.

thermal metamorphism Metamorphism resulting from the action of heat, involving chemical changes in the rock without the introduction of a material from elsewhere. Pressure is not significant. See **regional metamorphism**.

thermal resistance Resistance to the flow of heat. The unit of resistance is the *thermal ohm*, which requires a temperature difference of 1°C to drive heat at the rate of 1 watt. If the temperature difference is θ°C, the resistance S thermal ohms, and the rate of driving heat W watts, then $θ = SW$.

thermodynamics The mathematical treatment of the relation of heat to mechanical and other forms of energy. Its chief applications are to heat engines (steam engines and IC engines) and to chemical reactions. *Laws of thermodynamics*: *Zeroth law*: if two systems are each in thermal equilibrium with a third system then they are in thermal equilibrium with each other. This statement is tacitly assumed in every measurement of temperature. *First law*: the total energy of a thermodynamic system remains constant although it may be transformed from one form to another. This is a statement of the principle of the conservation of energy. *Second law*: heat can never pass spontaneously from a body at a lower temperature to one at a higher temperature (Clausius) *or* no process is possible whose only result is the abstraction of heat from a single heat reservoir and the performance of an equivalent amount of work (Kelvin-Planck). *Third law*: the entropy of a substance approaches zero as its temperature approaches absolute zero.

thermodynamic scale of temperature See **Kelvin thermodynamic scale of temperature**.

thermoluminescent material Material which, after radiation, releases light when subsequently heated in proportion to the radiation absorbed.

thermometer An instrument for measuring temperature. A thermometer can be based on any property of a substance which varies predictably with change of temperature. For instance, the *constant volume gas thermometer* is based on the pressure change of a fixed mass of gas with temperature, while the *platinum resistance thermometer* is based on a change of electrical resistance. The commonest form relies on the expansion of mercury or other suitable fluid with increase in temperature.

thermometric scales See **Centigrade scale, Fahrenheit scale, Kelvin thermodynamic scale of temperature, fixed points**.

thermometry The measurement of temperature.

thermoregulator Type of thermostat which keeps a bath at a constant temperature by regulating its supply of heat.

thermostat An apparatus which maintains a system at a constant temperature which may be preselected. Frequently incorporates a **bimetallic strip**.

thickener Apparatus in which water is removed from ore pulp by allowing solids to settle. To obtain continuous working the solids are worked towards a central hole in the bottom by means of revolving rakes.

thixotropy The property of some colloidal materials (particularly in geology, clays and fine sediments) to change from a gel to a sol when under stress and to revert to high viscosity at low stress. The clay thus strengthens when undisturbed.

tholeiite A silica-rich basalt abundant in mid-oceanic ridges and continental rifts.

thomsonite An orthorhombic zeolite; a hydrated silicate of calcium, aluminium and sodium, found in amygdales and crevices in basic igneous rocks.

thorianite Thorium dioxide; crystallizes in the cubic system and is found in pegmatites and gem gravel washings, as in Sri Lanka. It is an important source of thorium and uranium.

thorite Tetragonal thorium orthosilicate, found in syenites and syenitic pegmatites.

thorium A metallic radioactive element, dark-grey in colour. Symbol Th, at.no. 90, r.a.m. 232.0381, rel.d. 11.2, mp 1845°C. Its abundance in the Earth's crust is 8.1 ppm and there are few independent thorium minerals. Its commercial sources are *monazite*, which occurs widely in beach sands where it is derived from acid igneous rocks and pegmatites, and *thorite*. The thorium radioactive series starts with thorium of mass 232. It is fissile on capture of fast neutrons and is a fertile material, ^{233}U (fissile with slow neutrons) being formed from ^{232}Th by neutron capture and subsequent beta decay.

thorium series The series of nuclides which result from the decay of ^{232}Th. The mass numbers of the members of the series are given by $4n$, n being an integer. The series ends in the stable isotope ^{208}Pb. See **radioactive series**.

thoron Thorium emanation, an isotope of radon (^{220}Rn); half-life 54.5 s; a radioactive decay product of thorium. Symbol Tn.

throw (1) The amount of vertical displacement (*upthrow* or *downthrow*) of a particular rock, vein or stratum due to faulting. See **fault**. Cf. **lateral shift**. (2) The amplitude of shake of a concentrating table. (3) Deviation of a deep borehole from the planned path.

thrust plane A thrust plane, or thrust, is a *reversed fault* which dips at a low angle.

thulite A variety of zoisite, pink in colour due to small amounts of manganese.

thulium A metallic element, a member of the rare earth group. Symbol Tm, at.no. 69, r.a.m. 168.9342. One of the rarest elements, occurring in small quantities in euxenite, gadolinite, xenotime etc. Radioactive isotope emits 84 keV gamma-rays, frequently used in radiography.

thuringite A variety of **chlorite**.

Thurnian A stage in the Pleistocene. See **Quaternary**.

tiger's eye A form of silicified crocidolite stained yellow or brown by iron oxide.

tile ore The earthy brick-red variety of cuprite; often mixed with red oxide of iron.

till A poorly sorted mixture of unconsolidated sediment produced by glacial action. Also *boulder clay*.

tillite Consolidated and lithified till.

time scale A chronological sequence of geological events.

tin A soft, silvery-white metallic element, ductile and malleable, existing in three allotropic forms. Symbol Sn (L. *stannum*, tin), at.no. 50, r.a.m. 118.69, rel.d. at 20°C 7.3, mp 231.85°C. Not affected by air or water at ordinary temperatures. Electrical resistivity is 11.5×10^{-8} ohm metres at 20°C. The principal use is as a coating on steel in tinplate; also used as a constituent in alloys and with lead in low melting-point solders for electrical connections, and in tinned copper wire. Occurs very rarely as native metal and in independent tin minerals but dominantly in *cassiterite*, SnO$_2$. This occurs in lodes of tin associated with granite as in Cornwall, UK, and in placer deposits. 'Tin' cans are made of tin-plated steel. Non-toxic, but organic compounds of tin are toxic.

tincal The name given since early times to crude borax obtained from salt lakes, e.g. in Kashmir and Tibet. See **borax**.

tin pyrites See **stannite**.

tinstone See **cassiterite**.

tip See **dump**.

tipple Frame into which ore trucks are run, gripped and rotated to discharge contents.

tirodite A rare honey-yellow monoclinic amphibole, the name being used for the manganese-rich, magnesium-bearing end-member. See **dannemorite**.

titanaugite A titaniferous variety of the monoclinic pyroxene **augite**.

titaniferous iron ore See **ilmenite**.

titanite See **sphene**.

titanium A metallic element resembling iron. Symbol Ti, at.no. 22, r.a.m. 47.90, rel.d. at 20°C 4.5, mp 1850°C, bp above 2800°C. Manufactured commercially since 1948, it is characterized by strength, lightness and corrosion resistance. Widely used in aircraft manufacture, for corrosion resistance in some wet extraction processes, as a deoxidizer for special types of steel, in stainless steel to diminish susceptibility to inter-crystalline corrosion and as a carbide in cemented carbides. As dioxide, white pigment in paints. Abundance in the Earth's crust 0.6% (ninth commonest element). Occurs in accessory oxides and oxide minerals in igneous and metamorphic rocks (and in beach sands). The principal sources are *ilmenite* FeTiO$_3$, *rutile* TiO$_2$ and from other industrial processes.

toad's-eye tin A variety of **cassiterite** occurring in botryoidal or reniform shapes which show an internal concentric and fibrous structure. It is brownish in colour.

toadstone An old and local name for the basalts found in the Carboniferous Limestone of Derbyshire. The name may be derived from the rock's resemblance in appearance to a toad's skin, or from the fact that it weathers into shapes like a toad, or from the German word *Todstein* ('dead stone') in reference to the absence of lead ore.

Toarcian A stage in the Lower Jurassic. See **Mesozoic**.

todorokite Hydrated manganese oxide, enriched in other elements. It is one of the dominant minerals of the deep-sea **manganese nodules**.

tombolo A *bar* or *spit* of sand, shingle or gravel which joins an island to the mainland or to another island.

tonalite A coarse-grained igneous rock of dioritic composition carrying quartz as an essential constituent, usually with biotite and hornblend. See **plutonic rocks**.

toolpusher Field supervisor of drilling operations.

topaz Silicate of aluminium with fluorine, usually containing hydroxyl, which crystallizes in the orthorhombic system. It usually occurs in veins and druses in granites and granite-pegmatites. It is colourless, pale blue, or pale yellow in colour, and is used as a gemstone. Cf. **citrine, Oriental topaz, Scottish topaz, Spanish topaz.**

topazolite A variety of the calcium-iron garnet **andradite**, which has the honey-yellow colour and transparency of topaz.

top casing The topmost part of the casing of an oil well to which either the drilling gear or the flow-control gear is attached. Also *conductor, anchor string.*

topset beds Gently inclined strata deposited on the subaerial plain or the just-submerged part of a delta. They are succeeded seawards by the *foreset beds* and, in deep water, by the *bottomset beds.*

torbanite A variety of **boghead coal** or oil shale containing 70%–80% of carbonaceous matter, including an abundance of spores; dark brown in colour.

torque The turning moment exerted by a *tangential* force acting at a distance from the axis of rotation. It is measured by the product of the magnitude of the force and the distance. Unit newton-metre. See **moment.**

torr Unit of low pressure equal to head of 1 mm of mercury or 133.3 N/m^2.

Torridonian A succession of conglomerates and red sandstones and arkoses forming part of the Precambrian system in the north-west highlands of Scotland. They rest on the Lewisian schists and gneisses.

torsion balance A delicate device for measuring small forces such as those due to gravitation, magnetism, or electric charges. The force is caused to act at one end of a small horizontal rod, which is suspended at the end of a fine vertical fibre. The rod turns until the turning moment of the force is balanced by the torsional reaction of the twisted fibre, the deflection being measured by a lamp and scale using a small mirror fixed to the suspended rod.

Tortonian A stage in the Miocene. See **Tertiary.**

tossing The operation of raising the grade or purity of a concentrate by violent stirring, followed by packing in a *kieve.* Also *kieving.*

total internal reflection Complete reflection of incident wave at boundary with medium in which it travels faster, under conditions where Snell's law of refraction cannot be satisfied. The angle of incidence at which this occurs (corresponding to an angle of refraction of 90°) is known as the *critical angle.* See **gems and gemstones.**

touchstone See **Lydian stone.**

toughness Ability of mineral to withstand disruption, assessed empirically by comparison with standard minerals under controlled test conditions.

tourmaline A complex silicate of sodium, boron and aluminium, with, in addition, magnesium (*dravite*), iron (*schorl*), or lithium (*elbaite*), and fluorine in small amounts, which crystallizes in the trigonal system. It is usually found in granites or gneisses. The variously coloured and transparent varieties are used as gemstones, under the names *achroite* (colourless), *indicolite* (blue), *rubellite* (pink). The common black variety is *schorl.*

tourmalinization The process whereby minerals or rocks are replaced wholly or in part by tourmaline. See **pneumatolysis.**

Tournaisian The lowermost series of the Carboniferous of Europe. See **Palaeozoic.**

town gas Usually a mixture of coal gas and carburetted water gas; the energy density is about 500 Btu/ft^3 or 20 MJ/m^3, which is about half that of *natural gas* which in many areas has largely superseded town gas.

trace element A non-essential element (< 1%) in a mineral.

trace fossil Any sedimentary structure caused by the activity of a fossil organism during its life, e.g. trails, burrows.

tracers (1) Elements diffused as a **halo** around an ore body, identified by **geochemical prospecting.** These may be dilute concentrations of metals also present in the ore body, or elements found in association with ore bodies but not of primary interest themselves. (2) Minerals identified in soil and stream gravel which are common accessories to minerals of economic significance.

trachyandesite Fine-grained igneous rock, commonly occurring as lava flows, intermediate in composition between trachyte and andesite, that is, containing both orthoclase and plagioclase in approximately equal amounts.

trachybasalt A fine-grained igneous rock commonly occurring in lava flows and sharing the mineralogical characters of trachyte and basalt. The rock contains sanidine (characteristic of trachyte) and calcic plagioclase (characteristic of basalt).

trachyte A fine-grained igneous rock-type, of intermediate composition, in most specimens with little or no quartz, consisting largely of alkali-feldspars (sanidine or oligoclase) together with a small amount of coloured silicates such as diopside, hornblende, or mica. See **volcanic rocks**.

traction load That part of the load of solid material carried by a river which is rolled along the bed.

tram A small wagon, tub, cocoa-pan, corve, corf or hutch, for carrying mineral.

tramp iron Stray pieces of iron or steel, e.g. broken drills, mixed with ore from mine, which must be removed before crushing.

trans- In chemistry, the geometrical isomer in which like groups are on opposite sides of the bond with restricted rotation, or in which like ligands are on opposite sides of the central atom of a co-ordination compound.

transcurrent fault A strike-slip fault.

transform fault A strike-slip fault along which two plates slide past each other, e.g. *San Andreas fault*.

transgression The gradual submergence of land caused by a relative rise in sea level.

transition element See **transition metal**.

transition metal *Transition element*. Any one of a large group of elements in which the filling of the outermost electron shell to 8 electrons is interrupted to bring the penultimate shell (which can be used in bonding) from 8 to 18 or 32 electrons. This has profound effects upon the properties, and many metals of geological importance occur as transition elements.

translucent A mineral which is capable of transmitting light but through which no object can be seen.

transmittance Ratio of energy transmitted by a body to that incident on it. If scattered emergent energy is included in the ratio, it is termed *diffuse transmittance*, otherwise *specular transmittance*. Also *transmission*.

transmittivity Transmittance of unit thickness of non-scattering medium.

Transvaal jade Massive light green hydrogrossular garnet, used as a simulant for jade. Also *South African jade*.

trap (1) Imprecise and obsolete term for dark fine-grained volcanic and hypabyssal rocks, e.g. basalt. (2) Plateau baslts, e.g. the Deccan Traps of India. (3) A structure in which oil or gas may accumulate.

trappoid breccias A succession of breccias found near Nuneaton, Charnwood and Malvern, UK, consisting of angular blocks of rhyolite and feldspathic tuffs which are of Permian age. They probably represent fossil scree material.

trass A material similar to pozzuolana, found in the Eifel district of Germany; used to give additional strength to lime mortars and plasters.

traveling block The heavy duty block which supports the drill string while drilling and can lift it during a **round trip**.

travertine A variety of calcareous tufa of light colour, often concretionary and varying considerably in structure; some varieties are porous. A deposit characteristic of hot springs in volcanic regions.

trebles See **coal sizes**.

Tremadoc The oldest epoch of the Ordovician period, sometimes classed as Upper Cambrian. See **Palaeozoic**.

tremolite A monoclinic amphibole, hydrated calcium magnesium silicate. It is usually white or grey and occurs in bladed crystals or fibrous aggregates in metamorphic rocks. It differs from *actinolite* in having less iron; the name is used for the magnesium end-member. Also *grammatite*.

Triassic The geological period between Permian and Jurassic. It is the oldest period of the Mesozoic era and has a time span from approx. 245–210 million years. It was named by von Alberti from the three-fold division in Germany. The corresponding system of rocks. See **Mesozoic**.

triclinic system The lowest system of crystal symmetry containing crystals which possess only a centre of symmetry. Also *anorthic system*. See **Bravais lattices**.

tridymite A high-temperature form of silica, SiO_2, crystallizing in the orthorhombic system, but possessing pseudohexagonal symmetry. The stable form of silica from 870°C to 1470°C. Typically occurs in acid volcanic rocks.

trigonal system A style of crystal architecture characterized essentially by a principal axis of threefold symmetry; otherwise resembling the hexagonal system. Such important minerals as calcite, quartz and tourmaline crystallize in this system.

trilobites See panel on p. 232.

trip See **round trip**.

triphylite An orthorhombic lithium iron phosphate isomorphous with **lithiophilite**.

triple junction The focal point of three tectonic plates, e.g. at **convergence zones**, *divergence zones* or **transform faults**.

tripod drill Rock drill on heavy tripod.

Tripoli powder See **tripolite**.

tripolite A variety of opaline silica which is formed from the siliceous frustules of diatoms. It looks like earthy chalk or clay, but is harsh to the feel and scratches glass. When finely divided it is sometimes called *earthy tripolite*. Also *diatomite, infusorial earth, Tripoli powder*.

tritium The radioactive isotope of hydrogen, of half-life 12.5 years. Symbol T, r.a.m. 3.0221, mass no. 3. It is very rare, the abundance in natural hydrogen being one atom in

trilobites

Extinct marine arthropods that ranged from the Cambrian to the Permian. Trilobites (see figure) have a segmented oval body divided into a head or cephalon, a thorax and a pygidium or tail; in the cephalon and pygidium the segments are fused together. Longitudinal division into three parts (from which the name derives) is prominent. Trilobites are commonly from under 1 cm to well over 20 cm long, and rolled up specimens are often found. Trilobites were entirely marine, and most lived in or near the sea-bottom.

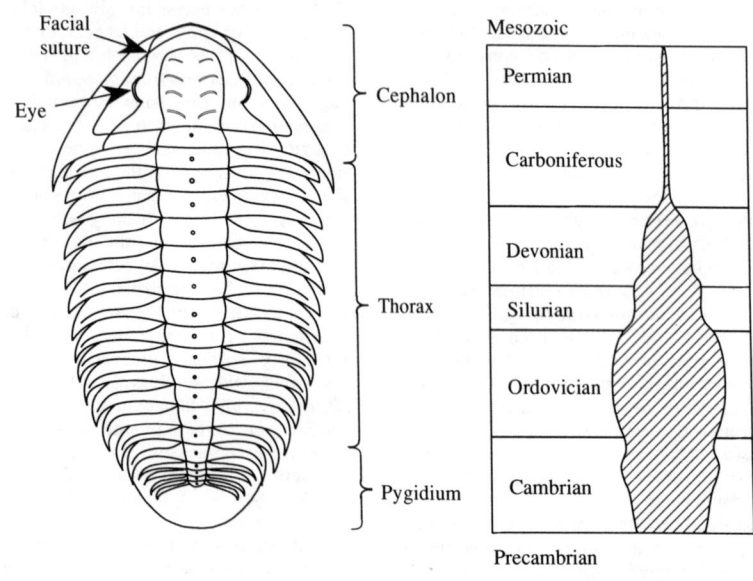

Trilobite sp. **Relative abundance**

Trilobites first appeared in the Lower Cambrian as fully developed and highly complex species, which suggests that there was a long earlier period of evolution in the Precambrian. They were the most important of the Cambrian faunas and reached their acme in the Ordovician Period, declining very markedly in the Carboniferous to extinction in the Permian.

10^{17} but tritium can be produced artificially by neutron absorption in lithium. Can be used to label any aqueous compound and consequently is of great importance in radiobiology.

troctolite A coarse-grained basic igneous rock, consisting essentially of olivine and plagioclase only. The former mineral occurs as dark spots against the feldspar, giving the rock a characteristic spotted appearance, whence the name *troutstone* (from the Gk.).

troilite A non-magnetic iron sulphide, FeS, which occurs mainly in meteorites.

trommel A cylindrical revolving sieve for sizing crushed ore or rock.

trondhjemite A coarse-grained leucocratic igneous rock consisting essentially of plagioclase (ranging from oligoclase to andesine), quartz and small quantities of biotite.

troutstone See **troctolite**.

truncated spur See **valleys**.

trunk conveyor In colliery, belt conveyor in main road.

tschermakite An end-member subspecies in the hornblende group of amphiboles, rich in aluminium and calcium.

tsunami A destructive sea wave caused by an earthquake or submarine eruption. Because of its very long wavelength it behaves as a 'shallow' *surface wave*. Its amplitude in

twinned crystals

Many crystals are found with two parts that are reversed on each other, but related in a definite geometrical manner, called *twinning*. Both parts are always of the same mineral species and re-entrant angles between them are common. Although the crystals have developed in this manner throughout their growth from a very early stage, the twinned form may be described as if it had resulted from a purely geometrical operation of symmetry acting on one individual to produce the other. If one part is thus considered to be produced from the other by rotation, it is usually through 180° about a *twin axis*. The plane of reflection when it occurs is the *twin plane* which is often identical with the *composition plane* along which the two parts are joined.

Simple twins have two component parts (Fig. 1, below) and *multiple twins* involve more than two individuals. When the individuals appear to penetrate each other they form an *interpenetration twin* as in Fig. 2, where the twin axis is a diagonal of the cube. In the interpenetration *Carlsbad twin* of the monoclinic feldspar, orthoclase, (Fig. 3) the twin axis is the vertical crystallographic axis and the composition plane is the *clinopinacoid*, i.e. parallel to the vertical and the clino axis.

A *geniculate twin* produces a knee- or elbow-shaped crystal, as in rutile (Fig. 4). Repeated twinning with parallel twin planes is *polysynthetic* twinning, as in albite (Fig. 5), the *twin lamellae* sometimes appearing only as fine lines on the crystal faces.

In quartz, Fig. 6, interpenetrant twinning is almost always present but usually difficult to detect. Quartz has two related forms, right- and left-handed, and twins may be right-handed or left-handed, twinned alone about the vertical axis, or interpenetrant right- and left-handed twins twinned about the vertical axis and reflected over the horizontal plane. This is known as the *Brazil* twin law and is revealed in etched basal sections as shown in Fig. 6.

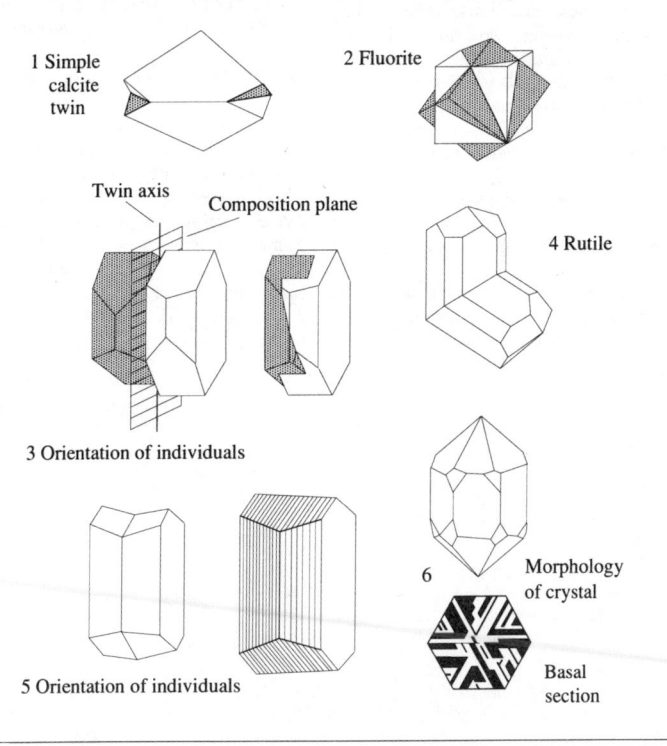

1 Simple calcite twin

2 Fluorite

Twin axis   Composition plane

4 Rutile

3 Orientation of individuals

5 Orientation of individuals

6 Morphology of crystal

Basal section

mid-ocean is very small; as it approaches land, the amplitude builds up and all the energy of the original disturbance is concentrated into a few wavelengths with devastating results. Erroneously *tidal wave*.

tub A tram, wagon, corf or corve.

tubbing The lining of a circular shaft, formed of timber or by steel segments.

tube mill Horizontal mill in which diameter/length ratio is usually high compared with that of standard ball mill, and which has high discharge.

tubing head See **casing head**.

tufa A porous, concretionary or compact form of calcium carbonate which is deposited from solution around springs, of which the dense variety is called tufa.

tuff A rock formed of compacted volcanic fragments, some of which can be distinguished by the naked eye. If the fragments are larger, then the rock grades into an agglomerate.

tundra A plain region characterized by water-logged soil underlain by permafrost.

tungsten Symbol W, at.no. 74, r.a.m. 183.85, rel.d. 19.1, mp 3370°C. A hard grey metal which is resistant to corrosion and is used in high-speed tool steel, cemented carbides for drills and grinding tools, and as wire in incandescent electric lamps. The main tungsten ores are wolfram (wolframite), $(MnFe)WO_4$, and scheelite, $CaWO_4$.

tungstic ochre, tungstite Hydrated oxide of tungsten. It is usually earthy and yellow or greenish in colour, and is a mineral of secondary origin, usually associated with wolframite.

turbidite A sediment deposited from a turbidity current, frequently poorly sorted as in a *greywacke*, it often shows **graded bedding**.

turbidity current A density flow of mixed water and sediment, capable of rapid movement downslope. See **turbidite**.

turbulent flow Fluid flow in which the particle motion at any point varies rapidly in magnitude and direction. This irregular eddying motion is characteristic of fluid motion at high Reynold's numbers. Gives rise to high drag, particularly in the **boundary layer** of aircraft. Also *turbulence*.

turgite See **hydrohematite**.

Turonian A stage in the Upper Cretaceous. See **Mesozoic**.

turquoise A hydrated phosphate of aluminium and copper which crystallizes in the triclinic system. It is a mineral of secondary origin, found in thin veins or small masses in rocks of various types, and used as a gemstone. The typical sky-blue colour often disappears when the mineral is dried. Much of the gem turquoise of old was fossil bone of organic origin and not true turquoise.

twinned crystals See panel on p. 233.

twinning Intergrowth of crystals of near-symmetry, such that (in quartz) the piezoelectric effect is not sufficiently determinate. See **twinned crystals**.

2E Apparent optic axial angle as measured in air.

2V The optic axial angle when measured in the mineral.

TWT, TWTT Abbrevs. for *two-way time*, *two-way travel time*. See **seismic prospecting**.

tye *Strake* in which a considerable thickness of low-grade concentrate is collected.

Tyler sieves Widely used series of laboratory screens in which mesh sizes are in $\sqrt{2}$ progression with respect to linear distance between wires.

Tyndall effect Scattering of light by very small particles of matter in the path of the light, the scattered light being mainly blue.

type locality The locality from which a rock, formation, fossil etc. has been named and described.

U V

U Symbol for **uranium**.
U Symbol for (1) potential difference; (2) tension; (3) internal energy.
ugrandite A group name for the *u*varovite, *gr*ossular and *and*radite garnets.
uintaite A variety of natural asphalt occurring in the Uinta Valley, Utah, as rounded masses of brilliant black solid hydrocarbon. Also *gilsonite*.
ulexite A hydrated borate of sodium and calcium occurring in borate deposits in arid regions, as in Chile and Nevada, where it forms rounded masses of extremely fine acicular white crystals. Also *cotton ball*.
ullmanite Nickel antimony sulphide. It crystallizes in the cubic system and occurs in hydrothermal veins.
ultrabasic rocks Igneous rocks containing less silica than the basic rocks (i.e. less than 45%), and characterized by a high content of mafic constituents, particularly olivine (in the peridotites) and amphiboles and pyroxenes (in the perknites and picrites). See **basic rocks**.
ultramafic rocks Igneous rocks containing more than 90% of dark minerals (olivine, pyroxene, amphiboles, micas, opaque minerals etc.). *Plutonic* rocks are classified according to the proportions of olivine, pyroxene (orthopyroxene and clinopyroxene) and hornblende, and include dunite, peridotite, pyroxenite, hornblendite etc. *Volcanic* ultramafic rocks are much rarer.
ultraviolet radiation Electromagnetic radiation in a wavelength range from 400 nm to 10 nm approximately, i.e. between the visible and X-ray regions of the spectrum. The *near* ultraviolet is from 400 to 300 nm, the *middle* from 300 to 200 nm and the *extreme* from 200 to 190 nm.
ultraviolet spectrometer An instrument similar to an optical spectrometer but employing non-visual detection and designed for use with ultraviolet radiation.
ulvöspinel An end-member species in the magnetite series, with composition ferrous and titanium oxide. First recognized in ore from Södra Ulvön, in northern Sweden.
umber Naturally occurring brown iron and manganese oxides or clays strongly coloured by oxides, formed by residual weathering and valued as a pigment. *Raw umber* has a greenish tinge; *burnt umber* that has been calcined is dark brown. See **ochre**.
unconformity A substantial break in the succession of stratified sedimentary rocks, following a period when deposition was not taking place. If the rocks below the break were folded or tilted before deposition was resumed, their angle of dip will differ from that of the overlying rocks, in which case the break is an *angular unconformity*.
underclay See **seat earth**.
underground gasification Technique for the remote extraction of hydrocarbons from coal by the deliberate combustion of coal seams in a controlled and restricted oxygen environment. Combustion and distillation products include carbon monoxide and volatile hydrocarbons, extracted via boreholes. Also *pyrolytic mining*.
underground volatilization Technique for remote extraction of hydrocarbons from coal. Solvents are introduced through boreholes and dissolved volatile hydrocarbons are extracted via a second set of boreholes.
underhand stopes Stopes in which excavation is carried downslope from access level.
underlay The departure of a vein or thin tabular deposit from the vertical; it may be measured in horizontal feet per fathom of inclined depth. Also *underlie*.
undersaturated Refers to an igneous rock in which there is a deficit of silica. This is normally shown by the presence of a feldspathoid. See **oversaturated**.
undersize See **riddle**.
uniaxial, uniaxial crystal A term for all those crystalline minerals in which there is only one direction of single refraction (parallel to the principal crystal axis and known as the optic axis). All minerals which crystallize in the tetragonal, trigonal, and hexagonal systems are *uniaxial*. Cf. **biaxial**. See **refractive index**.
uniformitarianism The concept that the processes that operate to modify the Earth today also operated in the geological past. In its more extreme form the concept also infers uniformity of rates as well as of processes.
unit One per cent of a specified element or compound in a parcel of ore, concentrates or metal being sold.
unit cell See panel on p. 236.
unit charge, unit quantity of electricity In SI units, 1 **coulomb**. In unrationalized MKS units, the electric charge which experiences a repulsive force of 1 newton when placed 1 metre from a like charge *in vacuo*. Similarly, in the CGS electrostatic units, the force is 1 dyne when 1 centimetre apart. See **SI units**.

unit cell

The smallest group of atoms, ions or molecules, whose repetition at regular intervals, in three dimensions, produces the lattice of a given crystal. There are four possible **crystal lattices** as shown in the diagram.

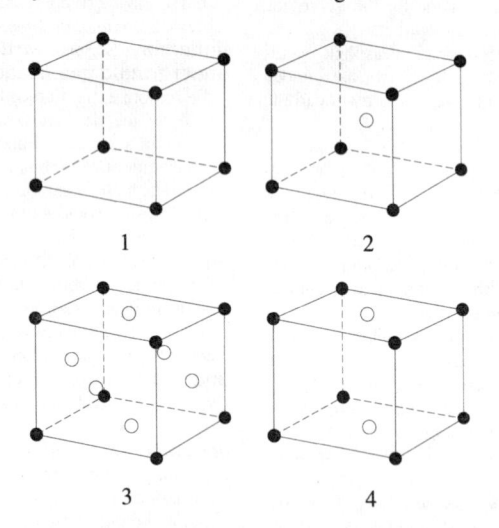

The four possible 3-dimensional unit cells

The primitive crystal lattice has lattice points only at the corners of the unit cell (P-lattice, 1 above).

The body-centred crystal lattice is one containing two lattice points in which the point at the centre, at the intersection of the four body diagonals, is identical with those at the corners (I-centred lattice, 2 above).

The face-centred lattice is a lattice in which the unit cell has a lattice point at the centre of each face as well as at each corner (F-centred lattice, 3 above).

A face-centred lattice may be face-centred on one face only (4 above), either an A-, B- or C-centred lattice, depending upon the face on which it is centred.

upper mantle The upper part of the mantle from the **Mohorovičić discontinuity** (the *Moho*) at the base of the crust to a depth of perhaps 1000 km. It is thought to be peridotitic in composition. See **mantle**.

Uralian emerald Not an emerald; green variety of **andradite** garnet (*demantoid*), occurring as nodules in ultra-basic rocks in the Urals; used as a semi-precious gemstone, though rather soft for this purpose.

uralite A bluish-green monoclinic amphibole, generally actinolitic in composition, resulting from the alteration of pyroxene.

uralitization A type of alteration of pyroxene-bearing rocks, involving the replacement of the original pyroxenes by fibrous amphiboles, as in some epidiorites.

uraninite See **uranium**.

uranite See **uranium**.

uranium See panel on p. 237.

uranium-lead dating A method of determining the age in years of geological material, based on the known decay rate of ^{238}U to ^{206}Pb and ^{235}U to ^{207}Pb.

uranium-radium series The series of radioactive isotopes which result from the decay of ^{238}U. The mass numbers of the members of the series are $4n+2$, where n is an integer. Series ends in the stable isotope ^{206}Pb. See **radioactive series**.

uranophane An ore of uranium, a yellow secondary hydrated calcium uranium silicate.

uranium

A hard grey metal with a number of isotopes. Symbol U, at.no. 92, r.a.m. 238.03, rel.d. 18.68, mp 1150°C. ^{235}U is the only naturally occurring readily fissile isotope and exists as one part in 140 of natural uranium. Uranium and all its compounds are radioactive. The primary mineral of uranium is *uraninite*, UO_2, which has the highest content of U of any uranium mineral. *Thorianite*, ThO_2, also contains UO_2. Uraninite contains lead, thorium, and the metals of the lanthanum and yttrium groups. It occurs as brownish to black cubic crystals and is an accessory mineral in granitic rocks and in metallic veins. When massive, and apparently amorphous, known as *pitchblende*.

There are many other complex uranium minerals. Many occur in pneumatolytic and hydrothermal deposits and in the alteration products associated with them. Numerous oxidized and hydrated salts occur, esp. phosphates, silicates and arsenates of uranium and other metals. Many have a spectacular appearance and are brightly coloured. Important minerals are *autunite*, $Ca(UO_2)_2(PO_4)_2.12H_2O$, of yellow colour, *torbernite*, $Cu(UO_2)_2(PO_4)_2.12H_2O$, a beautiful rich green colour and *carnotite*, $K_2(UO_2)_2(VO_4)_2.3H_2O$, a canary-yellow mineral. These and *coffinite*, $U(SiO_4)_{1-x}(OH)_{4x}$, are all worked as uranium ores.

The abundance of uranium in the Earth's crust is 2.3 ppm. Uranium deposits occur in most classes of rock, but dominantly in the Precambrian basement or in sediments immediately overlying the basement. Vein-type deposits occur in Proterozoic rocks, sometimes remobilized later, and in rocks associated with the Hercynian and Alpine orogenies. In conglomerates, there are substantial amounts of the world's uranium reserves, e.g. associated with gold in the Precambrian Witwatersrand deposits. Sandstone may be impregnated by carnotite and other minerals, as in the Colorado-Wyoming area of the US. Uranium has no known biological role: it is toxic and dangerous due to its radioactivity.

Synonym *uranotile*.

urtite An intrusive igneous rock composed mainly of nepheline.

U-shaped valley See **valleys**.

uvarovite A variety of garnet, of an attractive green colour; essentially silicate of calcium and chromium.

v Symbol for (1) **velocity**; (2) specific volume of a gas.

V Symbol for (1) **vanadium**; (2) **volt**.

V Symbol for (1) potential; (2) potential difference; (3) electromotive force; (4) **volume**.

vacancy Unoccupied site for ion or atom in a crystal.

vacuum Literally, a space totally devoid of any matter. Does not exist, but is approached in inter-stellar regions. On Earth, the best vacuums produced have a pressure of about 10^{-8} N/m^2. Used loosely for any pressure lower than atmospheric, e.g. train braking systems, 'vacuum' cleaners etc.

vadose zone The unsaturated zone between the water table and the surface of the ground.

Valanginian A stage in the Lower Cretaceous. See **Mesozoic**.

valentinite Antimony trioxide, Sb_2O_3, occurring as orthorhombic crystals or radiating aggregates. Snow-white when pure, it is formed by the decomposition of other ores of antimony.

valleys See panel on p. 238.

vanadinite Vanadate and chloride of lead, typically forming brilliant reddish hexagonal crystals or globular masses encrusting other minerals in lead mines.

vanadium A very hard, whitish metallic element. Symbol V. at.no. 23, r.a.m. 50.941, rel.d. at 20°C 6.11, mp 1710°C, electrical resistivity 22×10^{-8} ohm metres at 20°C. Its principal use is as a constituent of alloy steel, e.g. in chromium-vanadium, manganese-vanadium and high-speed steels. The principal ore minerals are *vanadinite*, *carnotite*, *patronite* and *roscoelite*. Vanadium also occurs in phosphate rock.

vanner See **frue vanner**.

vanning Rough estimate of cassiterite or other heavy mineral, made by washing finely-ground sample on a flat shovel, or in

valleys

Any hollow or low-lying tract of ground between hills or mountains, usually traversed by streams or rivers, which receive the natural drainage from the surrounding high ground. Usually valleys are developed by stream erosion; but in special cases, faulting may also have contributed, as in rift valleys. The valley usually has an outlet to another valley, a lake, an area of inland drainage or the sea. In a *V-shaped valley* the cross section resembles the letter V and is caused by a down-cutting stream. A *U-shaped valley* has a cross section resembling the letter U, as in the figure, with steep walls and a nearly flat floor. It is often caused by erosion by a valley glacier that has removed projecting spurs to produce *truncated spurs*, and produced *hanging valleys* where the main valley is deeper than the valleys of tributary streams. Waterfalls may descend from the hanging valleys. The longitudinal profile of a valley is often referred to as the *thalweg*.

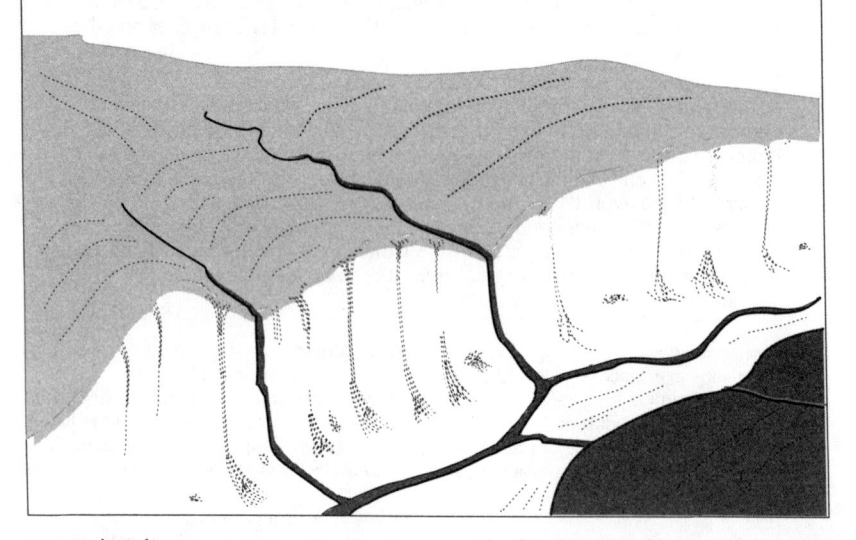

a vanning plaque.

vapour A gas which is at a temperature below its critical temperature and can therefore be liquefied by a suitable increase in pressure.

vapour pressure The pressure exerted by a vapour, either by itself or in a mixture of gases. The term is often taken to mean saturated vapour pressure, which is the vapour pressure of a vapour in contact with its liquid form. The saturated vapour pressure increases with rise of temperature.

variegated copper ore A popular name for **bornite**. So named from the characteristic tarnish that soon appears on the freshly fractured surface.

variolitic Said of a fine-grained igneous rock of basic composition containing small more or less spherical bodies (*varioles*), consisting of minute radiating fibres of feldspar, comparable with the more perfect spherulites in acid igneous rocks.

Variscan orogeny The late Palaeozoic mountain building period of Europe extending from the Carboniferous to the Permian; synonymous with *Hercynian orogeny*.

variscite A greenish hydrated phosphate of aluminium, $AlPO_4.2H_2O$, occurring as nodular masses.

varved clays Distinctly and finely stratified clays of glacial origin, deposited in lakes during the retreat stage of glaciation. The stratification is thought to be a seasonal banding, and its study enabled Baron de

Geer to work out the chronology of the Pleistocene Ice Age.

vaterite A less common polymorph of calcium carbonate, crystallizing in the hexagonal system; forms artificially. Cf. **calcite, aragonite.**

vegetation survey See **geobotanical surveying.**

vein A tabular or sheet-like body of rock, penetrating a different type of rock. Sometimes applied to particularly narrow igneous intrusions (dykes and sills), the term is more often applied to material deposited by solutions, such as quartz veins or calcite veins. Many ore deposits consist of veins in which the ore mineral is one of several constituents.

vein stuff The minerals occurring in veins or fissures.

velocity (1) The rate of change of displacement of a moving body with time; a vector expressing both magnitude and direction (cf. *speed* which is scalar). (2) For a wave, the distance travelled by a given phase divided by the time taken. Symbol v.

velocity of light See **speed of light.**

velocity of sound In dry air at s.t.p. 331.4 m/s (750 mi/h). In freshwater, 1410 m/s, and in seawater, 1540 m/s. The above values are used for sonar ranging but do not apply to explosive shock waves. They must be corrected for variations of temperature, humidity etc.

Vendian The uppermost part of the Proterozoic. See **Precambrian.**

vent See **volcanic vent.**

ventifact A wind-faceted pebble. See **dreikanter, zweikanter.**

ventube A flexible ventilating duct some distance away from the source of fresh air.

Venus' hair stone Variety of **rutile.** Also *Veneris crinis.* See **flèches d'amour.**

verdite A green rock, consisting chiefly of green mica (fuchsite) and clayey matter, occurring as large boulders in the North Kaap River, South Africa; used as an ornamental stone.

vermiculites A group of hydrated sheet silicates, closely related chemically to the chlorites, and structurally to talc. They occur as decomposition products of biotite mica. When slowly heated, they exfoliate and open into long worm-like threads, forming a very lightweight water-absorbent aggregate used in seed planting and, in building, as an insulating material.

Verneuil process The technique invented by the French chemist Verneuil for the manufacture of synthetic corundum and spinel by fusing pure precipitated alumina, to which had been added an amount of the appropriate oxide for colouring, in a vertical, inverted blow-pipe type of furnace.

vertebrates See panel on p. 240.

vesicle See **vesicular structure.**

vesicular structure A character exhibited by many extrusive igneous rocks, in which the expansion of gases has given rise to more or less spherical cavities (vesicles). The latter may become filled with such minerals as silica (chalcedony, agate, quartz), zeolites, chlorite, calcite etc.

vesuvianite Hydrated silicate of calcium and aluminium, with magnesium and iron, crystallizing in the tetragonal system. It occurs commonly in metamorphosed limestones. Also *idocrase.*

vicinal faces Facets modifying normal crystal faces, but themselves abnormal, as their indices cannot be expressed in small whole numbers; they usually lie nearly in the plane of the face they modify.

Vikoma Proprietary name for equipment used for recovering and cleaning up oil spills. See **boom.**

violane Massive violet-blue **diopside,** used as an ornamental stone.

viridine A green iron- and manganese-bearing variety of **andalusite.**

viscoelastic A solid or liquid which when deformed exhibits both viscous and elastic behaviour through the simultaneous dissipation and storage of mechanical energy. Shown typically by polymers.

viscometer An instrument for measuring viscosity. Many types of viscometer employ Poiseuille's formula for the rate of flow of a viscous fluid through a capillary tube.

viscosity The resistance of a fluid to shear forces, and hence to flow. Such shear resistance is proportional to the relative velocity between the two surfaces on either side of a layer of fluid, the area in shear, the **coefficient of viscosity** of the fluid and the reciprocal of the thickness of the layer of fluid. For comparing the viscosities of liquids, various scales have been devised, e.g. *Redwood No. 1 seconds* (UK), *Saybolt Universal seconds* (US), *Engler degrees* (Germany). See **kinematic viscosity.**

viscous flow A type of fluid flow in which fluid particles, considered to be aggregates of molecules, move along streamlines so that at any point in the fluid the velocity is constant or varies in a regular manner with respect to time, random motion being only of a molecular nature. The name is also used to describe *laminar flow* or *streamline flow.*

Viséan A series name in the Lower Carboniferous of Europe. See **Palaeozoic.**

vishnevite The sulphate-bearing equivalent

vertebrates

A subphylum of Chordata in which the notochord stops beneath the forebrain and a skull is always present. There are usually paired limbs. The brain is complex and associated with specialized sense organs, and there are at least ten pairs of cranial nerves. The pharynx is small and there are rarely more than seven gill slits. The heart has at least three chambers and the blood has corpuscles containing haemoglobin. The phylum Chordata includes the graptolites which existed only in the Palaeozoic (see panel on **graptolites**). There are several classes of vertebrates, of which the earliest was a group of fishes, the *agnathans*, a name meaning 'without jaws'. Originally these fishes were armoured but later lost this characteristic. They range from the Ordovician to the present day although most genera died out by the end of the Devonian, the lampreys are modern examples. Fossil agnathans are sometimes called ostracoderms (not to be confused with *ostracods* which are crustaceans). '*Pisces*' (fishes) is an old terminology class that loosely includes a number of aquatic vertebrates, including the *acanthodians* (Silurian to Permian), *placoderms*, heavily armoured fish which existed mainly in the Devonian and other classes. There are various classificatory groupings. During the Devonian the fishes became dominant. The cartilaginous fishes are the *chondrichthyans* (now represented by the sharks, rays and skates); the bony fishes, the *osteichthyans*, also first appeared in the Devonian in freshwater deposits and evolved into the marine environment. They became the most successful of aquatic vertebrates and are the dominant fishes of the present day. The lungfish, the *crossopterygians* and the *dipnoans* are classes that evolved in the Devonian.

Amphibians, members of the class Amphibia, are cold-blooded tetrapods which include frogs, newts and salamanders as living examples. They are semi-aquatic, breathing by means of gills in the early stages of life and lungs in the later stages, and probably evolved from crossopterygian fish in the Upper Devonian. They were dominant in the Carboniferous and Permian, the 'age of amphibians'.

Reptiles, vertebrates of the class Reptilia, are also cold-blooded tetrapods which first appeared in the late Carboniferous, were able to live completely on land and came to dominate life in the Mesozoic (the 'age of reptiles'). Present day representatives include snakes, lizards, crocodiles and turtles. The *dinosaurs* included the largest terrestrial animals (see panel on p. 60).

Birds (Aves) evolved from flying reptiles in the Jurassic, one of the first, *Archaeopteryx*, having reptilian features including teeth and a lizard-like tail.

The *mammals*, members of the class Mammalia, are warm-blooded, generally covered with hair. The young of most genera are born fully developed and initially nourished by milk. They evolved as small primitive furry, shrew-like animals in the Triassic during the dominance of the reptiles, but for about 150 million years remained as small animals with the class not expanding greatly until the Cenozoic (the 'age of mammals'). In the Tertiary, the modern groups of mammals (including horses, pigs, camels, elephants and primates) developed. The Primates are an order of Mammals including lemurs, monkeys, apes and Man. All living races of Man belong to the genus *Homo*, of the suborder Anthropoidea. *Homo sapiens*, modern man, appeared during the Pleistocene.

---

of **cancrinite**, found in nepheline syenite.

visible radiation Electromagnetic radiation which falls within the wavelength range of 780 to 380 nm, over which the normal eye is sensitive.

vitric tuff A tuff in which vitric (glassy) fragments are more abundant than lithic or crystal fragments.

vitrinite An oxygen-rich maceral that is found in coal.

volcanic rocks

Generally fine-grained crystalline or glassy igneous rocks that have resulted from volcanic action at or near the Earth's surface. They are named after the classical god Vulcan. Volcanic rocks are extruded through volcanoes as molten lava or ejected explosively to form *pyroclastic deposits*. Included with volcanic rocks are the associated minor intrusions of dykes and sills.

Basaltic rocks are by far the commonest type, forming e.g. the great modern shield volcanoes and lava flows of Hawaii, Iceland and the rocks of the sea bed, as well as those of former geological ages. Examples of these are the Tertiary lava flows of Northern Ireland and Western Scotland, and the Deccan Traps. These are formed from volcanic rocks of low viscosity that flow easily, some for long distances. When the magma is more viscous it builds up volcanic domes, as in the Puys of Central France. The composition of volcanic rocks varies considerably and there are many rock names relating to their mineralogical composition as well as textural characteristics.

When the minerals that compose the rocks can be determined, the classification broadly follows that of the **plutonic rocks**, as shown in the figure. Otherwise the classification must be chemically based. See **TAS**.

Pyroclastic rocks are formed of fragmented volcanic material expelled into the atmosphere, generally by explosions. Individual fragments are *pyroclasts*. *Tephra* are predominantly unconsolidated pyroclastic deposits. The fragments comprising the rocks are ash grade if < 2 mm in mean diameter, and make up rocks also known as *ashes*, or *ash tuffs* if consolidated. *Lapilli* are 2 – 64 mm in size and form beds of lapilli, or *lapilli tuffs* if consolidated. If the grain size is > 64 nm the pyroclasts are volcanic *bombs* or *blocks* forming unconsolidated deposits, or *agglomerates* or *pyroclastic breccia* if consolidated. *Nueés ardentes* (glowing clouds) are incandescent avalanches of particles that may flow at great speed for many miles, causing complete destruction in their paths. When they form consolidated deposits the fragments tend to weld together to form *welded tuffs* or ignimbrites. Pelé's hair (from the Hawaiian goddess of fire) is a mass of very fine threads of basaltic glass.

figure on the next page

vitriol See **blue-, green-, white-**.

vitroclastic structure The characteristic structure of volcanic ashes which have been produced by the disruption of highly vesicular glassy rocks, most of the component fragments thus having concave outlines.

vivianite Hydrated iron phosphate, $Fe_3P_2O_8 \cdot 8H_2O$. Monoclinic.

vogesite A hornblende-lamprophyre, the other essential constituent being orthoclase. Cf. **spessartite**.

volcanic ash The typical product of explosive volcanic eruptions, consisting of comminuted rock and **lava**, the fragments varying widely in size and in composition, and including deposits of the finest dust, lapilli and bombs. See **agglomerate, pyroclastic rocks, tuff**.

volcanic bomb A spherical or ovoid mass of lava, in some cases hollow, formed by the disruption of *molten* lava by explosions in an active volcanic vent. Its average diameter exceeds 64 mm. See **bread-crust bomb**.

volcanic muds, sands The products of explosive volcanic eruptions (*volcanic ash*) which have been deposited under water and have consequently been sorted and stratified, thus showing some of the characters of normal sediments, into which they grade. May also be the product of a mud flow down the side of a volcano.

volcanic neck A vertical plug-like body of igneous rock or volcanic ejectamenta, representing the feeding channel of a volcano.

volcanic rocks See panels on pp. 241–2.

volcanic vent The pipe which connects the crater with the source of magma below; it ultimately becomes choked with agglomerate or volcanic ash, or with consolidated lava.

volcano (1) A centre of volcanic eruption, having the form typically of a conical hill

volcanic rocks (contd)

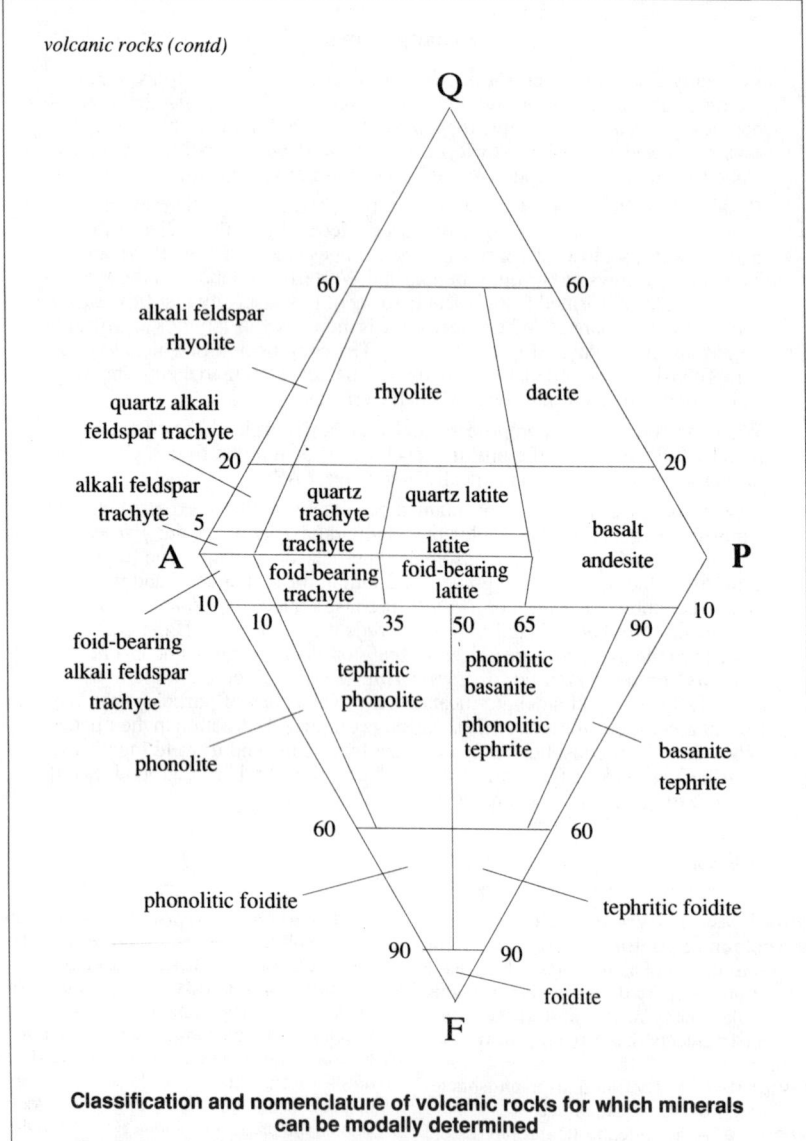

Classification and nomenclature of volcanic rocks for which minerals can be modally determined

Q = quartz; A = alkali feldspar; P = plagioclase feldspar; F = feldspathoids ('foids'). For either triangle, the quantities Q+A+P or F+A+P = 100%.

or mountain, built of ashes and/or lava flows, penetrated irregularly by dykes and veins of igneous rocks, with a central crater from which a pipe leads downwards to the source of magma beneath. Volcanoes may be active (periodically), dormant, or extinct; the eruptions may involve violent explosions (e.g. Krakatoa) or the relatively quiet outpouring of lava, particularly in those cases where the lava is basaltic (e.g. Hawaii). See **lava**. (2) A conical hill producing mud or sand. See **mud volcano**.

volt The SI unit of *potential difference*, electrical potential, or e.m.f., such that the potential difference across a conductor is 1 volt when 1 ampere of current in it dissipates 1 watt of power. Equivalent definition: if, in taking a charge of 1 coulomb between two points in an electric field, the work done on or by the charge is 1 joule, the potential difference between the points is 1 volt. Named after Count Alessandro Volta (1745-1827). See **SI units**.

volume The amount of space occupied by a body; measured in cubic units. Symbol V.

vough See vug.

vug, vough A cavity in rock or a lode, usually lined with crystals.

vulcanites A general name for igneous rocks of fine grain size, normally occurring as lava flows. Cf. **plutonites**.

W X Y Z

w Symbol for **work**.
W Symbol for **watt**.
W Symbol for (1) **weight**; (2) **work**.
wacke A sandstone in which the grains are poorly sorted with respect to size.
wad Bog manganese, hydrated oxide of manganese. See **asbolane**.
walking beam Rocking beam used for transmitting power, e.g. for actuating the cable in cable-drilling for oil.
walking dragline Large power shovel mounted on pads which are mechanically worked to manoeuvre it as required.
wall-rock Country rock to either side of a vein or lode.
wall-rock alteration Mineralogical and/or chemical alteration of the country rock adjacent to a mineralized vein or lode. Result either of diffusion of elements from the mineralizing fluid in the vein, or leaching of material from the country rock by the mineralizing fluid. See **halo**, **primary dispersion**.
Waltonian The lowest stage in the Pleistocene, above the Pliocene. Of long duration, it had a variable climate. See **Quaternary**.
wash The removal of impurities from a gas or vapour by passing it through a reactant solution or solvent that retains or dissolves the impurities. Cf. **scrubber**. Liquids can be washed using liquids or solutions that are immiscible with the product liquid.
wash box Box in which raw coal is jigged in coal washery.
wash gravel Alluvial sands worth exploitation for mineral values contained.
waste Waste rock, either host (enclosing) rock mined with the true lode, or ore too poor to warrant further treatment.
water A colourless, odourless, tasteless liquid, mp 0°C, bp 100°C. It is hydrogen oxide, H_2O, the liquid probably containing associated molecules, H_4O_2, H_6O_3 etc. On electrolysis it yields two volumes of hydrogen and one of oxygen. It forms a large proportion of the Earth's surface, occurs in all living organisms, and combines with many salts as water of crystallization. Water has its maximum density of 1000 kg/m^3 at a temperature of 4°C. This fact has an important bearing on the freezing of ponds and lakes in winter, since the water at 4°C sinks to the bottom and ice at 0°C forms on the surface. Besides being essential for life, water has a unique combination of solvent power, thermal capacity, chemical stability, permittivity and abundance. See **hydrogen**.

water blast A sudden escape of confined air due to water pressure, e.g. in rise workings.
water flooding Secondary oil recovery operation in which water is *injected* into a petroleum reservoir to displace additional oil and enhance production.
water gauge An instrument (e.g. *Pitot tube*) for measuring the difference in pressure produced by a ventilating fan or air current.
water of hydration, crystallization The water present in hydrated compounds. These compounds when crystallized from solution in water retain a definite amount of water, e.g. copper (II) sulphate, $CuSO_4.5H_2O$.
water sapphire See **saphir d'eau**.
water table The surface below which fissures and pores in the strata are saturated with water. It roughly conforms to the configuration of the ground, but is smoother. Where the water table rises above ground level a river, spring or lake is formed.
watt SI unit of power equal to 1 joule per second. Thus, 1 horse-power (hp) equals 745.70 watts. Symbol W.
watt-hour A unit of energy, being the work done by 1 watt acting for 1 hour, and thus equal to 3600 joules.
wave A time varying quantity which is also a function of position. The characteristic of a wave is to transfer energy from one point to another without any particle of the medium being permanently displaced; particles merely oscillate about their equilibrium positions. In electromagnetic waves it is the changes in electric and magnetic fields which represent the wave disturbance. The progress of the wave is described by the passage of a *waveform* through the medium with a certain velocity, the *phase* or *wave velocity*. The energy is transferred at the *group velocity* of the waves making the waveform.
waveform The shape, contour or profile of a wave; described by a phase relationship between successive particles in a medium. A waveform may be *periodic*, *transient* or *random*.
wavefront Imaginary surface joining points of constant phase in a wave propagated through a medium. The propagation of waves may conveniently be considered in terms of the advancing wavefront, which is often of simple shape, such as a plane, sphere or cylinder.
wavelength Symbol λ. (1) Distance, measured radially from the source, between

two successive points in free space at which an electromagnetic or acoustic wave has the same phase; for an electromagnetic wave it is equal in metres to c/f where c is the velocity of light (in m s^{-1}) and f is the frequency (in Hz). (2) Distance between two similar and successive points on a harmonic (sinusoidal) wave, e.g. between successive maxima or minima. (3) For electrons, neutrons and other particles in motion when considered as a *wave train*, $\lambda = h/p$, p is the momentum of the particle and h is Planck's constant.

wavelength of light The wavelength of visible light lies in the range from 400 to 700 nm approximately.

wavellite Orthorhombic hydrated phosphate of aluminium, occurring rarely in prismatic crystals, but commonly in flattened globular aggregates, showing a strongly developed internal radiating structure.

wave-particle duality Light and other electromagnetic radiations behave like a wave motion when being propagated, and like particles when interacting with matter. Interference, diffraction and polarization effects can be described in terms of waves. The photoelectric effect and the Compton effect can be described in terms of *photons*, quanta of energy $E = h\nu$, where h is Planck's constant and ν is the frequency.

wave theory Macroscopic explanation of diffraction, interference and optical phenomena as an electromagnetic wave, predicted by Maxwell and verified by Hertz for radio waves.

wax wall A wall of clay built round the gob or goaf, to prevent the entry of air or egress of gas.

way See **wind road**.

way up The upward direction of a succession of strata in an area of strong folding. The direction of *way up* is most commonly determined by bottom structures or by **cross bedding**. See **sole mark**.

weathering The processes of disintegration and decomposition effected in minerals and rocks as a consequence of exposure to the atmosphere and to the action of frost, rain and insolation. These effects are partly mechanical, partly chemical, partly organic and for their continuation depend upon the removal, by transportation, of the products of weathering. *Denudation* involves both weathering and transportation.

weber The SI unit of magnetic flux. An e.m.f. of 1 volt is induced in a circuit through which the flux is changing at a rate of 1 weber per second. 1 weber equals 1 volt-second equals 1 joule per ampere. Equivalent definition: 1 weber is the magnetic flux through a surface over which the integral of the normal component of the magnetic induction is 1 tesla m^2. See **SI units**.

websterite A coarse-grained ultramafic igneous rock, consisting essentially of hypersthene and diopside.

weddelite Hydrated calcium oxalate, CaC$_2$O$_4$.H$_2$O, crystallizing in the tetragonal system. It occurs uncommonly in the mineral world but freely in human *calculi*.

wedging Use of deflecting wedge near bottom of deep diamond boreholes, either to restore direction or to obtain further samples. Also *whipstocking*.

wedging crib, curb, ring A segmented steel ring on which shaft tubbing is built up and wedged in place.

Weg rescue apparatus Portable breathing apparatus with self-contained oxygen supply controlled automatically by wearer's breathing action.

weight The gravitational force acting on a body at the Earth's surface. The units of measurement are the newton, dyne or pound-force. Symbol W. Weight = mass × **acceleration due to gravity**, and must therefore be distinguished from *mass*, which is determined by quantity of material and measured in pounds, kilograms etc.

weightometer Device which automatically weighs and records the tonnage of ore in transit on a belt conveyor.

Weissenberg method A technique of X-ray analysis in which the crystal and photographic film are rotated in the beam of X-rays while the film is moved parallel to the axis of rotation.

welded tuff A **tuff** composed of glass fragments which have partially fused together as a result of being deposited while still at a high temperature. Also *ignimbrite*.

well head The top of the casing of a production well, with its control valves.

well logging The recording of the composition and physical properties of the rocks encountered in a borehole, particularly one drilled during petroleum exploration. Well logging includes a variety of techniques, e.g. resistivity log, gamma-ray log, neutron log, spontaneous or self-potential log, temperature log, calliper log, photoelectric log, acoustic velocity log etc.

Wenlock The middle series of the Silurian period. See **Palaeozoic**.

Wentworth scale The size of grains in a sediment or sedimentary rock. It is an extension of the *Udden grade scale*. See **particle size, phi grade scale**.

Wernerism See **neptunism**.

Westphalian The name of a stratigraphical series in the Carboniferous rocks of Europe, approximately corresponding to the Coal Measures in England and Wales. See **Palaeozoic**.

wet assay Qualitative or quantitative analysis of ores or their constituents in which dissolution and digestion with suitable solvents plays a part. See **dry assay**.

wheel-ore See **bournonite**.

whetstone See **hone**.

whewellite Hydrated calcium oxalate, $CaC_2O_4.H_2O$, crystallizing in the monoclinic system. It occurs uncommonly in the mineral world but is abundant in human calculi.

Whin Sill A sheet of intrusive quartz-dolerite or quartz-basalt, unique in Britain, as it is exposed almost continuously for over 300 km from the Farne Islands to Middleton-in-Teesdale. See **dykes and sills**.

whin, whinstone A popular term applied to doleritic intrusive igneous rock resembling that of the well-known Whin Sill.

whipstocking See **wedging**.

white copperas Goslarite.

white corundum See **white sapphire**.

white damp Carbon monoxide. Produced by the incomplete combustion of coal in a mine fire, or by gas or dust explosions. Invisible; very poisonous.

white iron pyrites See **marcasite** (1).

white lead ore See **cerussite**.

white light Light containing all wavelengths in the visible range at the same intensity. This is seen by the eye as white. The term is used, however, to cover a wide range of intensity distribution in the spectrum and is applied by extension to continuous spectra in other wavelength bands (e.g. *white noise*). Also *white radiation*.

white nickel A popular name for the cubic diarsenide of nickel, $NiAs_2$, **chloanthite**.

white oil (1) A term for oils that are substantially colourless and without bloom. Usually made from light lubricating oils by acid treatment or hydrogenation, they may be used medicinally (liquid paraffin) or in toilet preparations. (2) A term for oils that do not contaminate the tanker or other transport vehicle. Tankers are usually dedicated to one class of oil cargo. Cf. **black oil**.

white radiation See **white light**.

white sapphire More reasonably called *white corundum*, it is the colourless pure variety of crystallized corundum, Al_2O_3, free from those small amounts of impurities which give colour to the varieties 'ruby' and 'sapphire'; when cut and polished, it makes an attractive gemstone.

white smoker A plume of hydrothermal fluid, white with mineral precipitates, at the crest of an ocean ridge. Most of the mineral is barytes and silica. See **black smoker**, **chimney**.

white vitriol A popular name for **goslarite**, $ZnSO_4.7H_2O$.

whitlockite Trigonal calcium phosphate, occurring in sedimentary phosphate deposits.

wildcatting Prospecting at random, particularly in speculative boring for oil.

Wilfley table Flat rectangular desk, adjustable about the long axis for tilt, is given rapid but gentle throwing motion along this horizontal axis, while classified sands are washed across and down-tilt, against restraint imposed by horizontal riffles. Heavy minerals work across and progressively lighter ones gravitate down to separate discharge zones. Also *concentrating table, shaking table*.

willemite Orthosilicate of zinc, Zn_2SiO_4, occurring as massive, granular, or in trigonal prismatic crystals, white when pure but commonly red, brown, or green through manganese or iron in small quantities. In New Jersey and elsewhere it occurs in sufficient quantity to be mined as an ore of zinc. Noteworthy as the ore exhibits an intense bright yellow fluorescence in ultraviolet light.

window A closed outcrop of strata lying beneath a thrust plane and exposed by denudation. The strata above the thrust plane surround the 'window' on all sides.

wind road, way An underground passage used for ventilation.

winze Internal shaft, usually between two underground levels in plane of lode, used in exploration and subsequent extraction of valuable ore.

wireline tool Small tools or measuring instruments designed to be lowered into a well on a wire line.

withamite A mineral belonging to the epidote group, containing about 1% of manganese oxide.

witherite Barium carbonate, $BaCO_3$, crystallizing in the orthorhombic system as yellowish or greyish-white complex crystals of hexagonal appearance due to twinning; also massive. Occurs with galena in lead mines. Exploited as an important source of barium.

wolfram Synonym for **wolframite**.

wolframite Tungstate of iron and manganese (FeMn)WO_4, occurring as brownish-black monoclinic crystals, columnar aggregates, or granular masses. It forms a complete series from ferberite (FeWO$_4$) to hübnerite (MnWO$_4$). An important ore of tungsten. Also *wolfram*.

wollastonite A triclinic silicate of calcium, $CaSiO_3$, occurring as a common mineral in metamorphosed limestones and similar assemblages, resulting from the reaction of quartz and calcite. Also *tabular spar*.

Wolstonian A cold to glacial stage of the late Pleistocene. See **Quaternary**.

wood opal A form of common opal which has replaced pieces of wood entombed as fossils in sediments, in some cases retaining the original structure.

wood tin A botryoidal or colloform variety of cassiterite showing a concentric structure of brown, radiating, wood-like fibres.

work One manifestation of energy. The work done by a force is defined as the product of the force and the distance moved by its point of application along the line of action of the force. For example, a tensile force does work in increasing the length of a piece of wire; work is done by a gas when it expands against a hydrostatic pressure. As for all forms of energy, the SI unit of work is the *joule*, performed when a force of 1 newton moves its point of application through 1 metre along the line of action of the force. Alternatively, when 1 watt of power is expended for 1 second, 1 joule of work is done. See **joule**. Symbols W, w.

work function The minimum energy that must be supplied to remove an electron so that it can just exist outside a material under vacuum conditions. The energy can be supplied by heating the material (*thermionic* work function) or by illuminating it with radiation of sufficiently high energy (*photoelectric* work function). Also *electron affinity*.

work-over Term for any remedial operation on a producing well after completion, usually for maintenance or to increase production.

wulfenite Molybdate of lead, $PbMoO_4$, occurring as yellow tetragonal crystals in veins with other lead ores.

Würm The fourth and last stage of the Pleistocene epoch of the Alps. See **Quaternary**.

wurtzite Sulphide of zinc, ZnS, of the same composition as sphalerite, but crystallizing at higher temperatures and in the hexagonal system, in black hemimorphic, pyramidal crystals.

wyomingite An alkaline volcanic rock, composed of leucite, phlogopite and diopside.

xalostocite A pale rose-pink grossular which occurs embedded in white marble at Xalostoc in Mexico. Also *landerite*.

xanthophyllite A brittle mica, crystallizing in the monoclinic system; hydrated calcium, magnesium, aluminium silicate.

xenocryst A single crystal or mineral grain of extraneous origin which has been incorporated by magma during its uprise and which therefore occurs as an inclusion in igneous rocks, usually surrounded by reaction rims and more or less corroded by the magma. Cf. **xenolith**.

xenolith A fragment of rock of extraneous origin which has been incorporated in magma, either intrusive or extrusive, and occurs as an inclusion, often showing definite signs of modification by the magma.

xenomorphic A textural term implying that the minerals in a rock do not show their own characteristic shapes, but are without regular form by reason of mutual interference. See **granitoid texture**.

xenon A zero-valent element, one of the noble gases, present in the atmosphere in the proportion of 1:170 000 000 by volume. Symbol Xe, at.no. 54, r.a.m. 131.30, mp $-140°C$, bp $-106.9°C$, critical temp. $+16.6°C$, density at s.t.p. 5.89 g/dm^3.

xenotime Yttrium phosphate. YPO_4, often containing small quantities of cerium, erbium and thorium, closely resembling zircon in tetragonal crystal form and general appearance, and occurring in the same types of igneous rock, i.e. in granites and pegmatites as an accessory mineral. An important source of the rare elements named.

xonotlite A hydrous calcium silicate, of composition, $Ca_6Si_6O_{17}(OH)_2$.

Xmas tree See **Christmas tree**.

X-ray crystallography The main technique providing images of molecules in which the arrangement of atoms can be seen. Certain technical criteria must be fulfilled before X-ray crystallography can be applied, the most obvious of which is that the molecule or complex structure to be imaged must be crystallized. This involves the isolation in a highly pure form of adequate quantities of molecules, sometimes extremely rare. In an ordinary light microscope, image formation takes place in two stages. An incident light beam is scattered by the object and these scattered rays are collected by a lens and recombined to form an image. Unfortunately there is, at present, no lens able to refocus X-rays at an adequate resolution. This is because X-ray physics determines that their refractive index at most interfaces is very close to unity. The scattered rays can however be collected from the object, giving the *Fraunhofer diffraction pattern* which contains the *amplitudes* of the scattered rays but not their *phases*. Thus only half the information needed by a computer to calculate the image is available. See **X-ray microscope**. The phase problem has been solved in a number of ways including the replacement of specific atoms in the normal crystal with heavier ones. The techniques of X-ray diffraction are applied to *single crystals* or to *powders* of crystalline materials. See **powder photography**.

X-ray diffractometer An instrument containing a radiation detector used to record the X-ray diffraction patterns of crystals, powders or molecules. See **X-ray crystallography**.

X-ray fluorescence spectrography A method of chemical analysis in which the sample is bombarded by very hard X-rays or gamma-rays, and secondary radiations, characteristic of the elements present, are studied spectroscopically. Abbrev. *XRF*.

X-ray microscope Using soft X-rays and a Fresnel zone plate as the focusing device, a resolution of 200 nm has been obtained (1986) using a scanning technique. The zone plate is made using a scanning transmission electron microscope, and synchrotron radiation is used as the source of X-rays.

X-rays Electromagnetic waves of short wavelength (ca 10^{-3} to 10 nm) produced when high-speed electrons strike a solid target. Electrons passing near a nucleus in the target are accelerated and so emit a continuous spectrum of radiation (*bremsstrahlung*) ranging up from a *minimum wavelength*. In addition, the electrons may eject an electron from an inner shell of a target atom, and the resulting transition of an electron of a higher energy level to this level produces radiation of specific wavelengths. This is the *characteristic* X-ray spectrum of the target and is specific to the target element. X-rays may be detected photographically or by a counting device. They penetrate matter which is opaque to light; this makes X-rays a valuable tool for medical investigations and in X-ray crystallography.

X-ray spectrometer The name originally used for the *X-ray diffractometer*, but now abandoned in order to avoid confusion with *X-ray fluorescent spectrometry*.

X-ray unit kX, a unit in which X-ray wavelengths and cell dimensions were published up to about 1948. I kX unit = 1.002 02 Å.

XRF Abbrev. for **X-ray fluorescence spectrography**.

YAG Abbrev. for **Yttrium Aluminium Garnet**.

yellowcake A yellow powder which is the first stage in the production of uranium from ore concentrates, and is often produced at the mine itself. Approx. composition U_3O_8.

yellow ground Yellowish or buff-coloured, loose clay-rich material formed at the top of a **kimberlite** pipe by oxidation and alteration of **blue ground**.

yellow quartz See **citrine**.

yellow tellurium See **sylvanite**.

yield Tonnage extracted, or ratio of known tonnage, to that recoverable profitably.

yielding prop Support used just behind coal face, which shortens slightly under load, but can be reclaimed and reused.

ylem The basic substance from which it has been suggested that all known elements may have been derived through nucleogenesis (fusion of fundamental particles to form nuclei). It would have a density of 10^{16} kg/m^3, and would consist chiefly of neutrons.

yoderite A silicate of aluminium, iron and magnesium, crystallizing in the monoclinic system.

younging, direction of The direction in which a series of inclined sedimentary rocks becomes younger. See **way up**.

Young's equation Index of surface wettability, used in flotation research on minerals $\gamma_S = \gamma_{SL} + \gamma_L \cos\theta$, where θ is contact angle (that between water and air bubble adhering to mineral), γ is free energy per unit area, S and L are solid and liquid phases.

Young's modulus A modulus of elasticity applicable to the stretching of a wire (or thin rod), or to the bending of a beam. It is defined as the ratio

$$\frac{\text{tensile (or compressive) stress}}{\text{tensile (or compressive) strain}}$$

Symbol E. Known also as stretch or elongation modulus. Since **strain** is a numeric, its units are those of **stress**, viz. MN/m^2.

Ypresian A stage of the Eocene. See **Tertiary**.

ytterbium A metallic element, a member of the rare earth group. Oxide, Yb_2O_3, white, giving colourless salts. Symbol Yb, at.no. 70, r.a.m. 173.04, mp of metal about 1800°C. Obtained from *euxenite*.

yttrium A metallic element usually classed with the rare earths because of its chemical resemblance to them. Oxide, Y_2O_3, white, giving colourless salts. Symbol Y, at.no. 39, r.a.m. 88.9059, mp of metal 1250°C.

yttrium aluminium garnet A heavy colourless, isotropic artificial gemstone that is used as a simulant for diamond. Abbr.

YAG.

yttrocerite A massive, granular or earthy mineral, essentially a cerian fluorite, with the metals of the yttrium and cerium groups, commonly violet-blue in colour, and of rare occurrence.

yu-stone Yu or yu-shih, the Chinese name for the highly prized jade of gemstone quality.

Z The number of formula units per unit cell of a mineral.

Zanclian A stage in the Pliocene. See **Tertiary**.

zawn Cavern, natural or man-made in Cornwall, UK.

Zechstein The name of the upper series of the Permian. See **Palaeozoic**.

zeolites A group of alumino-silicates of sodium, potassium, calcium and barium, containing very loosely held water, which can be removed by heating and regained by exposure to a moist atmosphere, without destroying the crystal structure. They occur in geodes in igneous rocks, and as authigenic minerals in sediments, and include chabazite, natrolite, mesolite, stilbite, heulandite, harmotome, phillipsite, thomsonite etc.

zero-cut crystal Quartz crystal cut at such an angle to the axes as to have a zero frequency/temperature coefficient. Used for accurate frequency and time standards.

zinc A hard white metallic element with a bluish tinge. Symbol Zn, at.no. 30, r.a.m. 65.37, rel.d. at 20°C 7.12, mp 418°C, electrical resistivity 6.0×10^{-8} ohm metres. Because of its good resistance to atmospheric corrosion, zinc is used for protecting steel. It is also used in the form of sheet and as a constituent in alloys. Used as an electrode in a Daniell cell and in dry batteries. It is important nutritionally, trace amounts being present in many foods. Its principal ores are *sphalerite* (zinc blende) ZnS, and *smithsonite* ($ZnCO_3$) but there are many oxy-salts.

zinc blende A much-used name for **sphalerite**, the common sulphide of zinc.

zinc bloom A popular name for the massive basic zinc carbonate **hydrozincite**.

zinc dust Finely divided powder produced either by condensation of zinc vapour or by atomization of molten zinc. Once widely used to precipitate gold and silver from pregnant solution in the cyanidation of gold ore.

zincite Oxide of zinc, crystallizing in the hexagonal system and exhibiting polar symmetry; occurring rarely as crystals, usually as deep-red masses; an important ore of zinc. Also *red oxide of zinc*.

zinckenite A steel-grey mineral, essentially sulphide of lead and antimony, $PbSb_2S_4$, occurring as columnar hexagonal crystals, sometimes exceptionally thin, forming fibrous masses.

zinc spinel See **gahnite**.

zinnwaldite A mica related in composition to lepidolite (i.e. containing lithium and potassium) but including iron as an essential constituent; occurring in association with tinstones ores at Zinnwald in the Erzgebirge, in Cornwall and elsewhere.

zircon A tetragonal accessory mineral widely distributed in igneous, sedimentary, and metamorphic rocks. It varies in colour from brown to green, blue, red, golden-yellow, while colourless zircons make particularly brilliant stones when cut and polished. In composition, it is essentially silicate of zirconium, but often contains yttrium and thorium. A small amount of the rare element hafnium is present.

zirconium A metallic element. Symbol Zr, at.no. 40, r.a.m. 91.22, rel.d. 4.15, mp 2130°C. When purified from hafnium, its low neutron absorption and its retention of mechanical properties at high temperature make it useful for the construction of nuclear reactors. Also used as a refractory, as a lining for jet engines, and as a getter in the manufacture of vacuum tubes. The principal ores are *zircon* ($ZrSiO_4$) which is a very common accessory mineral of igneous rocks and concentrated in beach sands, and *baddeleyite* (ZrO_2).

zoisite Hydrated alumino-silicate of calcium crystallizing in the orthorhombic system and occurring chiefly in metamorphic schists; also a constituent of so-called saussurite. Clinozoisite has the same composition, but crystallizes in the monoclinic system.

zonal index See **zone**.

zone A stratigraphical unit with recognizable characteristics. The term has attracted many confusing definitions and is perhaps best used with an appropriate qualifier, e.g. *Dibunophyllum* zone, marine zone, ash zone. Also *zonal index*.

zone of cementation That 'shell' of the Earth's crust lying immediately below the zone of weathering, within which loose sediments are cemented by the addition of such minerals as calcite, introduced by percolating meteoric waters.

zone of weathering An 'Earth shell' comprising the exposed surface and that part which, through porosity, fracturing and jointing, is subject to the destructive action of the atmosphere, rain and frost. Soil develops in this zone.

zone plate A transparent plate divided into a series of zones by circles whose radii are in the ratio $\sqrt{1}:\sqrt{2}:\sqrt{3}:\sqrt{4}$ etc, the alternate zones being blacked. If a plane wave is incident normally on the plate, a maximum of light intensity is formed at a point on the axis as if the plate were acting as a lens of focal length f. Subsidiary focal points at $f/3$, $f/5$ etc. are also formed with progressively much weaker concentrations of light.

zoning Concentric layering parallel to the periphery of a crystalline mineral, shown by colour banding in such minerals as tourmaline, and by differences of the optical reactions to polarized light in colourless minerals like feldspars; it is due to the successive deposition of layers of materials differing slightly in composition.

zussmanite A pale green, tabular, trigonal hydrated silicate of iron, magnesium and potassium.

zweikanter A wind-faceted pebble with two curved surfaces intersecting at two sharp edges (whence its name, Ger. *zwei*, two). See **dreikanter, ventifact**.